Digital Image Processing

NATO ADVANCED STUDY INSTITUTES SERIES

*Proceedings of the Advanced Study Institute Programme, which aims
at the dissemination of advanced knowledge and
the formation of contacts among scientists from different countries*

The series is published by an international board of publishers in conjunction
with NATO Scientific Affairs Division

A	Life Sciences	Plenum Publishing Corporation
B	Physics	London and New York
C	Mathematical and	D. Reidel Publishing Company
	Physical Sciences	Dordrecht, Boston and London
D	Behavioural and	Sijthoff & Noordhoff International
	Social Sciences	Publishers
E	Applied Sciences	Alphen aan den Rijn and Germantown
		U.S.A.

Series C – Mathematical and Physical Sciences

Volume 77 – Digital Image Processing

Digital Image Processing

Proceedings of the NATO Advanced Study Institute
held at Bonas, France, June 23 - July 4, 1980

edited by

J. C. SIMON
Institut de Programmation,
Université Pierre et Marie Curie, Paris VI, France

and

R. M. HARALICK
Virginia Polytechnic Institute and State University, Blacksburg, U.S.A.

D. Reidel Publishing Company

Dordrecht : Holland / Boston : U.S.A. / London : England

Published in cooperation with NATO Scientific Affairs Division

Library of Congress Cataloging in Publication Data

NATO Advanced Study Institute (1980: Bonas, France)
    Digital image processing.

    (NATO Advanced Study Institutes Series. Series C, Mathematical and
Physical Sciences; v. 77)
    Includes index.
    1. Image processing-Digital techniques-Congresses. I. Simon, J. C.
(Jean Claude), 1923-   . II. Haralick, Robert M. III. Title. IV.
Series.
TA1632.N29   1980   621.36'7   81-11964
ISBN 90-277-1329-4   AACR2

CIP

Published by D. Reidel Publishing Company
P.O. Box 17, 3300 AA Dordrecht, Holland

Sold and distributed in the U.S.A. and Canada
by Kluwer Boston Inc.,
190 Old Derby Street, Hingham, MA 02043, U.S.A.

In all other countries, sold and distributed
by Kluwer Academic Publishers Group,
P.O. Box 322, 3300 AH Dordrecht, Holland

D. Reidel Publishing Company is a member of the Kluwer Group

TABLE OF CONTENTS

---

* Are Review Papers

PREFACE

        This book is the consequence of a NATO ASI held at the
Château de BONAS, from June 23 to July 4, 1980. It contains the
tutorial lectures and some papers presented at the Institute.
The book is divided in four sections:

        Issues of general interest. Some topics are broader than
the proper techniques of image processing, such as complexity,
clustering, topology, physiology; but they may be of interest...

        Feature detection and evaluation. The first level feature
detections are examined: edges and textures. Reorganization and
improvement of the results are obtained by relaxation and optimization process. Cooperative process are examined.

        Scenes and shapes concerns higher level problems, and
representation of images such as map and line-drawings.

        Applications in remote sensing, scene analysis, of one
or of a sequence of images.

        It is hoped that this book will serve to update a domain
in fast evolution.

Acknowledgment: This ASI, and this book, have been made possible
by the financial support of the NATO Scientific Affairs Division,
and the material support of INRIA and the Institut de Programmation
of the Université P. et M. Curie.

J. C. Simon and R. M. Haralick (eds.), Digital Image Processing, vii.

APPLICATION OF COMPLEXITY OF COMPUTATIONS
TO SIGNAL PROCESSING

S. Winograd

IBM Thomas J. Watson Research Center
Yorktown Heights, New York, U.S.A.

1. Introduction

Algebraic Complexity of Computations is concerned with determining the smallest number of arithmetic operations which are needed to perform a computation. In addition to this primary concern, this area is also concerned with the systematic development of minimal algorithms. In these lectures we will emphasize the results in Complexity of Computations which are applicable to those computations which are performed in digital signal processing. In particular we will discuss the applications to finite impulse response (FIR) filtering and to computing the discrete Fourier transform (DFT).

Even though the main objective of these lectures is the application to FIR filtering and the DFT, we will discuss the results of Complexity of Computations in a broader context. The reason for this generality is twofold:

(1) The algorithms which we will derive are a little intricate. It is only when they are described in their natural mathematical context that the simplicity of the ideas becomes clear.

(2) In order to extend the scope of the computations for which the Complexity results are applicable, it is important to know these results in as general a setting as possible.

In section 2 we will describe the relevant results in Complexity of Computations; in section 3 we will use them to obtain the algorithms for FIR filtering, and in section 4 we will apply the results of section 2 to DFT calculations.

2. Complexity of Computation

In this section we will summarize some of the results of Algebraic Complexi-

1

J. C. Simon and R. M. Haralick (eds.), Digital Image Processing, 1–17.

ty of Computation. We will concentrate on these results which will be used later in studying FIR filtering and DFT.

We will use G to denote the field of constants. In most of the uses, G will be the field of rational numbers, but for the time being we will leave it an arbitrary field of characteristics 0. Let F be another field which includes G. In FIR filtering, for example, F will be the field G extended by the variables (indeterminates) $x_0, x_1, ..., x_m$, while in DFT(n) F will be the field G extended by the primitive n^{th} root of unity $e^{2\pi j/n}$.

Consider the following t quantities to be computed

$$\psi_k = \sum_{i=1}^{n} f_{k,i} \, y_i, \quad k = 1, 2, ..., t,$$

where $f_{k,i} \epsilon F$, $1 \leq i \leq n$, $1 \leq k \leq t$ and y_i is a second set of indeterminates. Using matrix-vector notations we can write these t quantities as $A(F)y$, where $A(F)$ is the $t \times n$ matrix whose (k,i) entry is $f_{k,i}$, and y is the column vector $y = (y_1, y_2, ..., y_n)^T$. We will denote by $\mu(\psi_1, ..., \psi_t)$ the minimum number of multiplications needed to compute $\psi_1, ..., \psi_t$. In this number of multiplications we do not count multiplications by an element of $g \epsilon G$. Thus, if G is the field of rational numbers, and F is G extended by $x_1, x_2, ..., x_m$, then $(x_1 + x_2) \cdot (y_1 - y_3)$ is counted, while $17 \cdot (y_1 - y_3)$ or even $2/17 \cdot (y_1 - y_3/5)$ is not counted.

Definition   The row rank of $A(F)$, denoted by $\rho_r(A(F))$, is the largest number of rows of $A(F)$ such that no non-trivial G-linear combination of them yields a row all of whose elements are in G. The column rank of $A(F)$, denoted by $\rho_c(A(F))$, is the largest number of columns of $A(F)$ such that no non-trivial G-linear combination of them yields a column all of whose elements are in G.

We can use the definition to state the first theorem.

Theorem 1:   If $\psi_1, ..., \psi_t$ are given by $A(F)y$ then $\mu(\psi_1, ..., \psi_t) \geq \min(\rho_r(A(F)), \rho_c(A(F)))$.

Theorem 1 provides us with a bound on the number of multiplications. It does not tell us what $\mu(\psi_1, ..., \psi_t)$ is, nor does it tell us how to find a minimal algorithm. The second theorem does address itself to these two points.

Theorem 2:   Let $\psi_1, ..., \psi_t$ given by $A(F)y$. If $\rho_r(A(F)) = t$ then $\mu(\psi_1, ..., \psi_t) = t$ if and only if there exists a non-singular G-matrix C such that $\rho_r(cA(F)) = 1$, for every row c of C. Moreover, every minimal algorithm corresponds to such a non-singular matrix C.

Perhaps the best way of explaining the content of Theorem 2 is by an example. Consider ψ_0, ψ_1, ψ_2 given by

$$(\psi_0 + \psi_1 u + \psi_2 u^2) = (x_0 + x_1 u)(y_0 + y_1 u)$$

that is, $\psi_0 = x_0 y_0$, $\psi_1 = x_0 y_1 + x_1 y_0$, $\psi_2 = x_1 y_1$. We take G as the field of rational numbers, and F as G extended by x_0, x_1, x_2. Using the matrix notation we can write ψ_0, ψ_1, ψ_2 as

$$\begin{pmatrix} \psi_0 \\ \psi_1 \\ \psi_2 \end{pmatrix} = \begin{pmatrix} x_0 & 0 \\ x_1 & x_0 \\ 0 & x_1 \end{pmatrix} \begin{pmatrix} y_0 \\ y_1 \end{pmatrix} = A(F)y.$$

For which $c = (c_0,c_1,c_2)$ do we have $\rho_c(cA(F)) = 1$? Simple calculations show that this can happen in only two cases: Either

$c = c_0(1,\alpha, \alpha^2)$ in which case

$cA(F) = c_0(x_0 + \alpha x_1, \alpha x_0 + \alpha^2 x_1)$ and

$cA(F)y = c_0(x_0 + \alpha x_1)(y_0 + \alpha y_1)$ or

$c = c_0(0,0,1)$

in which case $cA(F) = c_0(0,x_1)$ and $cA(F)y = c_0 x_1 y_1$. We can construct the matrix C by taking two of its rows as c's of the first case, say with $\alpha = 0$ and $\alpha = -1$, and the third row as $(0,0,1)$. We use this C to obtain:

$$\begin{pmatrix} 1 & 0 & 0 \\ 1 & -1 & 1 \\ 0 & 0 & 1 \end{pmatrix} \begin{pmatrix} \psi_1 \\ \psi_2 \\ \psi_3 \end{pmatrix} = \begin{pmatrix} 1 & 0 & 0 \\ 1 & -1 & 1 \\ 0 & 0 & 1 \end{pmatrix} \begin{pmatrix} x_0 & 0 \\ x_1 & x_0 \\ 0 & x_1 \end{pmatrix} \begin{pmatrix} y_0 \\ y_1 \end{pmatrix}$$

$$= \begin{pmatrix} x_0 y_0 \\ (x_0 - x_1)(y_0 - y_1) \\ x_1 y_1 \end{pmatrix}.$$

Multiplying both sides by C^{-1} (which happened in this case to be the same as C) we obtain:

$$x_0 = x_0 y_0,$$
$$\psi_1 = x_0 y_0 - (x_0 - x_1) + x_1 y_1,$$
$$\psi_2 = x_1 y_1,$$

that is, an algorithm using 3 multiplications (and 4 additions). Theorem 2 guarantees that every algorithm which uses only 3 multiplications can be obtained by this procedure, and two different algorithms arise from different choices of C. A generalization of this procedure provides us with a useful corollary.

<u>Corollary:</u> If $\psi_0,\psi_1,...,\psi_{m+n}$ are given by

$$\sum_{k=0}^{m+n} \psi_k u^k = (\sum_{i=0}^{m} x_i u^i)(\sum_{j=0}^{n} y_j u^j)$$

then $\mu(\psi_0,...,\psi_{m+n}) = m + n + 1$, and all minimal algorithms can be classified.

The result of the corollary has one major drawback: It shows that when m and n are of even moderate size, every minimal algorithm needs a large number of additions. We cannot elaborate on it here, but the interested reader is invited to derive an algorithm for the case $m=n=3$.

The algorithm given in the example can be modified to yield an algorithm for a seemingly unrelated problem: that of multiplying two complex numbers. Let $(u_0 + ju_1) = (x_0 + jx_1)(y_0 + jy_1)$. Then $u_0 = x_0 y_0 - x_1 y_1$ and $u_1 = x_0 y_1 + x_1 y_0$. If ψ_0,ψ_1,ψ_2 are as in the example then $u_0 = \psi_0 - \psi_2$ and $u_1 = \psi_1$. If we use the algorithm of the example to compute ψ_0,ψ_1, and ψ_2 we obtain the following algorithm for u_0 and u_1:

$$u_0 = x_0 y_0 - x_1 y_1$$
$$u_1 = x_0 y_0 - (x_0 - x_1)(y_0 - y_1) + x_1 y_1$$

We will postpone, for a while, the question whether this new algorithm for u_0 and u_1 is minimal. Instead, we will develop another useful general technique for deriving new algorithms from old ones. Both ψ_0, ψ_1, and ψ_2 of the example and u_0 and u_1 are special, because they are *systems of bilinear forms*. In general, a system of bilinear forms is given by

$$\psi_k = \sum_{j=1}^{s} \sum_{i=1}^{r} g_{ijk} x_i y_j \quad k = 1,2,...,t$$

where $g_{ijk} \epsilon G$. Not only were the quantities to be computed special, even the algorithms which we derived for them were special: each multiplication was the product of a linear form of the x's by a linear form of the y's. We call such special algorithms *bilinear algorithms*. If a bilinear algorithm has N multiplications $m_1, m_2,...,m_N$ then

$$m_i = (\sum_{i=1}^{r} a_{\ell,i} x_i)(\sum_{j=1}^{s} b_{\ell,j} y_j) \quad \ell = 1,2,...,N.$$

(The $a_{\ell,i}$'s and $b_{\ell,j}$'s are, of course, elements of G.) Now, let $\psi_1,...,\psi_t$ be a system of bilinear forms, which can be computed by a bilinear algorithm with N multiplications. This can be described by

$$\psi_k = \sum_{j=1}^{s} \sum_{i=1}^{r} g_{ijk} x_i y_j$$

$$= \sum_{\ell=1}^{N} c_{\ell,k} (\sum_{i=1}^{r} a_{\ell,i} x_i)(\sum_{j=1}^{s} b_{\ell,j} y_j), \quad k = 1,2,...,t.$$

If we now multiply the equation above by z_k and sum over all k we get only one equation instead t of them, namely,

$$\sum_{k=1}^{t} \sum_{j=1}^{s} \sum_{i=1}^{r} g_{ijk} x_i y_j z_k$$

$$= \sum_{\ell=1}^{N} (\sum_{i=1}^{r} a_{\ell,i} x_i)(\sum_{j=1}^{s} b_{\ell,j} y_j)(\sum_{k=1}^{t} c_{\ell,k} z_k).$$

Let us now dismember this last equation by equating the coefficients of the x_i's. When we do that we obtain the r equations

$$\psi'_i = \sum_{k=1}^{t} \sum_{j=1}^{s} g_{ijk} y_j z_k$$

$$= \sum_{\ell=1}^{N} a_{\ell,i} (\sum_{j=1}^{s} b_{\ell,j} y_j)(\sum_{k=1}^{t} c_{\ell,k} z_k) \quad i = 1,2,...,r.$$

These r equations are recognized as a bilinear algorithm for the system of bilinear forms

$$\psi'_i = \sum_{k=1}^{t} \sum_{j=1}^{s} g_{ijk} \, y_j \, z_k \quad i = 1,2,...,r.$$

The new system of bilinear forms is called a *transpose* of the old one, and the new bilinear algorithm, which also has N multiplications, is called a *transpose* of the old one. Of course, we could have obtained another transpose system, and another transpose algorithm, by equating the coefficients of the y_j's.

Let us apply this construction to u_0 and u_1 of the complex multiplication. The two equations

$$x_0 y_0 - x_1 y_1 = x_0 y_0 - x_1 y_1$$
$$x_0 y_1 + x_1 y_0 = x_0 y_0 - (x_0 - x_1)(y_0 - y_1) + x_1 y_1$$

are summarized by the one equation

$$x_0 y_0 z_0 - x_1 y_1 z_0 + x_0 y_1 z_1 + x_1 y_0 z_1 =$$

$$= x_0 y_0 (z_0 + z_1) - (x_0 - x_1)(y_0 - y_1)z_1 + x_1 y_1 (z_1 - z_0).$$

Equating the coefficients of the x's we get

$$y_0 z_0 + y_1 z_1 = y_0 (z_0 + z_1) - (y_0 - y_1)z_1$$
$$y_0 z_1 - y_1 z_0 = y_1 (z_1 - z_0) + (y_0 - y_1)z_1$$

This new system of bilinear forms is recognized as the imaginary and real parts of $(y_0 + jy_1)(z_1 + jz_0)$. So the new system of bilinear forms is not new at all, *but the new algorithm is indeed new*.

Bilinear algorithms have another useful property which makes them of special interest: *They do not depend on the fact that multiplication is commutative.* The importance of this fact is that we can substitute matrices for the x's and matrices, or even vectors, for the y's and still get valid identities. For example, in the algorithm just derived we can substitute

$$Y_0 = \begin{pmatrix} \alpha_0 & -\alpha_1 \\ \alpha_1 & -\alpha_0 \end{pmatrix}, \quad Y_1 = \begin{pmatrix} \alpha_2 & -\alpha_3 \\ \alpha_3 & \alpha_2 \end{pmatrix}$$

$$Z_0 = \begin{pmatrix} \beta_0 \\ \beta_1 \end{pmatrix}, \quad Z_1 = \begin{pmatrix} \beta_2 \\ \beta_3 \end{pmatrix}$$

for y_0, y_1, z_0, and z_1, respectively. Having done that we obtain

$$Y_0 Z_1 - Y_1 Z_0 = Y_1 (Z_1 - Z_0) + (Y_0 - Y_1)Z_1$$
$$Y_0 Z_0 + Y_1 Z_1 = Y_0 (Z_0 + Z_1) - (Y_0 - Y_1)Z_1 .$$

But $Y_1(Z_1 - Z_0)$ is readily recognized as the real and imaginary part of $(\alpha_2 + j\alpha_3)((\beta_2 - \beta_0) + j(\beta_3 - \beta_1))$, and therefore its components can be calculated using only three multiplications. Similarly, both $Y_0(Z_0 + Z_1)$ and $(Y_0 - Y_1)Z_1$ are recognized as complex multiplications. The upshot of this substitution then, is a bilinear algorithm with 9 multiplications, for computing the system of bilinear forms

$$\begin{pmatrix} \alpha_0 & -\alpha_1 & -\alpha_2 & \alpha_3 \\ \alpha_1 & \alpha_0 & -\alpha_3 & -\alpha_2 \\ \alpha_2 & -\alpha_3 & \alpha_0 & -\alpha_1 \\ \alpha_3 & \alpha_2 & \alpha_1 & \alpha_0 \end{pmatrix} \begin{pmatrix} \beta_2 \\ \beta_3 \\ \beta_0 \\ \beta_1 \end{pmatrix}$$

The new system of bilinear forms is called the *tensor product* of the two old ones (in the example it was the tensor product of complex multiplication with itself), and the new algorithm is called the *tensor product* of the two algorithms. The reader should be warned that even though the two algorithms are minimal, their tensor product need not be minimal. The example above is such a case. We will see shortly that complex multiplication must involve three (real) products, but the tensor product of complex multiplication with itself can be done using only six products. Nonetheless, the tensor product construction is very useful in applications.

It is time to return to the question we postponed earlier: What is the complexity of complex multiplication? Before answering this question we will generalize it.

Our first algorithm for complex multiplication was derived from the algorithm for computing the coefficients of $(x_0 + x_1u)(y_0 + y_1u)$. This was not a coincidence! The two problems are intimately related. The minimal polynomial for $j = \sqrt{-1}$ is u^2+1, and the real and imaginary parts of $(x_0 + jx_1)(y_0 + jy_1)$ are the coefficients of the polynomial $(x_0 + x_1u)(y_0 + y_1u)$ mod $u^2 + 1$. Indeed, our original algorithm started by computing the coefficients of $(x_0 + x_1u)(y_0 + y_1u)$ and then reducing modulo u^2+1. Let us generalize this system by considering the system of bilinear forms which are the coefficients of $T(u)$ where

$$T(u) = (\sum_{i=0}^{n-1} x_i u^i)(\sum_{j=0}^{n-1} y_j u^j) \text{ mod } P(u)$$

where $P(u)$ is some monic polynomial of degree n with constant coefficients (that is, the coefficients of $P(u)$ are in G). We can generalize the problem even further by considering the system $\mathcal{T}$ of all the coefficients of all the polynomials,

$$T^\ell(u) = (\sum_{i=0}^{n(\ell)-1} x_i^{(\ell)} u^i)(\sum_{j=0}^{n(\ell)-1} y_j^{(\ell)} u^j) \text{ mod } P_\ell(u)$$

$\ell = 1,2,......,t$ where $P_\ell(u)$ is a monic polynomial of degree $n(\ell)$ with constant coefficients. (We do not insist that the $P_\ell(u)$ be different for different values of ℓ, but do demand that all the $x_i^{(\ell)}$'s and all the $y_j^{(\ell)}$'s be distinct). With this notation we can state the next theorem.

Theorem 3:   Let $\mathcal{T}$ be as above then

$$\mu(\mathcal{T}) = \sum_{\ell=1}^{t} 2n(\ell) - k(\ell),$$

where $k(\ell)$ is the number of distinct irreducible polynomials which divide $P_\ell(u)$, $\ell = 1,2,...,t$.

In particular, if we take $t=1$, $P(u) = u^2+1$, and G the field of rational numbers, we see that $\mu(x_0y_0-x_1y_1, x_0y_1 + x_1y_0) = 2 \cdot 2 - 1 = 3$.

It is important to add that all minimal algorithms have been classified in the special case that none of the polynomials $P_\ell(u)$ has a multiple root. We cannot describe the classification here, but can say that all minimal algorithms are derived from minimal algorithms for the systems of the corollary of Theorem 2. The interested reader is referred to (2) and (3). We must add, however, that one consequence of the classification is that if any of the $P_\ell(u)$'s has an irreducible component of even medium degree, the minimal algorithm will have a large number of additions. It is this practical difficulty which makes the tensor product construction, described earlier, so useful.

Having generalized the product of two complex numbers, let us now special-ize it. We saw that three multiplications are needed to compute the real and imaginary parts of $(x_0 + jx_1)(y_0 + jy_1)$, but what is the complexity of computing the real and imaginary parts of $(\sqrt{2} + j\sqrt{3})(y_0 + jy_1)$? Clearly three multiplica-tions will suffice, but can it be done with only two multiplications? More general-ly, let us substitute some specific numbers for the $x_i^{(\ell)}$'s and thus obtain a new system $\mathscr{S}(\mathscr{T})$ to be computed. If we substitute elements of G for the $x_i^{(\ell)}$'s then the resulting system can be computed without any multiplication. $((1 + 2j)(y_0 + jy_1)$ can be computed without any multiplication!) So we assume that we substitute some elements of a field F of the $x_i^{(\ell)}$'s when F is some field containing G. Let us denote by $\mathscr{S}(x_i^{(\ell)})$ the element of F which is substituted for $x_i^{(\ell)}$. We impose several conditions on the substitution $\mathscr{S}$:

1. For any $\ell=1,2,...,t$ no non-trivial G-linear combination of the $\mathscr{S}(x_i^{(\ell)})$'s yields an element of G.
2. If $P_\ell(u) = P_m(u)$ we allow $\mathscr{S}(x_i^{(\ell)}) = \mathscr{S}(x_i^{(m)})$, $i=1,2,...,n(\ell)-1$. In this case we say that $P_\ell(u)$ and $P_m(u)$ are strongly equal.
3. If $L = (\ell_1,\ell_2,...,\ell_n)$ are indices such that for any $\ell,m \in L$, $P_\ell(u)$ and $P_m(u)$ are not strongly equal, then no non-trivial G-linear combination of

$$\bigcup_{\ell \in L} \bigcup_{i=0}^{n(\ell)-1} \mathscr{S}(x_i^{(\ell)})$$

yields an element of G.

The reader may feel that these conditions are a bit artificial - and they are. The only justification I can give is that they replace the even stronger condition that all the $x_i^{(\ell)}$'s be distinct. More importantly, some important systems, like the DFT, satisfy these conditions.

<u>Theorem 4:</u> Let $\mathscr{S}$ satisfy the three conditions, and let $\mathscr{T}$, $n(\ell)$ and $k(\ell)$ be as in Theorem 3, then

$$\mu(\mathscr{S}(\mathscr{T})) = \sum_{\ell=1}^{t} 2n(\ell) - k(\ell).$$

The reader who is interested in the proof of Theorem 4, and its use in studying the DFT is referred to (4) and (5).

3. FIR Filtering

A FIR filter is described by the values $y_0, y_1, ..., y_{n-1}$ of its n taps. If $x_0, x_1, x_2, ...$ is the signal, then the output of the filter is the sequence $O_0, O_1,$. where

$$O_i = \sum_{j=0}^{n-1} x_{i+j} y_j.$$

We will denote by $F(m,n)$ the sequence $O_0, O_1, ..., O_{m-1}$ of m consecutive outputs. In this section we will use the results of the previous section to determine the complexity of $F(m,n)$, and to derive algorithms for computing these m quantities. Before describing the results we should emphasize that defining $F(m,n)$ as the first m outputs of the filter was quite arbitrary. In fact, an algorithm for computing the first m outputs of the filter can be used to compute any sequence of m consecutive outputs. This can be stated even more strongly: The computational process we have in mind uses the algorithm to compute the first m outputs, then uses the same algorithm to compute the second batch of m consecutive outputs, and so on. This computational process has an important consequence. Because every m consecutive output can be obtained from $F(m,n)$ by changing the indices of the x's and leaving the y's intact, any operation which involves only the values of the y's has to be done only once. If the number of outputs we wish to compute is much larger than m, and therefore we will use the same algorithm many times, we will not count those operations which involve only the values of the y's. Thus the computation of $3/7 \, y_1 + 2/5 \, y_2$ will not add to our count of the number of multiplications or additions.

We will start by assuming that the y_j's are distinct variables, and later in this section we will treat the case of a symmetric filter.

An immediate consequence of Theorem 1 is

Theorem 5:   Every algorithm for computing $F(m,n)$ uses at least $m+n-1$ multiplications.

Theorem 5 does not tell us what is the complexity of $F(m,n)$, nor how to find efficient algorithms for its computation. These are the questions which we will treat in the rest of this section.

The key to determining the complexity of $F(m,n)$, and more importantly to deriving efficient algorithms for computing $F(m,n)$, is that $F(m,n)$ is a transpose of the computation of the coefficients of a product of polynomials. More precisely, if we define the system of bilinear forms $t_0, t_1, ..., t_{m+n-2}$ by

$$\sum_{i=0}^{m+n-2} t_i \, u^i = \left(\sum_{i=0}^{m-1} z_i \, u^i \right) \left(\sum_{i=0}^{n-1} y_i \, u^i \right),$$

then $F(m,n)$ is a transpose of $t_0, t_1, ..., t_{m+n-2}$. In light of the corollary of Theorem 2 we have:

Theorem 6: The minimum number of multiplications needed to compute $F(m,n)$ is $m+n-1$.

We can now transpose algorithms for computing the coefficients of the product of polynomials to algorithms for computing $F(m,n)$. We will illustrate this process by transposing the algorithm of the example of the previous section. In that example we derived the algorithm

$$
\begin{aligned}
z_0 y_0 &= z_0 y_0 \\
z_1 y_0 + z_0 y_1 &= z_0 y_0 - (z_0 - z_1) \cdot (y_0 - y_1) - z_1 y_1 \\
z_1 y_1 &= z_1 y_1
\end{aligned}
$$

Multiplying the first equation by x_0, the second by x_1 and the third by x_2, and summing them we get

$$
\begin{aligned}
&x_0 z_0 y_0 + x_1 z_1 y_0 + x_1 z_0 y_1 + x_2 z_1 y_1 \\
&= (x_0 + x_1) z_0 y_0 - x_1 (z_0 - z_1)(y_0 - y_1) + (x_1 + x_2) z_1 y_1 .
\end{aligned}
$$

If we now equate the coefficients of the z's we get

$$
\begin{aligned}
x_0 y_0 + x_1 y_1 &= (x_0 + x_1) y_0 - x_1 (y_0 - y_1) \\
x_1 y_0 + x_2 y_1 &= (x_1 + x_2) y_1 + x_1 (y_0 - y_1).
\end{aligned}
$$

That is an algorithm for computing $F(2,2)$ which uses 3 multiplications and 4 additions (remember that we do not count $y_0 - y_1$ because it involves only the y's).

Perhaps a second example is in order. We start with an algorithm for computing the coefficients of $(z_0 + z_1 u + z_2 u^2)(y_0 + y_1 u)$. This algorithm, which can be derived by the procedure of the example of Section 2, is:

$$
\begin{aligned}
z_0 y_0 &= z_0 y_0 \\
z_1 y_0 + z_0 y_1 &= (z_0 + z_1 + z_2)(y_0 + y_1)/2 \\
&\quad - (z_0 - z_1 + z_2)(y_0 - y_1)/2 - z_2 y_1 \\
z_2 y_0 + z_1 y_1 &= (z_0 + z_1 + z_2)(y_0 + y_1)/2 \\
&\quad + (z_0 - z_1 + z_2)(y_0 - y_1)/2 - z_0 y_0 \\
z_2 y_1 &= z_2 y_1
\end{aligned}
$$

The reader can easily verify that the transpose of this algorithm is the algorithm for $F(3,2)$ given by:

$$
\begin{aligned}
x_0 y_0 + x_1 y_1 &= (x_0 - x_2) y_0 + (x_1 + x_2)(y_0 + y_1)/2 \\
&\quad - (x_1 - x_2)(y_0 - y_1)/2 \\
x_1 y_0 + x_2 y_1 &= (x_1 + x_2)(y_0 + y_1)/2 \\
&\quad + (x_1 - x_2)(y_0 - y_1)/2 \\
x_2 y_0 + x_3 y_1 &= (x_3 - x_1) y_1 + (x_1 + x_2)(y_0 + y_1)/2 \\
&\quad - (x_1 - x_2)(y_0 - y_1)/2
\end{aligned}
$$

This algorithm uses 4 multiplications and 8 additions. The reader should observe that in this algorithm we "pushed" the operation of division by 2 to that part of the algorithm which involves only the y's, that is, to that part in which no operation is counted. Clearly, this "pushing" can be done in general. We see that in FIR filtering we can justify the assumption that *division by a fixed integer should not be counted.*

Two-tap filters do not abound in practice. Yet if we try to follow the procedure used above to derive an algorithm for, say, $F(16,16)$ we will run into the problem that every minimal algorithm uses too many additions (and also lead to numerical inaccuracies and scaling problems). A way out of this dilemma is to derive an algorithm which does not use the minimum number of multiplications. It is in this derivation that the tensor product construction provides a powerful tool. Instead of describing the use of the tensor product construction "in general," I would like to use it in one small example. The reader should be able to provide the generalization.

Consider $F(4,4)$. This system of bilinear forms can be written as

$$\begin{pmatrix} O_0 \\ O_1 \\ O_2 \\ O_3 \end{pmatrix} = \begin{pmatrix} x_0 & x_1 & x_2 & x_3 \\ x_1 & x_2 & x_3 & x_4 \\ x_2 & x_3 & x_4 & x_5 \\ x_3 & x_4 & x_5 & x_6 \end{pmatrix} \begin{pmatrix} y_0 \\ y_1 \\ y_2 \\ y_3 \end{pmatrix}$$

If we denote

$$\begin{pmatrix} x_0 & x_1 \\ x_1 & x_2 \end{pmatrix} \text{ by } X_0, \begin{pmatrix} x_2 & x_3 \\ x_3 & x_4 \end{pmatrix} \text{ by } X_1$$

$$\begin{pmatrix} x_4 & x_5 \\ x_5 & x_6 \end{pmatrix} \text{ by } X_3, \begin{pmatrix} y_0 \\ y_1 \end{pmatrix} \text{ by } Y_0, \begin{pmatrix} y_2 \\ y_3 \end{pmatrix} \text{ by } Y_1,$$

$$\begin{pmatrix} O_0 \\ O_1 \end{pmatrix} \text{ by } O_0 \text{ and} \begin{pmatrix} O_2 \\ O_3 \end{pmatrix} \text{ by } O_1,$$

we can write the system as

$$\begin{pmatrix} O_1 \\ O_2 \end{pmatrix} = \begin{pmatrix} X_0 & X_1 \\ X_1 & X_2 \end{pmatrix} \begin{pmatrix} Y_0 \\ Y_1 \end{pmatrix}.$$

That is, $F(4,4)$ can be viewed as $F(2,2)$ with the x's replaced by matrices and the y's by vectors. We can use the algorithm for $F(2,2)$ and get

$$\begin{matrix} O_1 = (X_0 + X_1)Y_0 - X_1(Y_0 - Y_1) \\ O_2 = (X_1 + X_2)Y_1 + X_1(Y_0 - Y_1) \end{matrix}.$$

But $(X_0 + X_1)Y_0$ is an instance of $F(2,2)$ and therefore can be done by the $F(2,2)$ algorithm. Similarly $(X_1 + X_2)Y_1$ and $X_1(Y_0 - Y_1)$ are instances of $F(2,2)$. So altogether we derived an algorithm for $F(4,4)$ which uses 9 multiplications and 21 additions. Even though $F(4,4)$ can be done using only 7 multiplications, every such algorithm will use many more additions.

Using the tensor product construction we can derive, for example, an algorithm for $F(24,16)$ which uses 192 multiplications and 250 additions. That is, 8 multiplications per output and 10 5/12 additions per output. This algorithm, which was constructed from the $F(2,2)$ and $F(3,2)$ algorithms derived earlier, illustrate an interesting side effect of the construction. Even though both of the original algorithms reduced the number of multiplications at the expense of the number of additions, their combination reduced *both the number of multiplications and the number of additions*. Further discussion of constructing multiplicatively suboptimal algorithms can be found in (1).

The algorithm which we just derived may look new and unfamiliar. Yet when cast properly they can be seen as a generalization of a common method of FIR filtering - by using the DFT. Let me explain. Consider $F(m,n)$, and let N be $N = m+n-1$. We saw that a transpose of $F(m,n)$ is the problem of computing the coefficients of

$$(\sum_{i=0}^{m-1} z_i \, u^i)(\sum_{j=0}^{n-1} y_j \, u^j).$$

In the previous section we saw that every minimal algorithm for the product of the polynomials is described by a matrix C each of whose rows has the form $c(1,a,a^2,...,a^{N-1})$, with the possible exception of one row which is of the form $c(0,0,...,0,1)$. Now, if we choose C such that $c=1$ and the a's are all the Nth roots of unity, the resulting algorithm for $F(m,n)$ is the familiar method of FIR filtering via the DFT, "zero padding" and all.

This new derivation of the "Convolution Theorem" enables us to immediately extend it. It was found that it is almost always advantageous to use $(0,0,...,0,1)$ and $(1,0,0,...,0)$ as two rows of the matrix C. So let us choose these two rows and have the remaining $N-2$ rows be specified by choosing the a's to be the $(N-2)^{\text{nd}}$ roots of unity. The resulting algorithm will be related to using the DFT - but with modifications. A simple example, that of using a 3-tap filter, will explain the modifications. The reader can easily see how to extend and generalize the modifications.

Assume that we want to compute the outputs of a 3-tap filter using DFT(4) - the discrete Fourier transform of 4 points. To do that we "skd" the y's with zeros and "extend" the matrix to be cyclic. That is, we compute

$$\begin{pmatrix} x_0 & x_1 & x_2 & x_3 \\ x_1 & x_2 & x_3 & x_0 \\ x_2 & x_3 & x_0 & x_1 \\ x_3 & x_0 & x_1 & x_2 \end{pmatrix} \begin{pmatrix} y_0 \\ y_1 \\ y_2 \\ 0 \end{pmatrix}.$$

Of course, we can use only the first two outputs, the last two do not correspond to outputs of the filter. Consider now the following system

$$\begin{array}{l} (x_0 - x_4)y_0 + \\ \\ \\ (x_5 - x_1)y_2 + \end{array} \begin{pmatrix} x_4 & x_1 & x_2 & x_3 \\ x_1 & x_2 & x_3 & x_4 \\ x_2 & x_3 & x_4 & x_1 \\ x_3 & x_4 & x_1 & x_2 \end{pmatrix} \begin{pmatrix} y_0 \\ y_1 \\ y_2 \\ 0 \end{pmatrix}.$$

The matrix is still cyclic, so we can use DFT(4), but now all four outputs are usable. That is, by using modifications which involve two multiplications and four additions we obtained two more usable outputs. For very large filters the saving may be small, but at times the modifications may help in a tight situation. Assume that we need to compute exactly 36 outputs of a 31-tap filter. Using the 64 points, FFT yields 34 outputs, but what about the remaining two? The modifications just described may be just what the doctor ordered. This example is a bit artificial; but let me assure the reader that a similar situation, with different numbers, was encountered in practice.

Quite often we encounter symmetric filters, i.e., filters satisfying $y_i = y_{n-1-i}$ for all i. Besides their other advantages, like linear phase shift, symmetric filters lighten the computational load: Their straightforward implementation uses only half the number of multiplications of the unsymmetric ones. Let us denote by $F_S(m,n)$ the computation of m consecutive outputs of an n-tap symmetric filter.

Theorem 6:   The minimum number of multiplications needed to compute $F_S(m,n)$ is $m+n-1$ when n is odd and $m+n-2$ when n is even.

Theorem 6 is quite disappointing. It shows that as far as the minimum number of multiplications is concerned, symmetric filters do not enjoy a great advantage over unsymmetric ones. However, this first impression is a little misleading. The fact that the filter is symmetric does postpone the growth of the number of additions, and thus postpones the moment when multiplicatively suboptimal algorithms have to be used. I will not describe the derivation of algorithms for symmetric filters; they are very similar to the derivation for unsymmetric filters. (The reader is invited to consult (1) and (6)). I will end this section by giving the results of one such algorithm. But before doing that, I would like to mention one interesting side result coming out of the proof of Theorem 6. The proof shows how to convert an n-tap symmetric filter, when n is even, into an $(n-1)$-tap symmetric filter. The cost of the conversion: One addition per output. And now to the example.

Consider a 9-tap symmetric filter. More precisely, consider $F_S(10,9)$. It is possible to construct an algorithm for $F_S(10,9)$ which uses 24 multiplications and 106 additions. This algorithm is multiplicatively suboptimal - the minimum number of multiplications is 18 - but not by much. The use of only 2.4 multiplications per output instead of 5 (or even 4 if we normalize the center tap) seems attractive, but the cost of 10.6 additions per output, instead of 8, diminishes our enthusiasm. These 106 additions break down into 48 additions of the inputs and 58 additions performed after the products were computed. If we only had to compute the sum of the outputs of two or more such filters we could have reduced the cost of the additions. We would add the 24 products and then use the 58 "output additions" only once.

Computing the sum of the outputs of several filters is not as uncommon as one may think. A 19-tap symmetric filter with a 2:1 decimation is really the sum of two 9-tap symmetric filters. The new algorithm requires 18.8 additions per output and 4.8 multiplications per output, while the straightforward algorithm uses 18 additions per output and 9 multiplications per output. A 39-tap symmetric filter with 4:1 decimation is really the sum of four 9-tap symmetric filters. The new algorithm uses 36.2 additions and 9.6 multiplications per output, contrasted with 38 additions and 19 multiplications of the conventional algorithm. If the 39-tap symmetric filter has every fourth tap zero it is the sum of three 9-tap symmetric filters. The new algorithm uses 29 additions per output (contrasted with 30 of the conventional method) and 7.2 multiplications per output (contrasted with 15 of the conventional method). True, the control of the new algorithm is more complicated, but the potential for saving is there.

4.   Discrete Fourier Transform

The complexity results concerning the computation of the coefficients of the product of polynomials played a crucial role in the derivation of algorithms for $F(m,n)$. The related problem, of computing the coefficients of the product of two polynomials modulo a third one, plays a similar role in the derivation of algorithms for computing the DFT. Two basic observations enable us to use the complexity results to derive new algorithms for computing the DFT. The first observation is that cyclic convolution is a particular instance of computing the coefficients of the product of two polynomials modulo a third polynomial. The second observation is that the DFT, properly viewed, can be built up from cyclic convolutions.

If $x = (x_0, x_1, ..., x_{n-1})$ and $y = (y_0, y_1, ..., y_{n-1})$ are two n-dimensional vectors, then their cyclic convolution $x * y$ is the n-dimensional vector $z = (z_0, z_1, ..., z_{n-1})$ where

$$z_i = \sum_{j=0}^{n-1} x_{i+j} y_j \qquad i = 0, 1, ..., n-1$$

and the sum $i+j$ is taken modulo n. The cyclic convolution is often defined by $z_i = \sum_{i=0} x_{i-j} y_j$, where $i-j$ is taken modulo n. But these two definitions are intimately related: Replacing y_j by y_{n-j} replaces one definition by the other. Another relation, which is very useful in deriving algorithms, is that the two definitions are the transpose of each other. For the sake of definiteness we will use the first definition in this paper.

An alternative way of defining the z_i's is by observing that

$$\sum_{k=0}^{n-1} z_k u^k = (\sum_{i=0}^{n-1} x_i u^i)(\sum_{j=0}^{n-1} y_{n-j} u^j) \bmod (u^n - 1) .$$

This definition, which shows that cyclic convolution belongs to the class of bilinear forms of Theorem 3, enables us to deduce:

Theorem 7:   If the field of constants, G, is taken as the field of rational numbers, then the minimum number of multiplications needed to compute the cyclic convolution of two general n-dimensional vectors is $2n-d(n)$, where $d(n)$ is the number of divisors of n.

All that is needed to prove Theorem 7 is the result that the number of irreducible factors of u^n-1 is $d(n)$.

Computing the cyclic convolution may be of interest by itself. The results of Theorem 3, and the discussion following it, can be used to derive algorithms for cyclic convolution. This was done by Agarwal and Cooley (7). The reader who is interested in this topic is referred to their paper.

Casting the DFT as a cyclic convolution was done by Rader (8) for the DFT of a prime number of points. The key to this metamorphosis of the $DFT(p)$ is the isomorphism between the multiplicative group of integers modulo p, and the additive group of integers modulo $p-1$. This isomorphism implies that *after an appropriate renumbering of the indices*, $DFT(p)$ is turned into a cyclic convolution. Let me illustrate it by the small example of $DFT(5)$.

Using matrix-vector notation $DFT(5)$ can be written as

$$\begin{pmatrix} A_0 \\ A_1 \\ A_2 \\ A_3 \\ A_4 \end{pmatrix} = \begin{pmatrix} 1 & 1 & 1 & 1 & 1 \\ 1 & w & w^2 & w^3 & w^4 \\ 1 & w^2 & w^4 & w & w^3 \\ 1 & w^3 & w & w^4 & w^2 \\ 1 & w^4 & w^3 & w^2 & w \end{pmatrix} \begin{pmatrix} a_0 \\ a_1 \\ a_2 \\ a_3 \\ a_4 \end{pmatrix}$$

where $w = e^{2\pi j/5}$.

We now need the isomorphism between the group M_5 of multiplication modulo 5 and the group Z_4 of addition modulo 4. To avoid confusion, we will use $\{0,1,2,3\}$ to denote the elements of Z_4, and $\{1,2,3,4\}$ to denote the elements of M_5. Using this notation, the isomorphism is

$$0 - 1 , \quad 1 - 2 , \quad 2 - 4 , \quad 3 - 3 .$$

The reader is now invited to rewrite $DFT(5)$ *using the lexicographical order of Z_4.* We will "massage" $DFT(5)$ still further. Let us denote

$$\sum_{i=0}^{4} a_i \text{ by } a'_0,$$

and $A_i - A_0$ by A'_i $(i = 1,2,3,4)$. We further denote $w^i - 1$ by u_i $(i=1,2,3,4)$. Using this notation and the lexicographical order of Z_4, $DFT(5)$ can be written as

$$\begin{pmatrix} A_0 \\ A'_1 \\ A'_2 \\ A'_4 \\ A'_3 \end{pmatrix} = \begin{pmatrix} 1 & 0 & 0 & 0 & 0 \\ 0 & u_1 & u_2 & u_4 & u_3 \\ 0 & u_2 & u_4 & u_3 & u_1 \\ 0 & u_4 & u_3 & u_1 & u_2 \\ 0 & u_3 & u_1 & u_2 & u_4 \end{pmatrix} \begin{pmatrix} a'_0 \\ a_1 \\ a_2 \\ a_4 \\ a_3 \end{pmatrix}$$

This shows very clearly the cyclic nature of $DFT(5)$. This means that A'_1, A'_2, A'_3, A'_4 are given by

$$A'_1 + A'_2 u + A'_4 u^2 + A'_3 u^3 =$$
$$(u_1 + u_2 u + u_4 u^2 + u_3 u^3)(a_1 + a_3 u + a_4 u^2 + a_2 u^3) \bmod u^4 - 1.$$

Having gone thus far with this example, let us push it a little further, and break up the product modulo $u^4 - 1$ into products modulo the irreducible factors of $u^4 - 1$. The Chinese Remainder Theorem suggests the following change of variables. We define b_1, b_2, b_3, b_4 by $b_1 = a_1 + a_3 + a_4 + a_2$, $b_2 = a_1 - a_3 + a_4 - a_2$, $b_3 = a_1 - a_4$, $b_4 = a_3 - a_2$. We also implicitly define B_1, B_2, B_3, B_4 by $A'_1 = B_1 + B_2 + B_3$, $A'_2 = B_1 - B_2 + B_4$, $A'_4 = B_1 + B_2 - B_3$, $A'_3 = B_1 - B_2 - B_4$. In terms of the new variables, $DFT(5)$ can be written as

$$\begin{pmatrix} A_0 \\ B_1 \\ B_2 \\ B_3 \\ B_4 \end{pmatrix} = \begin{pmatrix} 1 & 0 & 0 & 0 & 0 \\ 0 & \alpha & 0 & 0 & 0 \\ 0 & 0 & \beta & 0 & 0 \\ 0 & 0 & 0 & \gamma & -\delta \\ 0 & 0 & 0 & \delta & \gamma \end{pmatrix} \begin{pmatrix} \alpha'_0 \\ b_1 \\ b_2 \\ b_3 \\ b_4 \end{pmatrix}$$

where $\alpha = -3/2$, $\beta = 1/2(cos2\pi/5 - cos4\pi/5)$, $\gamma = j\ sin2\pi/5$, $\delta = j\ sin4\pi/5$. We thus reduced DFT(5) to computing $\alpha \cdot b_1$, $\beta \cdot b_2$, and the real and imaginary parts of $(\gamma + j\delta)(b_3 + jb_4)$. This last computation can, of course, be done by the algorithm derived in Section 2.

The details of the resulting algorithm can be found in (9). In the same reference the reader can find algorithms derived for other DFT(p)'s, as well as an extension of this method to DFT(p^k). For a leisurely exposition of the mathematical background, we refer the reader to (10).

This method of converting DFT(p), or even DFT(p^k), into cyclic convolutions can be extended to DFT(n) for arbitrary n. We will not pursue this approach here, and turn instead to another way of deriving algorithms for DFT(n), when n is not the power of a prime number. It was noted by Good (11) that under certain circumstances we can convert a one-dimensional DFT into a multi-dimensional DFT. If m and n are two relatively prime numbers, then an appropriate renumbering of the indices transforms DFT($m \cdot n$) into DFT($m \times n$) - the two-dimensional DFT of $m \times n$ points. We will not illustrate this transformation, (the interested reader is referred to (9) and (11)) but instead describe an algorithmic consequence of this transformation.

Even a cursory examination of $W_{m \times n}$ - the matrix of DFT($m \times n$) - reveals that it is the tensor product of W_m (the matrix of DFT(m)) by W_n (the matrix of DFT(n)). This means that, once we have algorithms for DFT(m) and DFT(n), we can use the tensor product construction to derive an algorithm for DFT($m \times n$), and therefore an algorithm for DFT($m \cdot n$). Now, every integer n is the product of powers of prime numbers, and consequently every DFT(n) can be transformed into a multi-dimensional DFT *having a power of a prime number of points in each dimension*. The tensor product construction, starting from the algorithms for DFT(p^k), yields an algorithm for DFT(n) no matter what n is. The reader who is interested in these algorithms is referred to (9).

In the preceding paragraph we saw the implication of the transformation of one-dimensional DFT into a multi-dimensional one. A curious reversal of the process occurs when we study DFT($p \times p \times ... \times p$) - the k-dimensional DFT of p points in each dimension. As usual, p denotes a prime number. A recent result of Auslandner, Feig and the author, using finite field construction, shows how to transform DFT($p \times p \times ... \times p$) into a cyclic convolution. When this cyclic convolution is further broken down to its irreducible components the following result is obtained: *The k-dimensional DFT($p \times ... \times p$) is equivalent to $s = (p^k - 1)/(p - 1)$ copies of DFT(p)*. This result shows that an algorithm for DFT($p \times ... \times p$) can be built up from s copies of an algorithm for DFT(p). In fact, an immediate consequence of Theorem 4 states: *If the algorithm for DFT(p) is multiplicatively minimal, then the resulting algorithm for DFT($p \times p \times ... \times p$) is multiplicatively minimal as well*. To our regret, we cannot point the reader to a reference for these results. As this paper is being written, the results about DFT($p \times ... \times p$) are also in the process of being written.

We started this paper with results of theoretical interest, it is only fitting that we will end it with results of theoretical interest. Up to now we discussed the construction of algorithms for computing the DFT. We now turn to the question of the multiplicative complexity of this computation. In doing so we must remember that multiplicative complexity does not take into account the number of additions, which may be large, nor does it take into account multiplications by a fixed rational number which does not depend on the data - it simply ignores this kind of operation. To emphasize that we are again within the theoretical framework, we will denote the multiplicative complexity of the DFT by μ(DFT). These results, which follow from Theorem 4 are summarized in the following theorem.

Theorem 8:   If $d(n)$ denotes the number of divisors of n, and p denotes an odd prime number, then:

1. $\mu(\text{DFT}(p)) = 2p - d(p\text{-}1) - 3$
2. $\mu(\text{DFT}(p^n)) = 2p^n - d(p-1)(n^2+n)/2 - n - 2$
3. $\mu(\text{DFT}(2^n)) = 2^{n+1} - n^2 - k - 1$
4. $\mu(\text{DFT}(p \times p \times ... \times p)) = 2p^k - (\sum_{i=0} p^i)\ (1+d(p-1)) - 2$

The technique used to prove Theorem 8 is strong enough to determine the multiplicative complexity of $\text{DFT}(n_1 \times n_2 \times ... \times n_k)$ which satisfies the following condition: If p^2 divides n_i for some i, then p, does not divide n_j for any $j \neq i$. The general expression for $\mu(\text{DFT}(n_1 \times n_2 \times ... \times n_k))$ involves a complicated number theoretic function of $n_1, n_2, ..., n_k$. That is the reason why it was not included in Theorem 8. The only comment we should make is that $\mu(\text{DFT}(n_1 \times n_2 \times ... \times n_k))$ is bounded from above by $2 \prod_{i=1} n_i$.

5.   Conclusions

The main purpose of these lectures was to illustrate the relationship between the theory and its uses. As was mentioned briefly in the Introduction, one should not take the results presented here as the final results. Rather the bulk of these lectures should be taken as an indication of the kind of effect theoretical investigation can have on practical use. There are many computations in signal processing which were not discussed in these lectures. Some of them are amenable to the kind of analysis mentioned here, others will necessitate the development of new theoretical tools. These lectures are meant to stimulate and challenge as much as to inform. Most practical problems were not discussed here, and the challenge to the user is to subject his or her calculations to the analysis illustrated above. It is my hope that the reader will find the results of this kind of analysis rewarding.

References

1.   S. Winograd, "*Arithmetic Complexity of Computations,* to appear as CBMS monograph, SIAM.

2.   S. Winograd, *Some bilinear forms whose multiplicative complexity depends on the field of constants,* Math. Systems Theory, Vol.10, 1977, pp. 169-180.

3. S. Winograd, *On multiplication in algebraic extension fields,* Theoretical Comp. Sci., Vol.8, 1979, pp.359-377.

4. S. Winograd, *On the multiplicative complexity of the discrete Fourier transform,* Advances in Math., Vol.32, 1979, pp.83-117.

5. L. Auslander and S. Winograd, *The multiplicative complexity of certain semi-linear systems defined by polynomials,* to appear.

6. S. Winograd, *On the complexity of symmetric filters,* Proc. of 1979 International Symp. on Circuits and Systems, pp. 262-265.

7. R. C. Agarwal and J. W. Cooley, *New algorithms for digital convolution,* IEEE Trans. Acoustics, Speech and Sign. Processing, Vol.25, 1977, pp. 392-410.

8. C. M. Rader, *Discrete Fourier transforms when the number of data samples is prime,* Proc. IEEE, Vol.5, 1968, pp.1107-1108.

9. S. Winograd, *On computing the discrete Fourier transform,* Math. of Comp., Vol.32, 1978, pp.175-199.

10. J. H. McClellan and C. M. Rader, *Number Theory in Digital Signal Processing,* Prentice Hall, Englewood Cliffs, N.J., 1979.

11. I. J. Good, *The interaction of algorithms and practical Fourier series,* J. Roy. Stat. Soc., Ser.B, Vol.20, 1958, pp.361-372; Addendum, Vol.22, 1960, pp.372-375.

CLUSTERING IN PATTERN RECOGNITION

E. DIDAY[*] G. GOVAERT[**] Y. LECHEVALLIER[***] J. SIDI[****]

[*]   Université Paris IX-DAUPHINE, Paris - INRIA, Le Chesnay
[**]   Université Paris XII, 94 Créteil   - INRIA, Le Chesnay
[***]   INRIA, Le Chesnay
[****] INRIA, Le Chesnay

Abstract :

We present first the main basic choices which are preliminary to any clustering and then the dynamic clustering method which gives a solution to a family of optimization problems related to those choices. We show then how these choices interfere in pattern recognition using three approaches : the syntactic approach, the logical approach and the numerical approach. For each approach we present a practical application.

J. C. Simon and R. M. Haralick (eds.), Digital Image Processing, 19–58.
Copyright © 1981 by D. Reidel Publishing Company.

1. INTRODUCTION

The clustering problems consist in finding and ordering groups of objects, using the relation defined by the variables which characterize those objects.

During the last years "clustering" has known great development in theory as well as in application. It is not possible to make a complete survey in this paper and we would advise the reader to study surveys such as the one by Cormach [11] or Dorofeyuk [19] and more recently the one by Diday and Simon [18] and Diday [15].

"Clustering" is often considered as a set of techniques whose sole merit is to give clusters. In fact looking further into those techniques it appears that :

1) they are all dependent of some basic choices.

2) they cover different problems, generally formulable in terms of criterion to optimize (see Diday [14]).

In the different studies recently performed in image processing see Simon [52], clustering methods appear at all the stages :
- on the "thresholding" level - In multispectral image analysis, at each pixel is associated one or several continuous measures, or more precisely a measure which can take a great number of different values (for example a scale of greys). An important stage is the transformation of this measure in a measure which can only take a very few number of values. This coding is obtained by cutting the set of initial values in clusters and by transforming each value of a cluster in a unique mean value. Rosenfeld [49] shows that a way to do this is to use clustering methods. Otsu Nubuyaki [44] goes further : "more properly, the thresholding is a kind of clustering in one dimensional scale of a grey level ...". Besides he propose an algorithm optimizing a criterion of the within group sum of squares. Pratt [46] approached also this problem and presents a clustering algorithm: "the recursive thresholding technique is really a special kind of recursive classification by clustering". The images obtained by the LANDSAT are often dealt with thay way : Swain [56] uses ISODATA methods [5], Roche [47] dynamic clustering and Lowitz [39] proposes a similar method in order to perform data reductions.

- on the segmentation level - One of the next stages is the search for zones which are geographically close and close in frequence. Coleman [10] defines region segmentation as a problem of non supervised clustering. He proposes a method of segmentation

using the K-mean algorithm [41]. Do-Tu and Installe [20] use
ISODATA method to segmentalize on LANDSAT data. Fukada [26],
Kasvand [34], Backer [4] propose their own clustering algorithm.
Yokoya et al [61] construct a partition of their image,
depending on two parameters and show that in making those
parameters vary they obtain a hierarchy of clusters. We can also
quote the use of clustering for segmentazing contours : Haralick
and Shapiro [31] for the decomposition of polygonal patterns,
Charles and Lechevallier [8] for geographical maps.

- at higher levels - Fu et al [23-25,40], Lechevallier and
Suen [38] use clustering for character recognition. Those two
examples will be explicited later on.

The main characteristics often required from clustering methods
in image processing are :
- fast computing
- taking in account connexity constraints
- the possibility of using supplementary information often very
  important in image processing.

1.1. Basic choices

Before giving some details, let us give a list of the
main basic choices.
  1)   choice of objects
  2)   choice of variables
  3)   choice of the encoding of the variables
  4)   choice of the measure of similarity
  5)   choice of the structure of class
  6)   choice of the structure of representation
  7)   choice of the function of representation
  8)   choice of the function of allocation
  9)   choice of a criterion to validate the structure of
       class obtained
  10) choice of an algorithm.

After we have defined the set of objects E, we have to
choose the variables which characterize them, discrete or
continuous, quantitative or qualitative, ordinal or nominal;
next, there is the problem of encoding these variables and
defining a measure of similarity between objects. Various authors
have studied the problems related to these choices.

The problem of the determination of the most classifying
variables has been considered by Aivazian [2]. Once the sets of
objects and variables are well defined, another important
problem is how to encode the variables; for instance how to
transform a quantitative variable into a qualitative one, (see
Anderberg [3], Lechevallier [37]). The problem of an adapting

encoding has been studied by Taleng in [57].

A list of several measures of similarity may be found in Diday and Simon [18]: the problem of adapting the measure of similarity to the form of the clusters to determine, has also been studied by several authors (Rohlf [48], Chernoff[19], Diday and Govaert [17]).

After having chosen the set of objects, the variables, the encoding and the measures of ressemblance, the next basic choices are : the structure of classes S and the structure of representation. Let us give the definition of these two notions.

Définition of the interclass structure

A set E is provided with a structure of class : if we associate to it a set S of which each element, s, is a set of subsets of E, satisfying the following properties :

1) $\forall s = \{P_1, \ldots, P_k\} \in S$ we have : $\underset{i}{U} P_i \equiv E$

2) $\exists s \in S : E \in s$

3) $\exists s \in S : \forall e \in E \quad \{e\} \in s$

Many kinds of such a structure may be imagined, the most employed at present are :

$S = \mathbb{P}$ the set of partitions of E

$S = H$ the set of hierarchies (i.e. $h \in H$ verifies the following properties : $E \in H$; $E \in h$; $h_1, h_2 \in h$ $\Rightarrow \{h_1 \cap h_2 = \emptyset, h_1 \subset h_2 \text{ or } h_2 \subset h_1\}$).

$S = \theta$ the set of overlapping clusters of E.

An example of non classical interclass structure

S is a set in which each element $s = \{P_1, \ldots, P_k\}$ admits n groups of ordered parts of E : $s_1, \ldots, s_n$ which satisfies the following properties :

$S_1 = \{E\}$

$\left.\begin{array}{l} A \in s_i, B \in s_j \\ \\ i > j \end{array}\right\} \Rightarrow \left\{\begin{array}{l} A \cap B = \emptyset \\ \quad \text{or} \\ A \subset B \end{array}\right.$

$\exists \ell : \forall e \in E \quad \{e\} \in s_\ell .$

We deduce from this definition that :

. S satisfies the structure of class conditions 1), 2), 3).

. The clusters of a same level P_ℓ, P_k can be such that $P_\ell \cap P_k \neq \emptyset$ without necessarily $P_\ell \subset P_k$ or $P_k \subset P_\ell$ (contrary to what happens on a hierarchy). Let us call "ciel" each element of S.

The visual representation of a structure of class is a picture whose rules of visual interpretation satisfy ("at its best") the structure properties (see figure 1).

The space of representation

Definition : We say that E is provided with a representation structure if we associate it with a set $\mathbb{L}$ and an application D of $E \times \mathbb{L}$ in R^+

- $\mathbb{L}$ is called the space of representation

- each element of $\mathbb{L}$ is a representation

- if $(x,\ell) \in E \times \mathbb{L}$, then $D(x,\ell)$ is the measure of similarity of an object x to the representation ℓ.

Remark : We shall sometimes call (with too free a use or language but with no possible ambiguousness) "representation of the cover" $s = (s_1, \ldots, s_k) \in S$ all k-uple representations $L = (L_1, \ldots, L_k)$ obtained by associating a representation $L_i \in \mathbb{L}$ to each cluster s_i of the cover. We denote by L the set of those k-uples.

Examples of clusters representations :

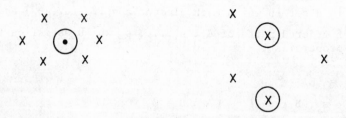

Center of mass $\mathbb{L} = \mathbb{R}^p$ The two points nearest to the centre of mass $\mathbb{L} = \mathbb{R}$

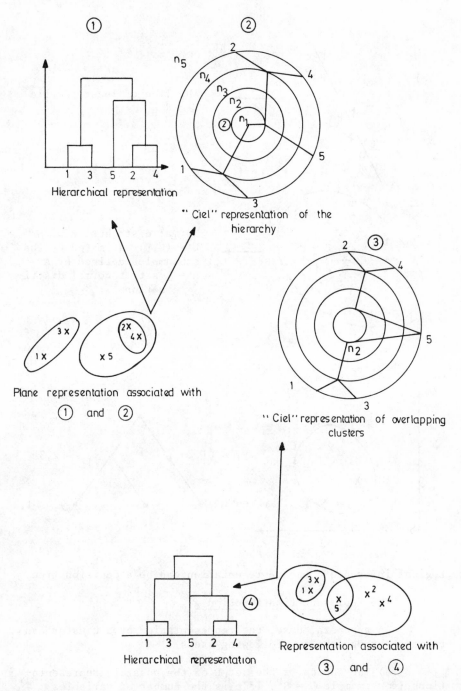

Figure 1

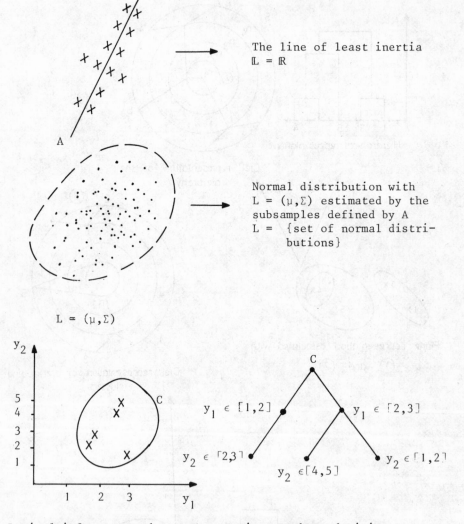

The line of least inertia
$\mathbb{L} = \mathbb{R}$

Normal distribution with
$L = (\mu, \Sigma)$ estimated by the
subsamples defined by A
$L = \{$set of normal distri-
butions$\}$

$L = (\mu, \Sigma)$

Logical inference : the representation can be a decision tree.

The representation function g

As a general case, the representation of a cluster can be made through the two following steps :

a) The choice of the set $\mathbb{L}$ of the possible representations (for example $\mathbb{L} = \mathbb{R}^p$, if p is the number of variables characterizing the objects).

b) The choice of a map g called "representation function": $\mathbb{P}_k \to \mathbb{L}$ which associates to each cluster of objects a unique representation (for example g(A) = G, the center of gravity of the set of objects of cluster A, considered as points in $\mathbb{R}^p$).

The allocation function f

Several clusters of objects being constituted, we then have to define a way to allocate an object to one of these clusters. This can be done in two steps :

a) The choice of a measure of consonnance D(x,a) between an object x and the representation of a cluster A.

b) The choice of an allocation function in one of the clusters.

These two steps can be summarized in the map f : $\mathbb{L}_k \to \mathbb{P}_k$ which associates to k representations L = $(L_1, \ldots, L_k)$ a partition P = $(P_1, \ldots, P_k)$, that is f(L) = P. The map f may be defined in the following manner :

$P_i = \{x \in E \mid D(x,L_i) \leq D(x,L_j)\}$ and in case of equality, x is allocated to the cluster of smaller index.

1.2. - The dynamic clustering method

The major steps of this method are the following :

1) Choice of a structure of class S.

2) Choice of a space of representation $\mathcal{L}$.

3) Choice of a criterion W which measures the fit between s $\in$ S and its representation L $\in \mathcal{L}$.

4) Definition of an optimization problem of the following type :
   Min W(s,L) where S' $\subset$ S, $\mathcal{L}' \subset \mathcal{L}$.
   s∈S'
   L∈$\mathcal{L}$'

5) Definition of an algorithm based on the choice of two functions g and f, characterizing the representation and allocation processes. If g and f are properly chosen, it can be proved that the criterion W improves at each iteration and converges to a stable position.

The main interest of the dynamic clustering method consists in :

1) Formulating classification problems in terms of optimizing the fit between a "class structure" and a "representation".

2) Defining a frame in which simple algorithms may be used to solve these problems.

3) It allows the unification of a large variety of problems as well as it gives algorithms leading to a (locally optimal) solution.

Let us now give several examples which illustrate this approach.

2. THE SYNTACTIC APPROACH

2.1. The problem

- in pattern recognition

Often in image processing, the first stage is extracting elementary patterns called primitives. The image is then represented by a more or less complicated gathering called string.

We can quote two examples :

1) Primitive extraction using borders or skeleton in the analysis of bubble chamber-photographs, in the analysis of chromosomes, in identification of finger-prints ...

2) Primitive extraction using the region notion (Pavlidis [45], Rosenfeld and Strong [50], Brice and Fennema [7]).

The next step is the analysis of those strings. The pattern recognition problem is then dealt in two stages :

1) Constructions of grammars (grammatical inference) on samples,

2) Construction of syntactic parsers associated to those grammars which will then allow accepting or rejecting of a new pattern.

- in clustering

Having a set of strings the aim is to form clusters of "homogeneous" strings; this problem has already been approached :

- by Fu et al [23-25,40] : they define distances between a
string and a grammar or between a string and a string, then
they define clustering algorithms using the method of assigning
to the nearest neighbor.

- by Miclet and Guerin : multiple inference without teacher [43]:
they propose a method allowing to find a covering of clusters
and a set of grammars associated with each cluster such as the
number of confusion (a string of class i recognized by a
grammar of class j; j ≠ i) will be minimum.

2.2. The similarity measures used

2.2.1. Definition of pattern grammar

Definition 1 : A grammar is a 4-uple
$G = (V_N, V_T, P, S)$ where V_N is a set of non terminal subpatterns,
V_T a set of terminal pattern primitives, $S \in V_N$ is the start
symbol and P a set of production rules of the type :

$$\alpha \to \beta \quad \text{where } \alpha \in (V_N \cup V_T)^* \times V_N \times (V_N \cup V_T)^*$$

$$\beta \in (V_N \cup V_T)^*$$

$$(V^* = V \cup \{\emptyset\} \, \forall V)$$

Remark : According to the form taken by the set
of P rules, the grammars will be "context-sensitive", "context-
free", regular, ... We shall note L(G) the set of strings which
can be generated by the grammar G.

2.2.2. Distance between two strings

For more details, see Aho and Peterson [1] and
Fu and al [23-25,40].

Let H be a set of strings and $H^* = H \cup \{\emptyset\}$. Three
transformations are defined on H :

Definition 2 : Let w_1 and $w_2 \in H^*$

substituting transformation : $\forall$ a and b $\in$ H $w_1 \, a \, w_2 \xrightarrow{T_s} w_1 \, b \, w_2$

deleting transformation : $\forall$ a $\in$ H $\quad w_1 \, a \, w_2 \xrightarrow{T_d} w_1 \, w_2$

inserting transformation : $\forall$ a $\in$ H $\quad w_1 \, w_2 \xrightarrow{T_i} w_1 \, a \, w_2$

We can now define the following distance :

Definition 3 : The distance between two strings
x, y $\in$ H* is the smaller number of needed transformations to go
from x to y.

It is also called Levenshtein distance. It can be
easily shown that the distances axioms are verified. Several
weighted distances can be defined either by assigning to the
three types of transformation the weights α_s, α_d, α_i or like Fu
et al [23-25,40] proposed,by taking weights which depend on the
elements of H which occur in the transformation $\alpha_s(a,b)$, $\alpha_d(a)$
and $\alpha_i(a)$. Let J be a sequence of transformations needed to go
from x to y, let $|J|$ be the sum of the weights associated,
then

$$d(x,y) = \min_{J} |J|.$$

An algorithm based on Wagner and Fisher's [59]
allows the calculation of such a distance.

2.2.3. Distance between a string and a grammar

First formulation : Let G be a grammar and x a
string, d the distance between two strings (definition 3). We
define :

$$D(x,G) = \min_{y \in L(G)} d(x,y)$$

Second formulation : We can show that it is
equivalent to finding the minimum number of transformation type
production rules to add to the grammar G so that x belongs to
the language of that grammar. To find those rules, there exist
parsers which are called "minimum-distance error-correcting
parsers".

2.3. A few clustering methods using dynamic clustering

2.3.1. The representation associated to each cluster is
       a grammar

Let V_T be a set of primitives, E is a finite set
of strings defined on V_T, G a set of grammars defined on V_T, D
a distance $E \times G$, $\mathbb{P}_k$ the set of partitions of E in k classes
and $\mathbb{L}_k = G^k$ the space of representation.

We shall note P = $(P_1,\ldots, P_k)$ an element of $\mathbb{P}_k$
        and   L = $(L_1,\ldots, L_k)$ an element of $\mathbb{L}_k$.

We define the two following functions :

- <u>representation function g</u> : $\mathbb{P}_k \xrightarrow{g} \mathbb{L}_k$

  $L = (G_1,\ldots, G_k) = g(P)$ is defined by

  $$\sum_{x \in P_i} D(x,G_i) = \min_{G \in G} \sum_{x \in P_i} D(x,G) \qquad \text{for all } i = 1,\ldots, k$$

- <u>assignment function f</u> : $\mathbb{L}_k \xrightarrow{f} \mathbb{P}_k$

  $P = f(L)$ is defined by

  $$P_i = \{x \in E \ / \ D(x,G_i) \leq D(x,G_j) \quad \forall j \neq i\}$$

  We define then the algorithm in the usual manner : a sequence $(P^{(n)}, L^{(n)})$ of $\mathbb{P}_k \times \mathbb{L}_k$ is constructed from a starting element $(P^{(1)}, L^{(1)})$ applying the two functions f and g.

  We can show the following proposition :

  <u>Proposition 1</u> : *The sequence* $(P^{(n)}, L^{(n)})$ *converges through a decrease of the following criterion* w :

  $$w(P,L) = \sum_{i=1}^{k} \sum_{x \in P_i} D(x,G_i)$$

  <u>The problems</u> : They lie mainly in building the representation function. The set G must be very limited. As a matter of fact, if we can associate to each cluster a grammar infered by the elements of the cluster the methods converge immediately starting from any initial partition.

  2.3.2. <u>The representation associated to each cluster is a string</u>

- We keep V_T, E, $\mathbb{P}_k$.

- $\mathbb{L}_k = \mathbb{L} \times \ldots \times \mathbb{L}$ where $\mathbb{L}$ is a finite set of strings defined over V_T (i.e. E itself).

- d a distance over the set of strings built from V_T.

- <u>representation function g</u> : $\mathbb{P}_k \rightarrow \mathbb{L}_k$

  $L = (\lambda_1,\ldots, \lambda_k) = g(P)$ is defined by

  $$\text{for all } i = 1,\ldots, k \quad \sum_{x \in P_i} d(\lambda_i,x) = \min_{\lambda \in \mathbb{L}} \sum_{x \in P_i} d(\lambda,x)$$

- the assignment function f and the algorithm are then defined as
above.

         We can show the two following propositions :

         Proposition 2 : The sequence $(P^{(n)}, L^{(n)})$
converges through a decrease of the criterion :
$$w(P,L) = \sum_{i=1}^{k} \sum_{x \in P_i} d(\lambda_i, x)$$

         The optimized criterion measures the number of
transformations separating each string of E from a "mean" string
associated to each cluster.

         Proposition 3 :
$$\sum_{i=1}^{k} \sum_{x,y \in P_i} d(x,y) \leq 2 \, w(P, g(P))$$
         The method therefore aims for clusters such as
the sum of distances between its strings be minimum.

   2.4. - Clustering methods suggested by Fu et al [23-25,40]
          and one illustrative example

         One of the methods proposed uses the distance between
a grammar and a string. We can summarize this method : at each
stage a supplementary string is classified :

- either in one of the already existing clusters if the distance
  between the grammar associated to this cluster and the string
  is the smallest and moreover if it is less than a threshold t.

- either in a new cluster if it is not the case.

         The grammar of the cluster thus modified or created is
then recalculated (inference). We shall then start again until
the last string. A partition of strings, an inferred grammar for
each cluster, will be available at the end.

- Another method uses distances between strings. A distance
  between cluster of strings and a string is then associated by
  taking the sum of the distances from the k-nearest neighbor
  of the cluster. The method is then the same as the former.

- An example of application
  The above two methods have been used over a set of 51 alphabetic
  characters (cf. figure 2). After a first treatment each
  character is coded by a string (cf. figure 2). The set to

cluster consists of the set of the 51 strings thus obtained.

The results obtained are quite stable and near ideal clustering. Computing time is very long and it seems difficult for the moment to use those methods on very large data sets.

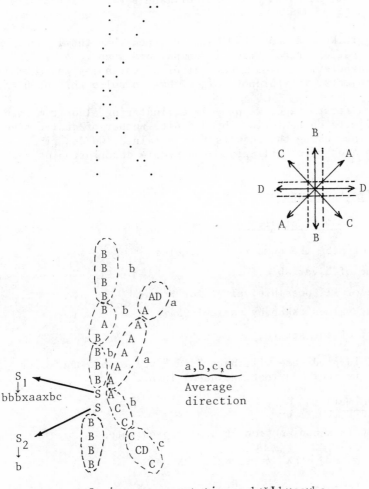

String representation : b+bbb×aa×bc

Primitive extraction of character K

Figure 2

3. THE LOGICAL APPROACH

3.1. Introduction

The problem we are interested in has been defined by Hayes-Roth [32]: "what kind of logical descriptions actually underlie cognitive processes and how they can be inferred directly from examples", and formalized by Simon et al [51].

Several approaches have been proposed. In particular : Haralick's [30], Hayes-Roth's [32], Ullmann's [58], Michalski's [42], Sidi's [53].

Zagoruiko and Lbov [62] have outlined that the logical approach in pattern recognition is compulsory when we deal with problems characterized by a great number of variables and a small number of objects. They propose algorithms to solve this problem.

In a first stage, we propose a clustering algorithm for a set of objects characterized by a finite number of qualitative variables, then using the results thus obtained, we describe a logical operator which will give us a simple description of each cluster.

3.2. Definitions and notations

Atomic predicate $[X^j \# A^j]$

Let p be the number of variables

X^j the name of a variable

M^j the set of all possible values for X^j

A^j a set of values taken by variable $X^j (A^j \subseteq M^j)$

$\#$ a symbol of the set $\{\subset, \not\subset\}$

This predicate will have "true" value when the value taken by X^j is included (not included) in A^j.

- formula :

It is a conjunction of atomic predicates :

$$[X^1 \# A^1] \wedge [X^2 \# A^2] \wedge \ldots \wedge [X^p \# A^p].$$

A formula will take "true" value when all the atomic predicates have the "true" value.

- We note F the set of all possible formulas

$F \in F$ a formula

$F^j \in F$ an atomic predicate over variable X^j.

- Elementary formula

It is an instantiation of a formula that is to say for all $j \in [1,p]$:

$[X^j = x^j]$ where x^j is a value belonging to M^j.

- Each object x of data set E which is to be clustered will be described by formula.

Remark : doing this allows us to deal with a data array characterized by a different number of variables for each object. That is to say that for some variables we will have $A^J = \{\emptyset\}$.

- We note
$P(E)$: the power set of E.
$\mathbb{P}_k$: the set of partitions of E in k clusters.

$\mathbb{L}'$: the set of non empty subsets of E $\mathbb{L}' = P(E) - \{\emptyset\}$

$\mathbb{L}'_k$: the set of k- tuples$(\lambda_1,\ldots, \lambda_k)$ where $\lambda_i \in \mathbb{L}'$.

D : a similarity measure precised later on.

$\mathbb{A}_k$: the set of k- tuples$(d_1,\ldots, d_k)$ $d_i \in D$

$\mathbb{L}_k$: the set of k- tuples couples : $\mathbb{L}_k = \mathbb{L}'_k \times \mathbb{A}_k$

3.3. - Dynamic clustering methods using adaptive distances

We use here an algorithm described by Govaert in [27].

3.3.1. The criterion

We introduce here the notion of formula F_1 included in Formula F_2 : it is expressed by the relation

$$F_1 \Rightarrow F_2$$

We shall say that $F_1 \Rightarrow F_2$ if for all $j \in [1,p]$, $F_1^j \Rightarrow F_2^j$. If F_1^j is expressed $[X^j = A^j]$ and F_2^j is expressed $[X^j = B^j]$, we shall measure the notion of F_1^j included in F_2^j by the measure d

$d\ (F_1^j,\ F_2^j) = 0$ if $A^j \subseteq B^j <=> F_1^j => F_2^j$

$d\ (F_1^j,\ F_2^j) \in\ ]0,1] \subset R^+$ otherwise

We note that if for F_1 the variable $\dot{X}^j$ is not relevant then $d\ (F_1^j,\ F_2^j) = 0$.

Let $\mathcal{D}$ be the set of similarity measures defined as follows.

Let $P = (P_1,\ldots,\ P_k) \in \mathbb{P}_k$

$x = (x^1,\ldots,\ x^p) \in E$

$y = (y^1,\ldots,\ y^p) \in E$

$c = (c^1,\ldots,\ c^p)$ the a priori costs of each variable

$\forall \ell \quad d_\ell \in \mathcal{D}$

$\Delta = (d_1,\ldots,\ d_k) \in \mathbb{A}_k$

$$d_\ell(x,y) = \sum_{j=1}^{p} \delta_\ell^j\ d\ (x^j,y^i)c^j$$

with $\delta_\ell^j \in \mathbb{R}^+$ and $\prod_{i=1}^{p} \delta_\ell^i = 1$

δ_ℓ^j represents the valuation of the jth variable in cluster ℓ. At the convergence of the algorithm a high value of δ_ℓ^j will correspond to a variable whose atomic predicate F^j will be important for the recognition of an object to that cluster ℓ. This weight defines our adaptive distance.

We define the following functions :

$D : E \times \mathbb{L} \rightarrow \mathbb{R}^+$

$x \in E,\ \lambda \in \mathbb{L}',\ d_\ell \in \mathcal{D}$ (i.e. $(\lambda,d_\ell) \in \mathbb{L}$)

$$D(x,(\lambda,d_\ell)) = \frac{1}{card(\lambda)} \sum_{y \in \lambda} d_\ell(x,y)$$

$I : P(E) \times \mathbb{L} \rightarrow \mathbb{R}^+$

$B \in P(E),\ \lambda \in \mathbb{L}',\ d_\ell \in \mathcal{D}$ (i.e. $(\lambda,d_\ell) \in \mathbb{L}$)

$$I(B,(\lambda,d_\ell)) = \sum_{x \in B} D(x,(\lambda,d_\ell))$$

Let $L = (L_1, \ldots, L_k) \in \mathbb{L}_k$

$\quad L' = (L'_1, \ldots, L'_k) \in \mathbb{L}'_k$

$\qquad$ The criterion to optimize is written as follows :

$$w(P,L) = w(P,(L',\Delta)) = \sum_{\ell=1}^{k} I(P_\ell(\lambda_\ell, d_\ell))$$

$$= \sum_{\ell=1}^{k} \sum_{x \in P_\ell} \frac{1}{\text{card}(\lambda_\ell)} \sum_{y \in \lambda_\ell} \sum_{j=1}^{p} \delta_\ell^j \, d\,(x^j, y^j) c^j$$

3.3.2. The optimization problem

$\qquad$ It lies in searching for the elements P^* of $\mathbb{P}_k$, $L^* = (L'^*, \Delta^*)$ of $\quad \mathbb{L}_k = \mathbb{L}'_k \times \Delta_k \quad$ such that :

$$w(P^*, (L'^*, \Delta^*)) = w(P^*, L^*)$$

$$= \min_{(P,L) \in P_k \times L_k} w(P,L)$$

$\qquad$ In that way, we are trying to gather elements which have in common the most atomic predicates in such a way that the representation of each cluster summarizes at its best the cluster. The adaptive distance by valuating the logical expression allows us to get out of the logical field, which is a necessary step (see Watanabe [60]) in order to obtain a good representation.

3.3.3. The algorithm

$\qquad$ It uses the allocation function f and a representation function g :

- the allocation function f

$\quad f : \mathbb{L}_k \rightarrow \mathbb{P}_k$ (i.e.) $\mathbb{L}'_k \times \Delta_k \rightarrow \mathbb{P}_k$

$\qquad f(L', \Delta) = P = (P_1, \ldots, P_k)$

where $P_\ell = \{x \in E \ / D(x, (\lambda_\ell, d_\ell)) \leq D(x, (\lambda_i, d_i)) \quad \forall \, i \in \,]0, k]\}$

- the representation function g is decomposed into two functions g_1 and g_2

$\quad$. Function $g_1 : \Delta_k \times \mathbb{P}_k \rightarrow \mathbb{L}'_k$

$\qquad g_1(\Delta, P) = L' = (\lambda_1, \ldots, \lambda_k)$

where λ_ℓ is defined by :

$$I(P_\ell,(\lambda_\ell,d_\ell)) = \min_{\lambda \in \mathbb{L}'} I(P_\ell,(\lambda,d_\ell))$$

by compelling card (λ) to be maximum (i.e. we want λ to be the set of all possible elements of E that minimize $I(P_\ell,(\lambda, d_\ell))$) we obtain :

$$\lambda_\ell = \{y \mid y \in E \text{ and } I(P_\ell,\{y\}, d_\ell)) = \min_{z \in E} I(P_\ell, (\{z\}, d_\ell))\}$$

. Function $g_2 : \mathbb{L}'_k \times \mathbb{P}_k \to \Delta_k$

$$g_2(L',P) = \Delta = (d_1,\ldots, d_k)$$

where $d_\ell \in \mathcal{D}$ is defined by :

$$I(P_\ell,(\lambda_\ell,d_\ell)) = \min_{d \in \mathcal{D}} I(P_\ell,(\lambda_\ell,d))$$

with $I(P_\ell,(\lambda_\ell,d_\ell)) = \sum_{j=1}^{P} \delta_\ell^i (\sum_{x \in P_\ell} \frac{1}{\text{card}(\lambda_\ell)} \sum_{y \in \lambda_\ell} d(x^j,y^j)c^j)$

Letting ϕ_ℓ^j be the expression in parenthesis, we obtain :

$$\delta_\ell^j = \frac{(\prod_{h=1}^{P} \phi_\ell^h)^{1/P}}{\phi_\ell^j}$$

The algorithm consists starting from an initial solution $\nu_o = (P^{(o)}, L^{(o)}) = (P^{(o)}, (L'^{(o)}, \Delta^{(o)}))$ in computing the different terms of the sequence $\nu_n = (P^{(n)}, L^{(n)}) \in \mathbb{P}_k \times \mathbb{L}_k$ up to the convergence which will occur when $w(\nu_n) = w(\nu_{n+1})$. This is proved by Diday in [16].

3.4. Describing the clusters

We are concerned here in associating to a cluster P_ℓ a set of formulas $F_\ell \subset F$ of minimum cardinality for which whatever $x \in P_\ell$ and $y \in E-P_\ell$ there exists a formula $F \in F_\ell$ which will be true for x, and there does not exist a formula $F \in F_\ell$ true for y.

The algorithm we present here under gives a solution to this problem.

We must first introduce supplementary notations and definitions.

3.4.1. Notations and definitions

. E   the set of given data

. $\mathcal{E}$ the set of all elementary formulas ($\mathcal{E} \subset F$)

. m^j the number of possible values for variable X^j

. $mf(F^j)$ = let $F^j = [X^j \neq A^j]$

   then $mf(F^j)$ = card (A^j)

. density of a formula : $F \to R^+$

   density $(F) = \dfrac{\text{number of objects of E for which F is "true"}}{\text{number of objects of } \mathcal{E} \text{ for which F is "true"}}$

. Complexity of a formula : $F \to \mathbb{R}^+$

   $$\text{complexity } (F) = \sum_{j=1}^{P} \text{Complexity } (F^j)$$

with Complexity $(F^j) \in [0,1]$
for more details see Sidi [53]

. Measure of similarity :

In the case of clusters obtained by the algorithm described in § 3.3.3. we shall use the adaptive distance obtained at convergence.

. Internal union : $\uplus$

   $F = F_1 \uplus F_2$

$\forall j$ we have $F^j = F_1^j \uplus F_2^j$

if $F_1^j = [X^j \subset A_1^j]$ and $F_2^j = [X^j \subset A_2^j]$

we define $F^j = [X^j \subset A]$ by $A = [A_1^j \cup A_2^j]$

3.4.2. The algorithm

We use an agglomerative type of algorithm as we wish to agglomerate as much as possible objects of the same cluster.

Starting from an initial formula $F_\ell^{(o)}$, we define in an iterative way $F_\ell^{(n)} = F_\ell^{(n-1)} \cup x$, $x \in P_\ell$. The x will be chosen in such a way as to optimize the following criterion taken in the lexicographic order (see Annexe 1) :

distance (F_ℓ, x) to be minimized

density ($F_\ell \cup$ x) to be maximized

complexity ($F_\ell \cup$ x) to be minimized

with the constraint $y \not\Rightarrow (F_\ell \cup x)$ $\forall y \in E-P_\ell$.

We thus obtain for each cluster a set of formulas describing the cluster as precisely as possible.

However in pattern recognition, it is useful to have a recognition operator which has a minimum number of elementary predicates.

To obtain this for each $F^j = [X^j \subset A^j]$, we have to increase the cardinality of A^j so as to minimize complexity as much as possible while respecting the constraint.

3.5. Applications

As an artificial example we tested our algorithm on the capital and small letters printed by our computer. There are 9 ordinal variables which are the number of stars on each line starting from the bottom.

After clustering on two classes we obtain for w=47 capital letters on class 1 and small letters in class 2, we obtained the following results :

2 formulas where needed by each class for characterization and for recognition operators this is due to the way the variables where choosen. The main problem being the small letters bdh beeing very high.

In annexe 2 we presents it data and the results.

This method has produced good results on data involving between 24 and 68 variables and 30 to 40 objects. One formula was sufficient to characterize each cluster.

4. NUMERICAL APPROACH

4.1. Introduction

In this chapter are described the different means of introducing various constraints in clustering problems encountered in pattern recognition. We introduce three types of constraints.

a) constraints on the maximum size of a cluster (§ 4.3.)

b) a priori classification of a subset of E (§ 4.4.)

c) spatial proximity constraints between points of E (§ 4.5.).

The use of classical clustering algorithms is often a critical problem (see Dubes, Jain in [21]).

It is necessary to introduce in classical clustering algorithms constraints which allow to respect the structure of the patterns which are to be classified. For example in the case of spatial proximity constraints, see Jarvis [33], Gowda [29] and De Falguerolles [12].

4.2. The problem

In the family of problems optimizing the representation, the introduction of the set of constraints C over the set E is done at the level of the structure of class S (§ 4.3.) and the space of representation $\mathcal{L}$ of that structure of class (§ 4.4.). This determines a subset S' of the structure of class S consistent with the set of constraints C, and a new space of representation $\mathcal{L}' \subset \mathcal{L}$ also consistent with the problem constraints. Thus the problem is reduced in finding the couple (s^*, L^*) of

$S' \times \mathcal{L}'$ such as

$$w(s^*, L^*) = \min_{\substack{s \in S' \subset S \\ L \in \mathcal{L}' \subset \mathcal{L}}} w(s,L)$$

Another possibility of introducing constraints in this optimization problem is to express the constraints in the criterion (§ 4.5).

In this section the structure of class chosen is the structure of partitionning, that is to say that S is the set of all the possible partitions of E. In this case, generally the optimum solution is trivial place, it is the finest partition of E; so that the first constraint is to define a maximum number k of clusters. The set $\mathbb{P}_k$ of partitions of E in k clusters being now chosen, we further assume that the criterion w over $S \times \mathscr{L}$ decomposed so :

$$w(P,L) = \sum_{\ell=1}^{k} D(P_\ell, L_\ell)$$

with $P \in \mathbb{P}_k \subset S$

and $L_\ell \in \mathbb{L}_k \ \forall \ell = 1,\ldots, k$ in other ways :

$$\mathscr{L} \equiv \mathbb{L}_k = \mathbb{L} \times \mathbb{L} \times \ldots \times \mathbb{L} \quad k \text{ times}$$

and $P = (P_1,\ldots, P_k)$ with $P_i \in P(E)$

thereby the admissible constraints in this optimization problem are the constraints which determine the following two non-empty subsets $\mathscr{L}'$ and S' : $\mathscr{L}' \subset \mathscr{L}$ and $S' \subset \mathbb{P}_k \subset S$.

4.3. <u>Introducing constraints on the structure of class</u>

4.3.1. <u>Constraints on the size of classes</u>

The constraints on the size of classes determine a subset $S' = S_\theta$ of $\mathbb{P}_k$ which can be for example :

$$S_\theta = \{P \in \mathbb{P}_k \ / \ \forall i = 1,\ldots, k \quad card(P_i) \leq \theta\}$$

this set is the subset of $\mathbb{P}_k$ consistent with the size constraint θ.

4.3.2. <u>Neighbourhood constraints</u>

It can happen that the user wishes to assign to some elements of E either to be in the same cluster, or in two separate clusters.

Those constraints lead the association of E to two graphs $u = (E, \Gamma)$ and $u' = (E, \Gamma')$ whose vertices are elements of E and :

$\Gamma = \{(x,y) \in E \times E \: / \: x \text{ and } y \text{ must be in the same cluster}\}$

we note $\gamma(x,y) = 1$ if $(x,y) \in \Gamma$
$\gamma(x,y) = 0$ otherwise

and

$\Gamma' = \{(x,y) \in E \times E \: / \: x \text{ and } y \text{ must be in two separate clusters}\}$

we note $\gamma'(x,y) = 1$ if $(x,y) \in \Gamma'$
$\gamma'(x,y) = 0$ otherwise

We can define a subset $S' \equiv S_c$ of $\mathbb{P}_k$

$S_c = \{P \in \mathbb{P}_k \quad \forall i = 1, \ldots, k, \: \forall x \in P_i, \: \forall y \notin P_i \: \gamma(x,y) = 0$

and $\forall x \in P_i, \: \forall y \in P_i \: \gamma'(x,y) = 0\}$

4.4. Introducing constraints in the space of representation

4.4.1. Example 1

In some clustering problems one wishes to associate to each cluster of the partition a subset of $P(E)$ of fixed cardinal as space of representation. This choice is often compulsory (see [36]) because without this size constraint the representation associated to a cluster is reduced to an unique element of E.

In defining as space of representation a subset $\mathbb{L}_\alpha$ of $P(E)$ such as (see [13]) :

$\mathbb{L}_\alpha = \{L \in \mathbb{L} \: / \: card(L) = \alpha\}$

The space of representation associated to the structure of class is $\mathcal{L}' = \mathbb{L}_\alpha \times \ldots \times L_\alpha \quad k \text{ times}$

4.4.2. Example 2

In this case the constraints may be the incomplete or full knowledge of an a priori classification of E in k clusters $C_1, \ldots, C_k$ verifying

$$C_i \in P(E) \quad C_i \cap C_j = \emptyset \text{ if } i \neq j$$

$$\text{and} \quad \bigcup_{i=1}^{k} C_i \subseteq E$$

Here the space of representation on $L' = \mathbb{L}^1 \times \ldots \times \mathbb{L}^k$ is so that :

$$\mathbb{L}^j = \{A \in P(E) \,/\, \forall a \in A, \, a \in C_j \text{ and } Card(A) = \alpha_j\}$$

In these two examples, the optimization problem is to find the couple (s^*, L^*) of $\mathbb{P}_k \times L$ so that :

$$w(s^*, L^*) = \min_{\substack{s \in \mathbb{P}_k \subset S \\ L \in \mathscr{L}' \subset \mathscr{L}}} w(s, L)$$

4.5. Introducing constraints in the criterion

4.5.1. The problem

Another way of introducing constraints is to associate to the set E a graph characterizing the constraints.

Let d be a distance over $E \times E$. We assume that :

$$0 \leq d(x,y) \leq 1 \qquad \forall x, y \in E$$

Let U be a set of edges.

Let $G = (E,U)$ be a graph whose incidence table is q
with $q(x,y) = 1$ if $(x,y) \in U$
$q(x,y) = 0$ otherwise.

This graph $G = (E,U)$ represents spatial neighbourhood of element of E (see [29]). Thus to two elements x and y of E spatially near is associated the value $q(x,y) = 1$ likewise to two elements x and z of E spatially far away is associated the value $q(x,z) = 0$.

The coherence of a cluster C of E is usually measured by the value of

$$\sum_{(x,y) \in C \times C} d(x,y)$$

The introduction of spatial constraints compels us to redefine the coherence of a cluster. It is now measured by :

$$\sum_{(x,y)\in C\times C} [1 - q(x,y)]\, d(x,y)$$

Thus the cluster C is penalized of the value $2 \times d(x,y)$ if the two elements x and y of C are not spatially near. If the notion of nearness is not symetric, three valuations of the coherence between two points x and y are possible.

- If $q(x,y) = 0$ and $q(y,x) = 0$ then the coherence is $2 \times d(x,y)$, in that case the two points are mutually distant.

- If $q(x,y) = 0$ and $q(y,x) = 1$ or the reverse, then the coherence is $d(x,y)$.

- If $q(x,y) = 1$ and $q(y,x) = 1$ then the coherence equals zero, in that case the two points x and y are mutually near.

But the coherence of a cluster C cannot measure the following case :

x is neighbour of y (i.e. $q(x,y) = 1$)

but x and y have not been classified in the same cluster. It is compulsory to penalize this clustering since two near points have been put in two different clusters; thus this penalization can be written :

$$\sum_{(x,y)\in C\times \bar{C}} q(x,y)\, [1 - d(x,y)]$$

with $\bar{C} = E - C$.

The coherence of a partition P in k clusters $(P_1,\ldots, P_k)$, according to the distance d and the matrix Q representing the spatial constraints is thus measured by the following criterion (cf [12]) :

$$w(P) = \alpha \sum_{i=1}^{k} \sum_{x\in P_i} \sum_{y\in P_i} [1 - q(x,y)]\, d(x,y) +$$

$$(1 - \alpha) \sum_{i=1}^{k} \sum_{x\in P_i} \sum_{y\notin P_i} q(x,y)\, [1 - d(x,y)]$$

(E1)

with $\alpha \in [0,1]$.

α weights the action of the two parts of the criterion.

4.5.2. An example with the k nearest neighbours

One of the possible examples is to use in this
method the method of the k nearest neighbours with a clustering
method minimizing inertia. The k nearest neighbours allows the
construction of the graph incidence matrix (see [28]).

Hence Q is thus defined :

q(x,y) = 1 if y is one of the nearest neighbours
q(x,y) = 0 otherwise

and this verifies :

$$\sum_{y \in E} q(x,y) = k \quad \forall \ x \in E$$

Two elements x and y are mutual neighbours
have :

q(x,y) = 1 and q(y,x) = 1

Thus the incidence table is symetric if we assign
that :

q(x,y) = 1 if x and y are mutual neighbours
q(x,y) = 0 otherwise

and this method can be applied to this new incidence table.

Thus, with this new incidence table, the partition
created from connected parts of the associated graph is an optimal
solution in the sense of w. A partition where the number of
clusters is greater than the number of connected parts of the graph
defined by function q can only be looked for through a compromise
between the cost of deleting edges of the graph and the gain in
creating supplementary clusters considering the inertia criterion
over d. In that case, this method allows finding a locally optimum
solution.

4.6. Clustering algorithms under constraints

In the different cases of clustering under constraints
that we have presented, the two algorithms hereunder defined
allow the criterion to decrease until convergence.

4.6.1. Algorithm number 1

This algorithm allows the introduction of some
constraints associated to the space of representation and the
structure of class. Suppose E and d a distance over E × E and

$C_1, \ldots, C_k$ an a priori partition of a subset C of E. We search for a partition P of E in k clusters $(P_1, \ldots, P_k)$ whose space of representation $\mathbb{L}_i$ associated to each cluster P_i is defined as follows :

$$\mathbb{L}_i = \{L \in P(E) \: / \: \forall x \in L; \: x \in C_i \text{ and } Card(L) = \alpha_i\}$$

and the criterion w is defined :

$$w(P,L) = \sum_{\ell=1}^{k} \sum_{x \in P_\ell} \sum_{y \in L_\ell} d(x,y)$$

The algorithm is decomposed in two stages. The first one lies in finding the best representative of each cluster and the second one generates a new partition of E in k clusters. In the algorithm described hereabove we suppose, to simplify that $\alpha_i = 1 \quad \forall i = 1, \ldots, k$ (i.e. Card $(L_i) = 1 \quad \forall i = 1, \ldots, k$).

Having d as the distance over E × E we define the proximity measure between $P(E)$ and L_i by :

$$A \in P(E), \: b \in \mathbb{L}_i \quad D(a,b) = \sum_{a \in A} \mu(a) \: d(a,b)$$

i)   <u>initialization</u>

We initialize by chosing $a_i \in C_i$ the best representative of the a priori cluster C_i such that $\sum_{y \in C_i} d(a_i, y)$ will be minimum.

ii)   <u>allocation</u>

We construct the partition P from the representatives of each cluster

$$P_i = \{x \in E \: / \: d(x, a_i) \le d(x, a_j) \quad i \ne j \text{ and } j = 1, \ldots, k\}$$

iii)   <u>representation</u>

We note $\bar{P}_i = E - P_i$ and $\bar{C}_i = \underset{\substack{j \ne i \\ j=1,\ldots,k}}{\cup} C_j = C - C_i$

We associate to the couple formed by P_i the obtained cluster and C_i the a priori cluster, three sets characterizing the bindings between P_i and C_i :

$P_i \cap C_i$: the set of elements of C_i well classified at this stage.

$\overline{P_i} \cap C_i$: the set of elements of C wrongly classified at this stage

$P_i \cap \overline{C_i}$: the set of elements of the other a priori clusters classified in this cluster

from which we search for $x \in P_i \cap C_i$ such that $\sum\limits_{y \in P_i} d(x,y)$ will be minimum.

iv)   We verify that the set of representatives is modified
        if it is go to (iii)
        if it is not we stop the algorithm.

Remark :In several clustering problems under the a priori cluster constraint, in addition to the problem of optimizing the representation we wish to find an element of $P_i \cap C_i$ which will allow at the next stage the transfer of elements of $P_i \cap \overline{C_i}$ to other clusters and will bring to this cluster elements of $\overline{P_i} \cap C_i$. Thus at the convergence of the algorithm if we wish to improve the recognition rate, we can use the following two stages :

a)   agglomeration stage

        We look for an element $x \in P_i \cap C_i$ such that $\sum\limits_{y \in C_i \cap \overline{P_i}} d(x,y)$ will be minimum.

b)   separation stage

        We look for an element $x \in P_i \cap C_i$ such that $\sum\limits_{y \in P_i \cap \overline{C_i}} d(x,y)$ will be minimum.

Having thus obtained a new representative we can use the algorithm number 1 in view of decreasing the criterion w.

4.6.2. Algorithm number 2 associated to constraints in the criterion

According to $c_\alpha(x,y)$, this algorithm allows the optimization of the cost of regrouping x and y. The expression (E1) can be expressed as follows :

$$w(P) = \sum_{i=1}^{k} \sum_{x \in P_i} \sum_{y \in P_i} c_\alpha(x,y)$$

where P is a partition in k clusters $P = (P_1, \ldots, P_k)$ and k a constant.

The kernel-partition method allows optimizing this criterion. The space of representation $\mathbb{L}$ is $P(E)$, set of all the clusters of E. The measure of fit between $P(E)$ and the space of representation is equal to :

$$D(A,L) = \sum_{x \in A} \sum_{y \in L} c_\alpha(x,y)$$

with $A \in P(E)$ and $L \in \mathbb{L} = P(E)$ thus the criterion is written :

$$V(P,Q) = \sum_{i=1}^{k} D(P_i, Q_i)$$

with $P = (P_1, \ldots, P_k)$ and $Q = (Q_1, \ldots, Q_k)$.

4.7. Application in pattern recognition : searching for character models set

The solution of this problem is to look for a set of k clusters so that all the elements of a same cluster belong to the same a priori cluster and that two elements of different clusters are compulsory in two separate a priori clusters. This set of k character clusters verifying the properties hereabove stated will be called a character models set. Algorithm number 1 will allow finding a solution locally optimal for this problem. This procedure is dependent only of the set E, the k a priori clusters $C_1, \ldots, C_k$ and a distance over E × E.

The set of letters to analyse includes the 26 letters of the alphabet and the 10 figures. To each letter is associated several characters (we use the word character to indicate the design of a letter). The aim of our analysis is to find among this set of characters the character which best represents its letter but which is most distant from the character of the other letters. Thus determined, this set of characters subsequently allows a good recognition of new characters. This set of characters is given by C.Y Suen [54] and [55].

This analysis can be split in two stages. The first is a description stage, which lies on looking for relevant features allowing the differentiation of these different families of characters. It associates to each character a description vector and determines a distance (see [39], [35]). The second stage consists in using algorithm number 1 over a set of 121 matrix point characters of 5 × 7 types.

For example, for the letter A we have four patterns :

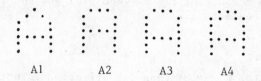

A1   A2   A3   A4

and for the letter G we have nine patterns :

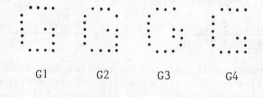

G1   G2   G3   G4

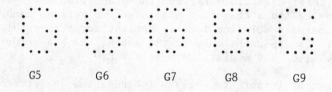

G5   G6   G7   G8   G9

This clustering method has allowed the selection of a set
of characters. With the allocation function, this set allowed a
good classification of 105 character out of the 121 of our
training set. Thus the allocation function can be used as
discrimination and decision function for other characters.

CONCLUSION

We have shown that several pattern recognition problems
can be set in terms of "clustering". Those problems need the same
basic choices and uses the same concepts : Among others the one
of "structure of class" and "representation" for which we have
given an exact mathematical definition. Numerous extension prospects and applications are still open for example in the three
proposed approaches (syntactic, numerical and logical) by
considering other structure of class and other kinds of
representation.

ANNEX 1

LEXICOGRAPHIC ORDER

In general we define the lexicographic order for s attributes as :

$$
\begin{pmatrix} a_1 \\ a_2 \\ \vdots \\ a_m \\ \vdots \\ a_s \end{pmatrix} < \begin{pmatrix} b_1 \\ b_2 \\ \vdots \\ b_m \\ \vdots \\ b_s \end{pmatrix}
$$

$\Longleftrightarrow$ $\exists\ m,\ s \geq m > i$ $a_i = b_i$ and $a_m < b_m$

ANNEX 2

THE BIGLETTERS

CARACTERIZATION OF THE CLASSES

Results for class 1

formula 1 [[ligne-1 = {0,1}].[ligne-2 = {2,4,6,7,8}].[ligne-3 = {2,3,4,5}].[ligne-4 = {2,4,
6}].[ligne-5={2,3,4,5,6}].[ligne-6 = {2,3,4,5,6,8}].[ligne-7 = {2,3,4,5,6}]...
[ligne-8 = {2,3,4,5,6}].[ligne-9 = {2,4,6,8}]]...

formula 2 [[ligne-1 = {0}].[ligne-2 = {2,8}].[ligne-3 = {2}].[ligne-4 = {2}].[ligne-5 = {2}]...
].[ligne-6 = {7}].[ligne-7 = {2}].[ligne-8 = {2}].[ligne-9 = {8}]]...

Results for class 2

formula 1 [[ligne-1 = {0,3,4,5}].[ligne-2 = {2,3,4,6,7}].[ligne-3 = {2,3,4,6,7}].[ligne-4 =
{2,4,6,7}].[ligne-5 = {2,3,4,5,6}].[ligne-6 = {0,2,4,5,6,7}].[ligne-7 = {0,2,3}...
4,7}].[ligne-8 = {0,2,3}].[ligne-9 = {0}]]...

formula 2 [[ligne-1 = {0}].[ligne-2 = {4,7}].[ligne-3 = {4,7}].[ligne-4 = {4}].[ligne-5 = {4}]...
].[ligne-6 = {7}].[ligne-7 = {2}].[ligne-8 = {2}].[ligne-9 = {2}]]...

RECOGNITION OPERATORS

Results for class 1

formula 1 [ligne-6 = {0,1,2,3,4,5,6,8}].[ligne-9 = {1,2,3,4,5,6,7,8}]]

formula 2 [ligne-3 = {0,1,2,3,5,6,7,8}].[ligne-8 = {1,2,4,5,6,7,8}]]

Results for class 2

formula 1 [ligne-9 = {0,1,3,5,7}]]

formula 2 [ligne-5 = {0,1,3,4,5,6,7,8}].[ligne-9 = {0,1,2,3,5,7,8}]]

BIBLIOGRAPHY

[1]   AHO A.V., PETERSON T.G. (1972) "A minimum distance
error-correcting parser for context-free languages".
SIAM J. Comput. Vol. 1, n° 4, p. 305.

[2]   AIVAZIAN S. (1976) "Les méthodes statistiques d'étude de
dépendances entre variables classifiantes".
T.S.E.M.I. Moscou, URSS.

[3]   ANDERBERG M.R. (1973) "Cluster analysis for applications".
Academic Press.

[4]   BACKER E. (1978) "Cluster analysis by optimal decomposition of induced fuzzy-sets". (Delft : Delftse Universi-
taire Pers.).

[5]   BALL G.H., HALL D. (1967) "A clustering technique for
summarizing variate data". Behavorial Science 12, n° 12,
p. 153-155.

[6]   BERTHOD M. (1980) "A new predicative learning scheme
and applications". 5 th Int. Conf. Pat. Rec. MIAMI, 1980.

[7]   BRICE C.R., FENNEMA C.L. (1970) "Artificial intelligence"
Vol. 1, n° 3, p. 205.

[8]   CHARLES C., LECHEVALLIER Y. (1979) "Pattern recognition
by a piecewise polynomial approximation with variable
points". Rapport IRIA Laboria n° 338.

[9]   CHERNOFF M. (1970) "Metric consideration in cluster
analysis". Proc. 6 th. Berkeley Symposium on Math. Stat.
and Prob.

[10]   COLEMAN G.B. (1977) "Scene segmentation by clustering"
University Southern California. Image Processing Institute. Report USCIPI.

[11]   CORMARK R.M. (1971) "A review of classification".
J. Royal Statist. Soc., serie A, p. 134-321.

[12]   DE FAGUEROLLES A. (1977) "Classification automatique :
un critère et des algorithmes d'échange". Séminaires IRIA

[13]   DIDAY E. (1972) "Nouvelles méthodes et nouveaux concepts
en classification automatique et reconnaissance des
formes". Thèse d'Etat, Paris VI.

[14]   DIDAY E. (1978) "Problems of clustering and recent
         advances". 11 th Congress of statistics, Oslo.

[15]   DIDAY E. (1978) "Selection of variables and clustering"
         4 th Int. Conf. Pat. Rec. Kyoto, Japan.

[16]   DIDAY E. ET AL. (1980) "Optimisation en classification
         automatique", (ed. INRIA).

[17]   DIDAY E., GOVAERT G. (1976) "Apprentissage et mesures
         de ressemblances adaptatives". Computer oriented
         learning processes. Nato Advances Study Institute,
         serie E, n° 14.

[18]   DIDAY E., SIMON J.C. (1976) "Clustering analysis".
         Communication and Cybernetics 10. Digital Pattern
         Recognition. (Ed. Springer-Verlag).

[19]   DOROFEYUK A. (1966) "Pattern Recognition machine
         learning without reward" Question of Technical Cyberne-
         tics, Nauka.

[20]   DO-TU H., INSTALLE M. (1978) "A fast algorithm procedure
         based on ISODATA algorithm with application to remote
         sensing". 4 th Int. Conf. Pat. Rec.. Kyoto, Japan,
         p. 326.

[21]   DUBES R., JAIN A.K. (1976) "Clustering techniques. The
         users dilemma". Pattern recognition, Vol. 8, n° 4,
         p. 247.

[22]   DURAN B.S., ODELL P.L. "Cluster analysis". Lectures notes
         in Economics and Mathematical Systems. Econometric 100
         (Ed. Springer-Verlag).

[23]   FU K.S. and LU S.Y. (1977) "A clustering procedure for
         syntactic patterns". IEEE Transactions on systems, man
         and Cybernetics, Vol. SMC-7, n° 10, p. 734.

[24]   FU K.S. and BOOTH T.L. (1975) "Grammatical Inference
         Introduction and Survey-Part I". IEEE Transactions on
         systems, man and Cybernetics, vol. SMC-5 n° 1, p. 95.

[25]   FU K.S. and BOOTH T.L. (1975) "Grammatical Inference
         Introduction and Survey-Part II". IEEE Transactions on
         systems, man and Cybernetics, Vol. SMC-5 n° 4, p. 409.

[26]   FUKADA Y. (1978) "Spatial clustering procedures for
         region analysis". 4 th Int. Conf. Pat. Rec. Kyoto, Japan,
         p. 329.

[27]   GOVAERT G. (1975) "Classification automatique et dis-
         tances adaptatives", Thèse de 3° cycle, Paris VI.

[28]   GOVAERT G., LECHEVALLIER Y. (1978) "Classification uti-
         lisant la notion de voisinage". Premières Journées
         Nationales sur la classification, Vannes.

[29]   GOWDA K.C., KRISHNA G. (1978) "Agglomerative clustering
         using the concept of mutual nearest neighbourhood".
         Pattern recognition, vol. 10, p. 305.

[30]   HARALICK R.M. (1978) "Scene matching problems". IEEE
         SMC-8, p. 600.

[31]   HARALICK R.M., SHAPIRO L. (1977) "Decomposition of
         polygonal shapes by clustering". Proc. IEEE Conf. On
         pattern Recognition and Image processing Troy. N.Y. (USA)
         p. 183.

[32]   HAYES-ROTH F. (1976) "Representation of structured
         events and efficient procedures for their recognition".
         Pattern Recognition, vol. 8, n° 3, p. 141.

[33]   JARVIS R.A., PATRICK E.A. (1973) "Clustering using a
         similarity measure based on shared near neighbours".
         IEEE Trans. Comp. Vol. C 22 n° 11, p. 1025.

[34]   KASVAND T. (1978) "Scene Segmentations and segment
         clustering experiments". 4 th. Int. Conf. Pat. Rec.
         Kyoto, Japan, p. 426.

[35]   KWAN C.C. (1977) "A study of the selection and recogni-
         tion of hand-printed characters". Thesis in Dept.
         Computer Science, Concordia University, Montreal Canada.

[36]   LECHEVALLIER Y. (1974) "Optimisation de quelques cri-
         tères en classification automatique". Thèse de 3° cycle,
         Paris VI.

[37]   LECHEVALLIER Y. (1976) "Classification automatique sous
         contrainte d'ordre total". Rapport de recherche IRIA-
         LABORIA n° 200.

[38]   LECHEVALLIER Y., SUEN C.C. (1978) "Sélection d'un jeu
         de caractères par une méthode de classification automa-
         tique". Congrès AFCET-IRIA, p. 687.

[39]   LOWITZ G.E. (1978) "Compression des données par recon-
         naissance des formes et clustering". Congrès AFCET-IRIA,
         p. 699.

[40] LU S.Y., FU K.S. (1978) "A sentence to sentence clustering procedure for pattern analysis". IEEE Transaction on systems man, and cybernetics, Vol. SMC-8 n° 5, p. 381.

[41] MAC QUEEN J. (1967) "Some methods for classification analysis of multivariate observations". 5th Berkeley Symposium on Mathematics, Statistics and probabilities, vol. 1, n° 281.

[42] MICHALSKI R.S. (1975) "Synthesis of optimal and quasi-optimal variable valued logic formula". Int. Symp. on multiple valued logic. Indiana University Bloomington.

[43] MICLET L., GUERIN C. (1978) "Inférence multiple en grammaires régulières". Congrès AFCET-IRIA, p. 73.

[44] OTSU N. (1978) "Discriminant and least squares threshold selection". 4 th Int. Conf. Pat. Rec. Kyoto. Japan, p. 592.

[45] PAVLIDIS T. (1978) "Algorithms for shape analysis of contours and waveforms". 4 th. Int. Conf. Pat. Rec. Kyoto. Japan, p. 70.

[46] PRATT W.K. (1978) "Quantitative approches to image feature extraction and segmentation". Congrès AFCET-IRIA p. 897.

[47] ROCHE C., REBUFFET M. (1978) "Méthodes de classification d'imagerie multi-spectrale". Congrès AFCET-IRIA, p. 715.

[48] ROHLF F.J. (1970) "Adaptative hierarchical clustering schemas". Systematic Zoology, vol. 18, p. 58-82.

[49] ROSENFELD A. (1979) "Some recent developments in texture Analysis". Proc. of Pattern Recognition and Image Processing, Chicago.

[50] ROSENFELD A., STRONG J.(1971) "A grammar for Maps". In Software engineering. Vol. 2, ed. by J.T. TOU (Academic Press, New-York).

[51] SIMON J.C., BACKER E., SALLENTIN J. (1980) "A structural approach of Pattern recognition". Signal Processing Vol. 2 n° 1.

[52] SIMON J.C. (1979) "Clustering and digital image analysis". Inst. Phys. Conf. Sc. n° 44 : chapter 1.

[53]   SIDI J. (1980) "L'approche logique en classification au-
        tomatique et reconnaissance des formes". Thèse de 3° cycle.

[54]   SUEN C.C., SHIAU R., SHINGAL R., KWAN C.C. (1976)
        "Reliable recognition of hand printed characters".
        Joint workshop on Pattern Recognition and Artificial
        Intelligence.

[55]   SUEN C.C., SHINGHAL R. (1977) "Assessment on handprint
        quality based on dispersion measurements". IFIP Cong.

[56]   SWAIN P.H. (1979) "Image data Analysis in remote
        sensing". Digital Image Processing and Analysis, ed.
        R.M. HARALICK and J.C. SIMON (Leyder : Nordhoff).

[57]   TALENG (1980) "Codage optimal adaptatif". Optimisation
        en classification automatique, Diday et al. Ed. INRIA.

[58]   ULLMANN (1978) "A relational view of text image
        processing" NATO BONAS.

[59]   WAGNER R.A., FISHER M.J. (1974) "The string to string
        correction problem". J. Ass. Comput. Mach. Vol. 21
        Janv.

[60]   WATANABE S. (1969) "Knowing and Guessing". John Wiley
        and Sons INC.

[61]   YOKOYA N., KITAHASHY T., TANAKA K., ASANO T. "Image
        segmentation schema based on a concept of relative
        similarity". 4 th Int. Conf. Pat. Rec. Kyoto. Japan.
        p. 645.

[62]   ZAGORUIKO N.G., LBOV G.S. "Algorithms of pattern
        recognition in a package of applied programs" Oteks"".
        4 th Int. Conf. Pat. Rec. Kyoto. Japan. p. 1100.

TOPOLOGIES ON DISCRETE SPACES

J.M.CHASSERY and M.I.CHENIN

Laboratoire IMAG - BP 53X - 38041 GRENOBLE CEDEX - FRANCE

Abstract
        To define regions in digital image we use topological
notion of connectivity. This notion refers to the existence
and the description of a topology.
Our problem is to prove for different tessallations of the
support, the existence and unicity and eventually to describe
such topology for which the notion of connectivity will be
the same we use in image analysis.

Introduction
        In Digital Picture Processing, after the sampling
problem we have the step of segmentation.
        Segmentation decomposes a picture into regions using
concepts of connected components. Each region will correspond
to isolated objects or background and next step will consist
in evaluation of some parameters on these connected components.
        Our problem is to study if this notion of connected
component is related to an usual topology.
        After description of current definitions used in
image analysis we shall enumerate properties to caracterize
such topologies. In a third part we shall consider examples of
tessallations currently used in digital picture processing.

I - Definitions and Properties
        Let E a set of points (I,J) on the discrete space
ZxZ which generally correspond to the regular support of the
digital picture.
        For every point x belonging to E, we associate
a subset P_x satisfying the hypothesis :

59

J. C. Simon and R. M. Haralick (eds.), Digital Image Processing, 59–66.
Copyright © 1981 by D. Reidel Publishing Company.

$$\forall x \in E \qquad Card \ P_x < + \infty$$

$$\forall x \in E \qquad x \in P_x$$

$$x \in P_y \Longleftrightarrow y \in P_x$$

$$P_x = P_y \Longleftrightarrow x = y$$

Remark : P_x will correspond to the notion of 4-neighborhood or 8-neighborhood currently used in digital pictures.

4-neighborhood   8-neighborhood

I.1 Definition

Let x and y two points into E.

By a P-path from x to y we mean a sequence of points $x_0 = x$, x_1 , x_2 ... $x_n = y$ such that $x_i \in P_{x_{i-1}}$ $i = 1 \to n$

We say that x and y are P-connected in a subset S of E if there exists a P-path from x to y included into S.

The notion of "P-connectivity in S" is an equivalence relation and the equivalence classes defined by it are defined as the P-connected components of S.

I.2 Properties

We have evident properties.

Property 1

$\emptyset$ and E are P-connected components.
{x} is a P-connected component.

Property 2

The subsest {x,y} is a P-connected component $\Longleftrightarrow$ $y \in P_x$ (or $x \in P_y$).

Property 3

$\forall x \in E$ P_x is a connected component.

Property 4

if A and B are P-connected components} $\Rightarrow$ A∪B is a
if A ∩ B ≠ $\emptyset$ P-connected
                                                          component in E.

For further study we make a new hypothesis on E.

$\forall x \in E$, $E \setminus P_x$ has a finite number of P-connected components.

This hypothesis is always satisfied in Image Analysis where the support of picture is bounded.

I.3 Definition of Compatible Topology.

Let $\mathcal{T}$ a topology on E ($\mathcal{T}$ will be for example defined by a base of open sets).

We shall say that $\mathcal{T}$ is compatible with (E,P) if the P-connected subsets of E are the connected subsets of E for the topology $\mathcal{T}$.

We remind that A is a connected set for the topology $\mathcal{T}$ if there is no partition of A into open subsets.

II - Properties on compatible topologies

Let $\mathcal{T}$ a compatible topology with (E,P).

For every point x belonging to E we define the subset Θ_x as the intersection of all the open sets including x.

Our purpose is to define the topology by its base composed by the Θ_x subsets. From relations between Θ_x and P_x , we shall construct such compatible topologies from information about P_x . This comes from the next propositions.

II.1 Proposition 1

1) $\forall x \in E$ $\Theta_x \subseteq P_x$
    $\forall x \in E$ Θ_x is open

2) $\{\Theta_x \; ; \; x \in E\}$ and $\emptyset$ compose a topological base for $\mathcal{T}$

Prove :
1) Let us consider the set $(E \setminus P_x) \cup \{x\}$
   $\{x\}$ is a connected set.
   $E \setminus P_x$ has a finite number of P-connected components. By the last hypothesis, it can be considered as a closed set composed by a finite union of closed subsets.
   So $\{x\}$ considered as the complementary set of $\{E \setminus P_x\}$ into $(E \setminus P_x) \cup \{x\}$ is an open set.
   So there exists an open set Θ such that $\{x\} = \Theta \cap ((E \setminus P_x) \cup \{x\}\ )$.
This implies $\Theta \subseteq P_x$ and by definition of Θ_x we have $\Theta_x \subseteq P_x$.

If Θ is equal to Θ_x then Θ_x is open.
Else there is a point y belonging to Θ and not to Θ_x and an open set Θ' such that $x \in \Theta'$ and $y \notin \Theta'$.
Consider $\Theta \cap \Theta'$ which is open.
If $\Theta \cap \Theta' = \Theta_x$ then Θ_x is open.
Else we iterate. The number of iterations is finite (card $\Theta < +\infty$).

2) We can remark that we have two possibilities about $\Theta \in \mathcal{T}$

$$\Theta = \emptyset \quad \text{or} \quad \Theta = \bigcup_{x \in \Theta} \Theta_x$$

So $\{\Theta_x \; ; \; x \in E\} \cup \emptyset$ form a topological base for $\mathcal{T}$.

II.2 Proposition 2

Let $\mathcal{T}$ a compatible topology with (E, P). Then

1) $x \in \Theta_y \Longleftrightarrow \Theta_x \subseteq \Theta_y$

2) We have equivalent properties

    a) $\{x, y\}$ is a connected set for the topology
    b) $\{x, y\}$ is a P-connected component
    c) $x \in P_y$
    d) $y \in P_x$
    e) $x \in \Theta_y$ or $y \in \Theta_x$
    f) $\Theta_x \subseteq \Theta_y$ or $\Theta_y \subseteq \Theta_x$

3) $\Theta_x = \Theta_y \Longleftrightarrow x = y$

4) $(x \in \Theta_y$ and $x \neq y) \Longrightarrow y \notin \Theta_x$

Proves

1) evident

2) . $a \Longleftrightarrow b \Longleftrightarrow c \Longleftrightarrow d$
  . $e \Longrightarrow c$ by proposition 1
  . $\neg e = (x \notin \Theta_y$ and $y \notin \Theta_x)$
    $\left. \begin{array}{l} x \notin \Theta_y \Longrightarrow \{y\} = \{x, y\} \cap \Theta_y \\ y \notin \Theta_x \Longrightarrow \{x\} = \{x, y\} \cap \Theta_x \end{array} \right\} \Longrightarrow \{x, y\} = (\underbrace{\{x, y\} \cap \Theta_x}_{\text{open}}) \cup (\underbrace{\{x, y\} \cap \Theta_y}_{\text{open}})$

  . so we have $\neg e \Longrightarrow \neg a$ and by deductions $c \Longrightarrow e$
  . $e \Longleftrightarrow f$ evident using 1).

3) if $\Theta_x = \Theta_y$
  $z \in P_y \Longleftrightarrow \Theta_z \subseteq \Theta_y$ or $\Theta_y \subseteq \Theta_z \Longleftrightarrow \Theta_z \subseteq \Theta_x$ or $\Theta_x \subseteq \Theta_z$
    $\Longleftrightarrow z \in P_x$
  Then $\Theta_x = \Theta_y \Longrightarrow P_x = P_y \Longrightarrow x = y$

4) evident

II.3 Proposition 3

Let $x \in E$ and Θ_x the smallest open set including x.
We have
$$\forall y \in \Theta_x \; (z \in P_x \setminus P_y \Longrightarrow z \in \Theta_x)$$

Prove :

Suppose $z \in P_x \setminus P_y$ and $z \notin \Theta_x$

$$z \in P_x \xRightarrow{\text{Prop. 2}} \left. \begin{array}{l} x \in \Theta_z \text{ or } z \in \Theta_x \\ z \notin \Theta_x \end{array} \right\} \Rightarrow x \in \Theta_z$$

$$x \in \Theta_z \xLeftrightarrow{\text{Prop.2}} \left. \begin{array}{l} \Theta_x \subset \Theta_z \\ y \in \Theta_x \end{array} \right\} \Rightarrow y \in \Theta_z \Rightarrow y \in P_z \text{ and } z \in P_y \text{ contradiction.}$$

These different propositions will permit us now to construct compatible topologies on discrete spaces for a tassallation i.e. a definition of P_x for every point x.

III – Application for determination of compatible topologies on current tessallations

In image analysis for the square tessallation the subset P_x is defined from a metric d on ZxZ. To every point x into E we associate a d-neighborhood which corresponds to our P-connected set P_x .

For each example of P_x the previous propositions will permit us to caracterize all the possibilities of subsets Θ_x satisfying the compatibility condition.

Example 1

On E = ZxZ we define the metric d_1 by :
$d_1(x,y) = |I_x - I_y| + |J_x - J_y|$ where (I_x,J_x) and (I_y,J_y) are respectively the coordinates of points x and y.

We define for each point x :
$P_x = \{y \in E ; d_1(x,y) \leq 1 \}$

Property :

Suppose the existence of a compatible topology.
Then we have $\Theta_x = \{x\}$ or $\Theta_x = P_x$

Prove :

Suppose $\Theta_x \neq \{x\}$

. prop. 1 $\Rightarrow \Theta_x \subseteq P_x$

. Let $y \in \Theta_x$ with $y \neq x$
prop 3 $\Rightarrow P_x \setminus P_y \subset \Theta_x$
By definition of subsets P_x we have for $y \in P_x$
and $y \neq x$
$$P_x = \{y,x\} \cup (P_x \setminus P_y)$$
So $\left. \begin{array}{l} \{y,x\} \subset \Theta_x \\ P_x \setminus P_y \subset \Theta_x \end{array} \right\} \Rightarrow P_x \subseteq \Theta_x$

We conclude $\Theta_x = P_x$

Construction of the unique compatible topologies for $d = d_1$

The discrete topology ($\Theta_x = \{x\}$) is not compatible with (E,P) because two points x and y , d_1-neighbors constitue a P_1-connected component and a not-connected set for the discrete topology.

So there is at least one point x for which $\Theta_x = P_x$

Let $y \in \Theta_x$ with $y \neq x$
prop 2.4 $\Longrightarrow x \notin \Theta_y$
prop 2.2 $\Longrightarrow y \in \Theta_x \Longrightarrow x \in P_y$ $\Bigg\} \Longrightarrow \Theta_y \neq P_y$

Then by precedent property $\Theta_y = \{y\}$

We recognize current topologies used in image analysis [1,2]

$\mathcal{T}_1$
   $\Theta_x = P_x$ if $I_x + J_x$ is odd
      $= \{x\}$ else
or
$\mathcal{T}_2$
   $\Theta_x = \{x\}$ if $I_x + J_x$ is odd
      $= P_x$ else

More than the existence we prove now the unicity of such compatible topologies for the case of the d_1-neighborhood.

Example 2 [3]
   On E = ZxZ we define the metric d_∞ by
$d_\infty(x,y) = \text{Max} (|I_x - I_y|, |J_x - J_y|)$
where (I_x,J_x) and (I_y,J_y) are respectively the coordinates of points x and y.
   For each point x we define
   $P_x = \{y \in E ; d_\infty(x,y) \leq 1\}$

Property :
   Suppose the existence of a compatible topology then we have $\Theta_x = \{x\}$ or $\Theta_x = P_x$

Prove :
   It is the same demonstration than previously if we remark that we have
$P_x = \{x,y\} \cup (P_x \setminus P_y) \cup \bigcup_{z \in P_x \setminus P_y} (P_x \setminus P_z)$

Construction of compatible topology for $d = d_\infty$

Noting that Discrete Topology is not compatible, there exists at least a point x into E for which $\Theta_x = P_x$

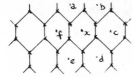

For every point y belonging to P_x , different from x we have $y \in \Theta_x$ and $y \neq x \Rightarrow x \notin \Theta_y$ (prop. 2.4)

$$\left.\begin{array}{c} x \in P_y \\ x \notin \Theta_y \end{array}\right\} \Rightarrow \Theta_y \neq P_y \Rightarrow \Theta_y = \{y\} \ .$$

Referrencing to the figure we have $\Theta_y = \{y\}$ and $\Theta_z = \{z\}$. The set $\{y,z\}$ is a P-connected component and not a connected set.

So we conclude that there is no compatible topology for this definition of P_x.

Example 3 Hexagonal tessallation

If we use as support a regular grid composed by hexagonal elements, we can define P_x by

$$P_x = \{x,a,b,c,d,e,f\}$$

Property :

Suppose the existence of a compatible topology, then we have $\Theta_x = \{x\}$ or $\Theta_x = P_x$

Prove :

It is the same demonstration than previously, if we remark that we have $P_x = \{x,y\} \cup (P_x \backslash P_y) \cup \underset{z \in P_x}{\cup} \backslash P_y \ (P_x \backslash P_z)$

Construction of compatible topology

As the Discrete Topology is not compatible, there exists at least a point x sucht that $\Theta_x = P_x$

As in the previous example we demonstrate (referrencing to the figure) that $\{b\}$ and $\{c\}$ are open sets and $\{b,c\}$ is a P-connected component and not a connected set.

So there is no compatible topology.

Example 4 Triangular tessallation

if we use as support a regular grid composed by triangular elements , w e can define P_x by

$$P_x = \{x,a,b,c\}$$

Property :

Suppose the existence of a compatible topology, then we have $\Theta_x = \{x\}$ or $\Theta_x = P_x$

Construction of the unique compatible topologies

As the Discrete Topology is not compatible we have at least one point x for which $\Theta_x = P_x$

Let $y \in \Theta_x$ and $y \neq x$

Then by proposition 2.4 we have $x \notin \Theta_y$

by proposition 2.2 we have $y \in \Theta_x \Rightarrow x \in P_y$ $\Big\} \Rightarrow \Theta_y \neq P_y$

Then for every point y belonging to Θ_x we have $\Theta_y = \{y\}$

So we have 2 compatible topologies corresponding to the configuration of the subset P_x .

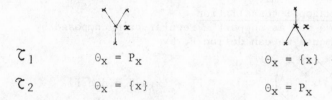

$\mathcal{T}_1$ $\Theta_x = P_x$ $\Theta_x = \{x\}$

$\mathcal{T}_2$ $\Theta_x = \{x\}$ $\Theta_x = P_x$

IV – Conclusion

In conclusion, this problem of compatible topology is solved and it would be more correct to use notion of consecutivity instead of connectivity.

In second conclusion this aspect of formulation has to be extended to other notions such as the notion of convexity. In this case, as in the problem of measuring parameters such as area or perimeter, we have problems of approximation and convergence in function of the step of digitization.

References

[1]   F.Wyse et al, Solution to Problem 5712. Amer Math.
        Monthly 77, 1970, 1119

[2]   A.Rosenfeld, Digital Topology. Amer Math.
        Monthly oct 1979, 621-630

[3]   J.M.Chassery, Connectivity and Consecutivity in Digital
        Pictures. Comp. Graphics and Image Processing,
        9, 1979, 294-300.

A MODEL OF THE RECEPTIVE FIELDS OF MAMMAL VISION

J.-P. CRETTEZ & J.-C. SIMON

Institut de Programmation, Université P. & M. Curie,
4 place Jussieu, 75230 Paris Cedex 05, France.

The retina, the lateral geniculate bodies, the visual cortex
are the three physiological levels of the visual pathway.
At each level, to a definite cell corresponds a "receptive field"
in the visual space. A multilayer model may explain the size
constancy effect in perception. Each layer is supposed to have an
homogeneous distribution of retinal ganglion receptive fields.
A simple transformation allows to model complex and hypercomplex
cells. Called a "synaptical heptarchy", this set of transforms
explains a number of features of the receptive fields:
symmetry, orientation, spatial frequency characteristics, and
complexity.

INTRODUCTION

The perception of an image is processed through the
visual pathway at three successive physiological levels: the
retina, the lateral geniculate bodies and the visual cortex.
The visual cortex is divided into three Brodmann's areas: the
striate area, the peristriate area, the parastriate area. What
is the signal processing done in this retino-geniculate-striate
pathway, in which the physiologists have made many tests?
    This pathway is made of neurone cells, connected by
synaptic contacts between dendrites and axons . To each cell
corresponds a receptive field projected in the visual space.
Such a visual field may be defined as the (usually small) region
of the visual space, which may influence the cell activity.
Let θ be the angular position and ψ be the eccentricity of a
field, which also has a defined size in square degrees, an ocular
dominance and a polarity, and many characteristics defined below.

J. C. Simon and R. M. Haralick (eds.), Digital Image Processing, 67–81.
Copyright © 1981 by D. Reidel Publishing Company.

ELECTROPHYSIOLOGICAL DATA

At the retina level, signal is processed by three types
of cells: receptive, bipolar and ganglion cells, and horizontal
and amacrine cells. The receptive fields of ganglion cells has a
circular symmetry, showing an antagonism between the center and
the peripheric region . FISCHER (1) has shown that the set of
receptive fields of the ganglion cells is not like a tessellation
of the visual space, but rather like a random overlapping of
fields of different size. If a spot of light falls on the retina,
approximately 35 ganglion cells will fire correspondingly. The
axones of the ganglion cells are the million and a half fibers
of the optical nerves. The receptive fields of the geniculate
cells are similar to those of the ganglion cells. They also overlap.

The visual cortex has a uniform thickness of two milli-
meters, and a surface of 1300 mm^2 in the macaque monkey, 6000 mm^2
for men (this at least is a proved superiority!). Usually six
layers, and even more are distinguished along a..perpendicular
to the surface. HUBEL and WIESEL (2) speak of four classes of
neurone cells, according to the complexity of their receptive
fields: the cells similar to the ganglion cells, the simple cells,
the complex and hypercomplex cells. The simple cells answer to
a line or an edge of light, with a definite position and orientation. The receptive field has an excitory and an inhibitory region,
with a summation effect in each region. The complex cells are
specific in orientation, but not in position. An increase in the
width of the light slit reduces the answer. The hypercomplex cells
have been observed commonly on several experiments. For instance,
GROSS et al. (10) describe the behaviour of a cell in the infero-
temporal cortex of a macaque, which would fire over a very wide
visual angle. But it fired most vigorously on seeing a stimulus,
which was "the shadow of a human or monkey hand". Even:"curiously,
fingers pointed downward elicited very little response as compared
to fingers pointed upward or laterally; the visual orientation in
which the animal would see hid own hand". Many physiologists have
shown that the striate cortex has an elaborate structure. HUBEL
and WIESEL (3), ALBUS (4). Penetrating the cortex along a perpendicular, different cells are met, with receptive fields overlapping
randomly around a central direction in the visual space. The orientation and ocular dominance are the same, but the size are different.
MAFFEI and FIORENTINI (5) found, penetrating the cortex in a different direction, that the orientation of the visual field would
turn clockwise or counterclockwise.

An extensive literature exists on these topics. For a
more detailed survey, oriented towards image processing specialists,
presented at the former NATO ASI of june 1976, the reader should
report to the very good paper by M. EDEN (12).

A MULTILAYER MODEL OF THE VISION FIELD

        In this section, we introduce a model of the retino-geni-
culate region, trying to explain some features. Let us make some
assumptions:
        - The visual space is divided into layers of receptive
fields, called RF layers.
        - In each of these RF layers, the receptive fields cor-
responding to geniculate cells are uniformly distributed in a sort
of regular tessellation.
        - Each RF layer has the same number of receptive fields,
i.e. concerns a same number of geniculate cells.
        - The different RF layers cover different total visual
fields, the first covering the fovea, the last the maximum angular
vision.

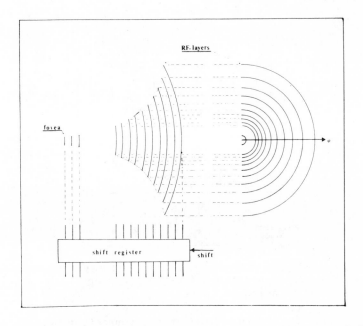

        Figure 1. Representation of the schema of the multilayer
        model, showing an expanded view of the N_f homogeneous layers.

        In a previous publication, J.P. CRETTEZ (6) has explained
how this organisation may model the size invariance of a moving
image. Usually called the law of EMMERT, this invariance may be
observed in after-images, and also explains the MULLER-LYER illusion. The stack of layers corresponds to 1,5 million ganglion cells.
Assuming 30.000 cells for each layer, the number of layers would
be of the order of 50; this in turn would take into account the
approximate number of 35 cells firing for a spot of light.

The rectangular box of Fig. 1 figures a "shift register" which would be controlled by the inner ear. The shifting of the recorded image would be done from a layer to its neighbooring one, according to the direction of movement. This suppose a rather elaborate connecting circuitry, which has not been found experimentally. We will propose a model of synaptic connections, which may however illustrate what we envision.

A SYNAPTIC NETWORK

A human brain counts about 10^{11} neurones and 10^{14} synaptic connections. BENZER (7) has pointed out that the connections could not be wired in only by genetic information, but that this genetic information could provide the planning of the connecting process, in an overall draft.

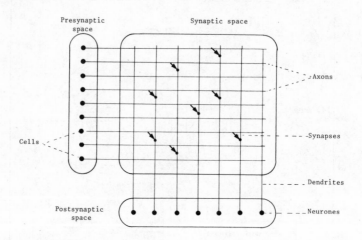

Fig. 2. Connection between a presynaptic and a postsynaptic region or space.

Fig. 2 displays a schema of the connecting network between for instance the geniculate region and the striate cortex. The dendrites of neurones of the synaptic network receive some optical information from the afferent axons of the geniculate cells. STEVENS (8) has shown the existence of two types of synaptic contacts: - excitory contact, which we code as +1,
            - inhibitory "   "   "   "   " -1,
we add a missing   "   "   "   "   " 0.
The synaptic network may be formalized as a matrix transform, the elements of which are only +1, -1, 0. Such a transform suggests the use of the HADAMARD transform...

A HIERARCHY OF CONNECTING NETWORKS.

A uniform distribution of the RF layers and the circular symmetry of ocular vision suggest a centered tessellation, which would minimize the gaps between the circular receptive fields of the ganglion cells. Let us suppose that seven ganglion cells displayed as a, b, ...g in Fig 3 are connected to seven neurones of the striate cortex, a',b',....g', through a connecting network represented by T.

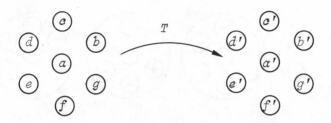

a, b,....g are the ganglion cells states.

a',b',...g' are the states of the connected neurones of the cortex.

Fig. 3. Inputs and outputs of a connecting network. The configurations are called "heptuplets" HPl.

Let us suppose that the cental activity may be broken in two, and let us apply the HADAMARD trasform of order 8. After some simplifications, we obtain the T transform of an HPl

$$
\begin{vmatrix} a' \\ b' \\ c' \\ d' \\ e' \\ f' \\ g \end{vmatrix} = \begin{vmatrix} 1 & 1 & 1 & 1 & 1 & 1 & 1 \\ 1 & -1 & 1 & -1 & -1 & 1 & -1 \\ 1 & 1 & -1 & -1 & 1 & -1 & -1 \\ 1 & -1 & -1 & 1 & -1 & -1 & 1 \\ 0 & 1 & 1 & 1 & -1 & -1 & -1 \\ 0 & 1 & -1 & -1 & -1 & 1 & 1 \\ 0 & -1 & -1 & 1 & 1 & 1 & -1 \end{vmatrix} \begin{vmatrix} a \\ b \\ c \\ d \\ e \\ f \\ g \end{vmatrix}
$$

The state of each of the cortical cell a',b'...g', is obtained as an algebraic sum of the states of the ganglion cells a, b,....g. Let us call "sign or synaptic configuration" SC, the set of signs on each line of the matrix T. An SC represents the polarity of the synapses joining seven cells of an HPl to a cortical neurone.

Seven settings are obtained, displayed in Fig. 4 in the corresponding RF layer.

- One called 'A' averages the seven ganglion states.
- Three symmetrical configuration 'B', 'C', 'D' may be interpreted as line detectors.
- Three antisymmetrical configurations 'E', 'F', 'G' may be interpreted as edge detectors.

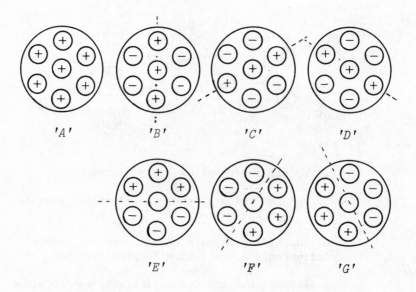

Fig. 4. The first level of the receptive field configurations corresponding to the first level HP1.

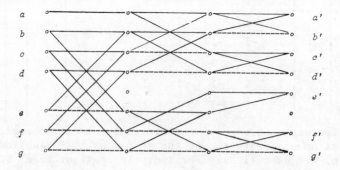

Fig. 5. The connections of the "fast" HADAMARD algorithm.

The HADAMARD transform may be implemented by wiring,
corresponding to a "fast transform". Such a process is shown by
the Fig. 5. It is interesting to remark that the disposition
introduces naturally layers, and reduces the number of additions
to be performed: 19 instead of 49, which would be necessary with
the matrix product. Also these operations may be performed in
parallel as in the nervous system.

A generalisation:

Let HP2 be an heptuplet of order 2, generated in a RF
layer from seven heptuplets HP1. Such an HP2 would cover 49 retinal ganglion receptive fields; thus in average a much larger angle
than a HP1........
More generally, let HPn be an heptuplet of order n, generated from seven heptuplet HPn-1. An HPn covers 7^n ganglion receptive fields. It is easy to realize that the angular field increases
with n, and of course the complexity of the setting in the receptive
field of vision. Let us call an "heptarchy" such a hierarchy of
successive transforms.

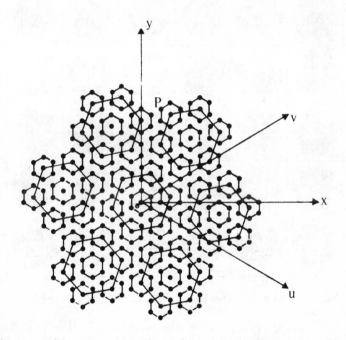

Fig. 6. Heptarchy on an hexagonal tessellation

In this organization, any point is defined by its heptal abcissa z, represented by a number of n digits in the base seven. For example the point P of the Figure 5 is the fourth element of an HP1, which is the fourth element of an HP2, which is the second element of an HP3.

Figure 7 displays the 64 sign-configurations of the HP2.

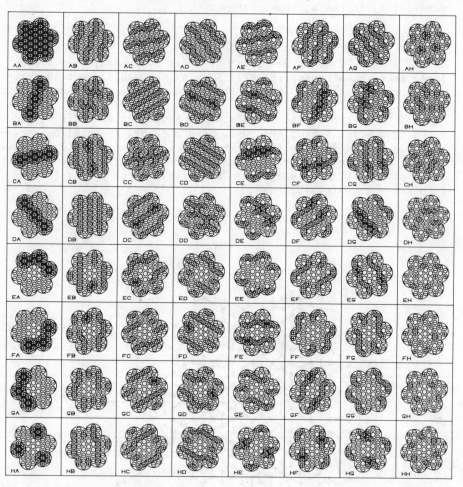

Fig. 7, the 64 possible configurations of HP2.

Let T_1, T_2,....., T_n be the matrices giving HP1, HP2,..HP$_n$ from the initial elements of the hexagonal tessellation. T_2 is built up from the kroneckerian product of two matrices T_1;... T_n from the kroneckerian product of a matrix T_{n-1} and a matrix T_1, or n matrices T_1. As the lines of T_i give the signs + or - , the sign configurations of HP2, HP3, ..., HPn can be obtained

by the kroneckerian product of matrices of inferior rank, in
particular of HP1. This process is to be compared to the following
statement by HUBEL and WIESEL (3): "It is enough to point out
that a complex (resp. hypercomplex) cell can most easily be understood by supposing that it receives its inputs from many simple
(resp. complex or more rarely simple) cells, all with the same
orientation preference".

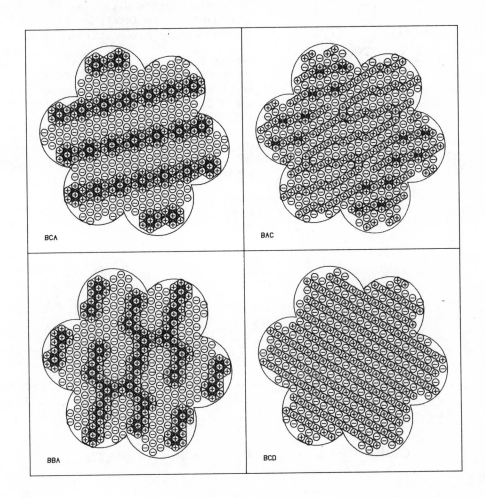

Fig 8: Four of the possible HP3 of different orientations.

It is possible to see on Fig. 8 some prefered orientations.
These HP have in fact a very interesting property in the Fourier
domain: they allow to sample this space in an homogenous way,
as it will be shown in the next paragraph.

The computational process, giving the successive HPn, and using kroneckerian products of matrices, can be visualized in the image plane.

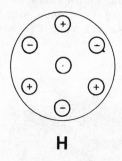

H

Fig. 9: H.

Let A, B,....,G, H; be the eight configurations of HP1. A is seven sign +. B to G are given by Fig. 4. The configuration H is linearly dependant of the former configurations and is given by the Fig 9.

Let us describe the process which allows to obtain HPn from HPn-1 and HP1. For this we show how HP2 can be obtained from HP1 and HP1:

i) Multiply by μ in scale one of the HP1, for instance C. $\mu = \sqrt{7}$
ii) Do the convolution with one of the HP1, for instance D.
iii) A configuration HP2 is obtained, denoted CD.

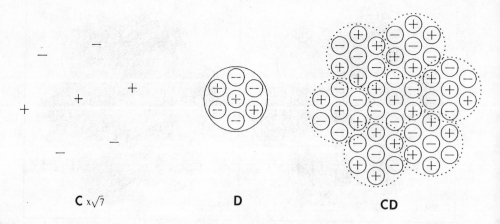

C x√7   D   CD

Fig. 10: On the left, the configuration C; in the center D; on the right, the HP2 configuration CD, result of the convolution of D and C scaled by $\mu = \sqrt{7} = 2,64$.

N.B. So to allow a tessellation without holes, C is also rotated clockwise of 19°10.

The configuration HPn is obtained in the same way, i.e. a convolution of a HPn-1, scaled by μ, with a HP1.

FOURIER TRANSFORM.

Let us recall the formulas of the discrete Fourier transform. See for instance ROSENFELD (11) § 2.2.

Let p,q and u,v be respectively the coordinates in the image plane and the Fourier plane.

Let p,q and u,v be respectively the unit vectors in the image plane and the Fourier plane.

A Fourier transform may be defined in a non orthogonal system of coordinates. The only requirements between p,q,u,v, is that the scalar products:

$$p.u = q.v = 1 \, , \qquad p.v = q.u = 0$$

In the image   plane a point is designated by

$$r = p\,p + q\,q \, ,$$

in the Fourier plane by

$$s = u\,u + v\,v.$$

The Fig. 11 shows the system of coordinates in the image plane of Fig. 10 and in the corresponding Fourier plane.

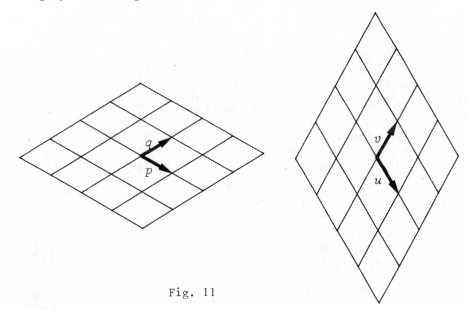

Fig. 11

Image plane   Fourier plane

The Fourier transform is given by F(u,v),

$$F(u,v) = \frac{1}{\lambda} \sum_{p} \sum_{q} f(p,q) \exp - \frac{2\pi i}{\lambda} (p u + q v), \quad \lambda \text{ is a coefficient}$$

This transformation is denoted

$$f(p,q) \xrightarrow{\text{TF}} F(u,v)$$

Convolution theorem

Let f and g be two discrete distributions. The convolution product is given by h,

$$h(k,1) = \beta \sum_{p} \sum_{q} f(p,q).g(k-p,1-q), \quad \beta \text{ is a coefficient.}$$

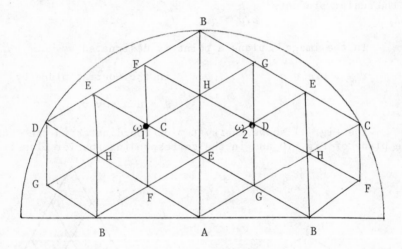

Fig. 12, peaks of the FT of the HP1.

The Fourier Transform of a discrete HP1 configuration is composed of sin. and cosin. functions. Each has three peaks, symmetrical of the points ω_1 and ω_2 in the Fourier plane.

The Fig. 12 gives the positions of these peaks. They sample the Fourier plane.

To obtain the spectrum of an HP2, we have to multiply the spectrum of an HP1, with the spectrum of an HP1 (not necessarily the same) scaled by μ and turned 19°10 counterclockwise.

$$f(\mu p, \mu q) \xrightarrow{\text{TF}} \frac{1}{\mu} F(\frac{u}{\mu}, \frac{v}{\mu})$$

The coordinate-axis of the spectrum of HP2 are deformed according to Fig 11.

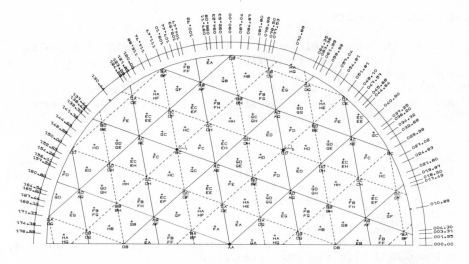

DEMI-SPECTRE DE FOURIER des configurations HP2

Fig. 13   The peaks of the spectrums of the HP2.(half plane)

F(u/μ,v/μ) covers the whole plane but with a periodicity
2,64 as fast as F(u,v). It is easy to see that again the plane is
uniformly sampled by the peaks of the HP2.
        A peak in the spectrum corresponds to a response maximum
to a grating of bands alternatively black and white, perpendicular
to the radius, and the spacing of which increasing with the value
of the radius. Again each HP2 has three spectrum maximums, symmetric versus the points ω_1 and ω_2.

        Physiologists have obtained experimental results for the
grating stimuli. See for example MAFFEI & FIORENTINI (5).
        POLLEN & RONNER (9) moved a narrow slit across the recep-
tive field of complex cells. They usually found five principal
peaks for the cells of area 17. This is most easily explained by
some of our HP2 configurations. Even they found a cell on the border of the area 17, which would give thirteen peaks in the response
curve. This is easily explained with a HP3 configuration such as
BCD of Fig. 8. *This similarity of results is rather encouraging
for the validity of our model.*

        A process similar to the former one allows to obtain the
spectrum of HPn from the spectrum of HP1 multiplied by the spectrum of HPn-1 scaled 1/μ Again the spectrum of an HPn is found
to have three peaks (except in the degenerate cases). These peaks
are symmetrical versus ω_1 and ω_2, and they sample the Fourier
plane uniformly.

CONCLUDING REMARKS

I)   As a model of mammal vision, the heptarchical organiza-
         tion of the visual field seems a rather suggestive system.
Many   experimental features are readily explained:
- The visual field is covered with receptive fields of different ψ
and angular definitions. The simple fields HP1 are usually more
selective that the complex HP2 or hypercomplex HP3 fields.
- The hexagonal tesselation is found in the retina of mammals.
- "Heptuplets" are found in the visual system of the fly (musca):
an omatidium is such an heptuplet. FRANCESCHINI has shown that,
in the center,two receptive cells receive the light, which is thus
parted in two as in our cell "a" of the HP1.
- Only excitory and inhibitotry linkings are used (or no linking).
- The kroneckerian process building up the heptarchy reminds of
the statement of HUBER & WIESEL (3) that a complex HP2 (resp. hypercomplex HP3) is made up of a combination of simple cells HP1 (resp.
HP2 and HP1) with the same orientation.
- The simple cells HP1 are "bar detector" or "edge detector".
- The response of complex cells of area 17 is similar to those of
HP2 or HP3:
         - To a slit of light: They answer better to an oriented
line stimulus, and an increase of the thickness of the line decreases the cell response.
         - To a grating of light stimuli, i.e. the frequency res-
ponse: our HP2 or HP3 seem to answer preferentially to three sets
of orientation and frequency. It is probable that they would answer to one only if the grating was moving.
         - POLLEN & RONNER found 5 or 13 peaks in the answer to
a properly oriented slit of light, as for an HP2 or an HP3.
         - Some of the HP3 have the feature of the "monkey paw"
in a proper orientation, as found by GROSS for an hypercomplex cell.
- The sampling of the Fourier plane i.e. of the spatial spectrum,
by the different HPi may explain many frequency responses.

II)   The former feature of uniform sampling of the spatial
         spectrum is a very attractive property of the sets of
digital filters HPi. Organized in an heptarchy, they may be a
very valuable mean for feature detection. The first experimental
results in this direction seem very promising.

REFERENCES

(1) FISCHER B. "Overlap of receptive field centers and represen-
         of the visual field in the cat's optic tract." Vision
         Research, 13, pp. 2113-2120, 1973.
(2) HUBEL D.H. & WIESEL T.N. "Receptive fields, binocular inter-
         action and functional architecture in the cat's visual
         cortex". Jr. Physiology, 160, pp. 106-154, 1962.

(3) HUBEL D.H. & WIESEL T.N. "Sequence regularity and geometry of
    orientation columns in the monkey striate cortex". Jr.
    comp. neur., 158, pp. 267-294, 1974.
(4) ALBUS K. "A quantitative study of the projection area of the
    central and paracental visual field in area 17 of the cat"
    Exp. Brain Res. 24, pp. 159-179, 1975.
(5) MAFFEI L. & FIORENTINI A. "Spatial frequency rows in the stri-
    ate visual cortex" Vision res., 17, pp. 257-264, 1977.
(6) CRETTEZ J.P., "A visual model based on the overlapping of the
    cell receptive fields in the visual pathway". Proc. of
    the first scandinavian conf. on image analysis, Linköping
    Sweden, jan. 1980.
(7) BENZER S., "From the gene to behaviour", Journ. of the Amer.
    Med. Assoc., 218, N°7, pp. 1015-1022, 1971.
(8) STEVENS Ch., "Interactions between intrisic membrane protein
    and electric field". Biology Jr., 22, n° 2, pp 295-306,
    1971.
(9) POLLEN D.A. & RONNER S.F., "Periodic excitability changes
    across the receptive fields of complex cells in the striate
    and parastriate cortex of the cat". Jr. Physiology, 245,
    pp. 667-697, 1975.
(10) GROSS C.G., ROCHA-MIRANDA C.E. & BENDER D.B., "Visual proper-
    ties of neurons in the inferotemporal cortex of the maca-
    que". Jr. Neurophysiol. 35, p. 95, 1972.
(11) ROSENFELD A. & KAK A.C., "Digital image processing". Academic
    Press, N.Y., 1976.
(12) EDEN M., "Visual image processing in animal and man" in SIMON
    J.C. (edit.) "Digital Image Processing and Analysis",
    published by Noordhoff, Leyden, Holland, 1977.

IMAGE CODING SYSTEM USING AN ADAPTIVE DPCM

S. CASTAN[*] L. MASSIP-PAILHES[*] M. BALAND[**]

[*] Laboratoire Cybernétique des Entreprises Reconnaissance
des Formes Intelligence Artificielle.
Université Paul Sabatier 118 route de Narbonne TOULOUSE

[**] Société SINTRA, 20 rue des Jardins 92600 ASNIERES

We are trying to build a real time transmitter for large scale
aerial digital image (5 000 x 5 000 pels) coded on 32 greys levels
(5 bits) making a bit rate reduction of the transmitted information for a human interpretation.

For financial and real time consideration we wish to use the least
memory as possible, that is the reason why the orthogonal transforms are not suitable and we have selected a differential pulse
code modulation image coding system.

1. SIMULATION.

All simulations have been made on images analysed and synthetized
on S.A.I.N.T. We have used only original images of 512 x 512 pels
and coded on 5 bits (1).

1.1. Adaptive DPCM

For a transmission using a Differential Code Modulation (DPCM), a
continuous image is spatially sampled and the difference between
an actual pel and its estimate is quantized and coded for transmission.

1.1.1. Predictive coder. The predictive signal $\hat{S}_o$ is formed
from a linear combination of several previously scanned pels along
a scan line and from the previous line (fig. 1).

J. C. Simon and R. M. Haralick (eds.), Digital Image Processing, 83–94.

$$\hat{S}_o = \sum_{i=1}^{n} A_i \, S_i$$

$$
\begin{array}{ll}
S_3 & S_2 \\
X & X \\
\\
\\
X & \\
S_1 & \quad\cdot\ S_o
\end{array}
$$

Figure 1

Spectral predictive image coding system where the A_i are prediction weighting constant determined by (2).

$$R_{oj} = \sum_{i=1}^{n} A_{ji} \, R_{ji} \quad \text{For all } j = 1, 2, \ldots, n$$

where $R_{ji} = E(S_j S_i)$ by experimental studies.

A first, second and third order predictors have been determinated and experimented on three aerial images.
The three predictors L_1, L_2, L_3 are :

$$L_1 : \hat{S}_o = A_1 S_1 \text{ with } A_1 = 1$$

$$L_2 = \hat{S}_o = A_1 S_1 + A_2 S_2 \text{ with } A_1 = A_2 = 0,5$$

$$L_3 : \hat{S}_o = A_1 S_1 + A_2 S_2 + A_3 S_3 \text{ with } A_1 = A_2 = 1 \quad A_3 = -1$$

The second order predictor give us a minimum entropy (table 1).

Table 1 : Entropy of the three aerial images

Type \ Image	I	II	III
Image Inital E	4,539	4,34	4,22
L_1	2,64	2,54	2,55
L_2	2,299	2,28	2,39
L_3	2,62	2,73	2,99

The entropy is given by :

$$H = - \sum_{i=1}^{n} P_{(i)} \cdot Log_2 (P_{(i)})$$

where $p_{(i)}$ denotes the probability of occurence of the level i for the n values of all the pels in the image.

1.1.2. Adaptive coder. To get a better reduction of the entropy we have to use an adaptive predictive coder. Usually the difference signal is quantized on 8 levels. We have determinated a coding system which permit us to use only three coding levels. For each predictive value (I $\in$ 0,31) the difference signal histogram is quantized in p runs. Each run is determined such a each population have to be nearest as possible of the total population divide by p (3).

Table 2 : Difference signal histograms between original and reconstruted image for the first aerial image with specific quantizer levels.

a \ b	8	7	7PD	6	5	4	3
- 10	72	30	0	50	25	170	62
- 9	53	32	0	72	39	217	72
- 8	60	36	2	76	58	342	74
- 7	52	62	48	98	74	524	164
- 6	76	86	78	181	99	614	261
- 5	85	105	98	305	166	380	432
- 4	118	186	87	497	240	1236	626
- 3	140	385	186	765	501	1863	1143
- 2	276	759	1437	1624	1106	4356	2396
- 1	1188	3183	19175	6246	33838	53997	85896
0	257593	249301	229387	244692	214543	182513	143790
1	1801	6068	10464	6893	9324	13463	21593
2	237	1621	1086	224	707	700	2339
3	117	61	79	80	356	356	1069
4	3	17	10	44	178	172	679
5	1	2	1	9	68	68	509
6	1	0	0	8	26	52	303
7	0	0	3	1	31	19	166
8	0	0	1	1	11	13	89
9	2	0	2	2	4	9	60
10	1	0	0	0	2	1	42

a : Quantizer levels   b : Difference signal

We have studied several quantization for the three images, and
table 2 show us the comparative results with the first image for
the levels 8, 7, 6, 5, 4, 3, and the Panter and Dite quantizer
for 7 levels.

Table 3 : a) Part of the difference signal histograms for a gene-
ralisated three quantizer levels.
b) Entropy and compression coefficient.

Image / Difference signal	I	II	III	
− 10	36	50	88	
− 9	60	100	43	
− 8	49	96	172	
− 7	55	107	168	
− 6	99	110	90	
− 5	200	122	120	
− 4	346	164	211	
− 3	531	325	179	
− 2	1134	1038	392	
− 1	15256	37387	15691	
0	129724	128790	123266	a
1	103567	87735	110379	
2	5681	3778	6893	
3	2159	1259	1563	
4	1182	439	553	
5	329	188	452	
6	531	53	350	
7	368	14	293	
8	184	3	289	
9	74	3	259	
10	43	0	211	

IMAGE	I	II	III	
Entropy	1,38	1,49	1,35	
Compression Coefficient	3,62	3,35	370	b

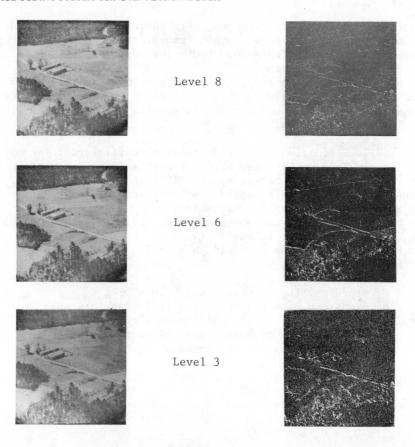

Level 8

Level 6

Level 3

Reconstructed images and difference signal images for 8, 6, 3
generalised quantized levels.

Figure 2

On table 3a we have the difference signal histogram for a generalized three quantizer levels rule ; the corresponding entropy and
compression coefficient are on table 3b.
Figure 2 show us reconstructed images and difference signal images
for generalized 8, 6 and 3 quantizer levels rules.

1.2. Run-length coding

1.2.1. Principle. Differantial coding on reconstructed pels :
In the part 1.1.2. the DPCM coder use likeness between neighbours
pels on original image. Now we are going to use the likeness between neighbour pels on the reconstructed image to increase the
compression coefficient.
In fact when the value of the reconstructed pel is equal, in

function of a likeness criterium, with the corresponding pel of
the previous, scan line, we do not transmit the coded difference
signal, but we decide to recopy the value of this previous pel,
and so on for each pel.
Differential coding on predictive pels : We use the likeness between the predictive pels and its original value. The difference
signal is transmited only when it is greater than the value of
the likeness criterium.

1.2.2. Estimation of the compression coefficient for grey
tone image. Thus we have two kinds of points : The I_d points coded
with an adaptive DPCM coding system (like in 1.1.2.) and the I_j
points which are to be recopied. For this last class we have to
use a run length coding for transmission.
For the estimation of the compression coeffcient, we have to use
the two sets of points I_d and I_j. (4,5).
Let h_{Jd} the entropy for a pel in this coding system.
$$h_{Jd} = P_d h_d + P_J h_j.$$

where P_d = probability to have a pel belonging at I_d
        h_d = entropy of a pel belonging at I_d (in this case we
consider that each pel is assimilated at a run lenght)
        P_J = probability to have a pel belonging at I_J
        h_J = entropy of a pel belonging at I_J
        $h_J = H_J/\bar{L}$
where H_J : Entropy of the I_J run-lenght
        L_J their means lenght.
If we use a Huffman code based on the mesured probabilities of the
run lenght and probabilities of the difference signal we have :
$$h_J \leq b_J \leq h_J + 1/L_J$$

$$h_d \leq b_d \leq h_d + 1$$

where b_j and b_d are the bit rate in bits per run.

For the image we obtain :
$$P_d h_d + P_J h_J \leq b_{Jd} \leq P_d h_d + P_d + P_j h_j + P_J/\bar{L}_J$$

$$h_{Jd} \leq b_{Jd} \leq h_{Jd} + P_d + P_J/\bar{L}_J$$

Then the compression coefficient if we suppose the two sets of
points like independant, is :

$$CL_M = \frac{N(\bar{L}_d + \bar{L}_J)}{H_J + H_d L_d}$$

where N : number of bits per pel. $\bar{L}_d$: means length of the I_d

1.2.3. Results. On table 4 we have the compression coefficients for the likeness criterium 0, 1, and 2, for 6, and 8 quantizer levels, for the differential coding on reconstructed pels.

Table 4 : Compression coefficients

Likeness criterium \ Quantizer levels	8	6
0	2,56	2,96
1	4,034	4,38
2	6,92	12,78

On table 5 we have the corresponding difference signal histogram.
The figure 3 show us the reconstructed image and the difference
signal multiplicated by 5 for a likeness criterium of one and two
for 6 quantizer levels.
On table 6, we have the compression coefficient for the differential coding on predictive pels.
On table 7, we have the corresponding difference signals histograms we see that the shape of the histograms is very different.
In the first case we have a flat histograms.
In the second case, the population is quasi completely determinated by the value of the likeness criterium which is like a filter
for the difference signals.
The figure 4 show us the reconstructed image and the difference
signal multiplicated by 5 for 6 quantizer levels and for a likeness criterium of one and two.

Table 6 : Compression coefficients

Likeness criterium \ Quantizer levels	8	6
0	2,986	3,097
1	4,994	5,094
2	8,104	8,2

This last method show us that it is possible to obtain reconstructed image good enough for usual inspection with a compression
coefficient about 8.

Table 5 : Part of the difference signal histograms between original image and reconstracted image for different likeness criterium and quantizer levels for image 1.

Difference signal	Likeness criterium	1		2	
	Quantizer levels	8	6	8	6
− 10		1	1	0	3995
− 9		0	6	0	3771
− 8		0	32	2	3657
− 7		2	69	8	4575
− 6		4	134	35	5963
− 5		17	317	110	7186
− 4		57	737	285	10334
− 3		202	1501	722	16823
− 2		518	3628	27492	34375
− 1		57006	59051	53729	51235
0		143943	134112	94055	61154
1		59593	61595	58117	36137
2		196	310	26785	15270
3		65	88	161	184
4		20	29	81	67
5		5	9	31	21
6		0	7	13	16
7		2	2	4	6
8		0	1	0	3
9		0	0	0	0
10		0	0	0	0

Likeness criterium
of one

Likeness criterium
of two

Figure 3 : Reconstructed image and difference signal for 6 quantizer levels (likeness criterium on reconstructed pels).

Likeness criterium
of one

Likeness criterium
of two

Figure 4 : Reconstructed image and difference signal for 6 quantizer levels (likeness criterium on predictive pels).

Table 7 : Part of the difference signal histograms between original image and reconstracted image for different likeness criterium and quantizer levels for image 1.

	EPS=1		EPS=2	
	8	6	8	6
− 10	0	0	0	0
− 9	1	1	0	0
− 8	0	0	0	1
− 7	0	0	1	0
− 6	0	0	0	0
− 5	1	2	1	3
− 4	3	22	0	8
− 3	22	76	5	33
− 2	87	209	11267	11265
− 1	40026	40928	30312	30637
0	142164	140083	88285	87345
1	79310	80274	80778	80958
2	7	27	80913	51327
3	1	4	12	1
4	2	2	14	13
5	2	2	2	7
6	3	1	14	16
7	2	0	13	9
8	0	0	6	3
9	0	0	1	1
10	1	0	3	4

2. REALISATION.

SINTRA has developed a line of photograph transmission and reception modules for remote use of data acquired during air reconnaissance missions, photographic data needed by the authorities for
identification tasks, data received by satellites, etc.
SINTRA photographic transmission equipment is based on the most
recent methods and technologies which make for high stability in
image quality and excellent safety in the transmission of data.
There are 6 to 20 scan lines per millimeter, on both axes.
The scale of gray values is from 2 to 32 levels.
Full transmission of an image format 250 x 250 mm takes from 1 to
10 minutes via a 64 Kbits/sec binary transmission channel, depen-
ding on image context and desired photometric parameters.
Analysis and inscription of image data involves the use of Helium
Neon lasers.
Redundancy reduction of digitalized video signals is obtained by
a micro-programmed processing system based on adaptable code.
Data reception is automatic and image recording on 3M 777 support
is obtained by controlled heating of the support during reception.

Optionally, the photograph transmission system may be completed
by one-line cripto decives, connected by HDLC process.
The transmission channel may have a capacity of between 2.4 and
128 K-bits/sec.
Fig. 5 show us the result after transmission and reception.

Fig. 5 : Reconstructed image by the SINTRA's system

REFERENCES.

1 - "Synthetiseur et analyseur d'images numériques de Toulouse
    S.A.I.N.T.".
    Patrice DALLE, Automatisme Mars-Avril 1976

2 - "Comparision of n the order D.P.C.M. encoder with linear
    transformations and block quantization techniques".
    ALI HABIBI, Member I.E.E.E.
    I.E.E.E. Transaction on communication technology. Vol. COM 19
    n° 6 Décembre 1971.

3 - "Un système de codage adaptatif en transmission d'image par MICD".
S. CASTAN, L. MASSIP-PAILHES. Congrès AFECT-IRIA Septembre 79.

4 - "Coding of two-tones images".
THOMAS S. HUANG. TR.EE. 10 March 1977

5 - "One dimensional coding of black and white facimile pictures".
ULF ROTHGORDT, Gilles ARRON, G. RENELT
Acta electronica 21.01.1978

LOOKING FOR PARALLELISM IN SEQUENTIAL ALGORITHM : AN ASSISTANCE IN HARDWARE DESIGN

B.Y. BRETAGNOLLE, C. RUBAT DU MERAC
Centre interuniversitaire de Calcul de Grenoble
BP 53 X - 38041 Grenoble Cedex - France

P. JUTIER
INSERM U 194
91 Bd de l'Hopital
75013   PARIS - France

INTRODUCTION

Two years ago we have been faced with the necessity of rebuilding our image processing facilities. We are operating in a large university computer center, and what we need is an equipment which can be used freely by scientific personnel for research and experimentation of processing algorithms.

Our customers, now well accustomed to time-sharing, want easy interaction excluding any specialized and complicated procedures. The image processing facility must be as straightforward to use as any other terminal.

The user wants to code easily (in normal programming fashion) a new algorithm, and see what happens. A few seconds response time is adequate. There is no need to work faster, as very large production is not asked for at the present time.

To sum up, our constraints were :
- be reasonably fast
- absolutely no limitation on algorithm
- avoid any exotic programming rules
- and, of course, keep cheap.

J. C. Simon and R. M. Haralick (eds.), Digital Image Processing, 95–103.

The answer was pretty obvious :
    - microprocessors : universal, not exotic nor expensive
    - several of them together to get enough power.

Then multiprocessing ? pipeline ? - better - parallel processing !

Hereunder we propose a few heuristics which may be of some help
for integrating both real world constraints of image processing
and technological opportunities.

I - ALGORITHMS AND PARALLELISM IN IMAGE PROCESSING

Images include a tremendous amount of information. There are reasons to believe that the human eye-brain system implements parallelism to a vast extend.

Is useful information to be found in current litterature on multi-
processors ? Flynn classification of data processing [1] is of
little use in our case, because it stops were we are starting to
be interested, at the multiprocessor level. Mazaré's taxonomy [2]
is very good as description, but of little help as a guide for
architectural choices. Let us look more specifically at parallelism.

1. Various types of parallelism :

A first approximation approach is to consider three main categories:
    - plain parallelism
    - functional parallelism
    - hidden parallelism

a) *Plain parallelism :*
is of the type "FOR ALL x DO action(x)".

b) *Functional parallelism :*
in other words, roughly, pipe-line : keeping as many copies of
the image as there are stages in the process. Only interesting
for dedicated architectures.

c) *Hidden parallelism (in sequential algorithms) :*
Our first assumption is that human expression abilities always
lead back to a sequential algorithm, at least at the programming level. So let it be our starting point.
Our second assumption is that a vast amount of parallelism is
embedded in all image processing problems.
Let us then dig out this hidden parallelism and construct the
kind of hardware which can face it. There are many different
ways of doing that but we guess that it is always possible to

apply the sort of strategies described hereunder.

2. A possible guide-line : Bernstein rules

Let us consider a sequential algorithm (W) as a succession of actions
   W = A1 ; A2 ; Ai;   ; An
   Each action Ai uses two data sets
   ORi : set of read objects
   OMi : set of modified objects.

Two actions Ai, Aj are said to be independent (meaning that they can be run "simultaneously") if
   OMi ∩ OMj = ∅
   OMi ∩ ORj = ∅
   ORi ∩ OMj = ∅

Bernstein proposed [3] these rules at a time -1966- when the problem was to run multiple tasks on a single processor, which guaranteed permanent disponibility of all OR to all tasks. He could disregard "Bus contention problems". This is no more possible in normal to-day practice. Hardware will have to be designed to cope with access problems, implicitely satisfying an additional rule (separation of data path) :
   ORi ∩ ORj = ∅

3. A trial and error methodology

Let us proceed as follows :
   a) We choose an obviously parallel problem.
   b) We construct the usual sequential algorithm.
   c) We apply our set of rules.
   d) We realise that they are not satisfied ! - hardly never in any practical case.
   e) We map out precisely the subsets of OR and OM which break the rules.
   f) We construct a machine (hardware and software) allowing when necessary the use of multiple objects instead of unique ones ; this beeing obtained both by duplication of objects and segmentation of data sets.

We tried that methodology on various examples of the three main categories of image processing algorithms :
   - pixel algorithms
   - neighbourhood algorithms
   - global algorithms.

A very common example of evidently parallel pixel algorithm is

thresholding. For an image represented by an intensity array I(k)
of n elements (k = 1 to n) the sequential thresholding algorithm
is :

```
    DO k = 1 TO n
      IF I(k) < T                    (threshold)
        THEN I(k) = B                B and C being two different
        ELSE I(k) = C                fixed values of intensity
    END
```

which performs n actions A (one on each element) ; let us consider two of these actions :

```
    Ai :  IF       I(i) < T
             THEN  I(i) = B
             ELSE  I(i) = C

    Aj :  IF       I(j) < T
             THEN  I(j) = B
             ELSE  I(j) = C
```

action Ai using ORi = I(i), T, B, C, code (program)
                OMi = I(i)
action Aj using ORj = I(j), T, B, C, code
                OMj = I(j)

These actions can be run in parallel if the set of rules is satisfied.

We find :   OMi ∩ OMj = ∅
                  ORi ∩ OMj = ∅
                  OMi ∩ ORj = ∅

Bernstein rules apply (!). Nevertheless, access problems do certainly arise for the common read objects : T, B, C, and code.
They have also to be kept in mind for array I ; I(i) and I(j)
while being different, may possibly be accessed by a common data
path.

Hereabove we defined exactly the two problems which hardware will
have to face in that particular case.

Instead of n actions A (meaning in practice n parallel processors)
we can consider m (= n/ℓ) actions working on m corresponding dis-
connected subsets of I. This type of partitioning, especially
when applied to less trivial algorithms as the one just considered,
leads to a very complicated formalism which cannot be included in
this short presentation.

This methodology is of particular interest in the case of neighbourhood algorithms, as it points out very clearly local conflicts

between modified points (OMi) and read points (ORi) and enables
to measure exactly the consequences of any given partitioning factor ℓ. An optimum value of ℓ will finally emerge. Note that we
already came to the same conclusion using a different (but fonda-
mentally equivalent) approach in a previous paper [4].

II - PARALLEL ARCHITECTURE AND DYNAMIC RECONFIGURATION

In practice, an image is very often a TV image, the TV camera
being a good sensor for various reasons :
- cost effectiveness - mass produced equipment for surveillance
  or hobbyists offer a very favorable cost/performance ratio,
- relative precision - the standard TV image is good enough
  for many scientific and technical usages,
- prevalence of the technique involved - no need for specialized personal.

1. Parallel organization of memory and processors :

The partitioning of the TV raster scan image is evident. Such an
image divides quite naturally in horizontal slices, containing
each a certain number of adjacent raster lines. The whole image
can be loaded into a matrix organized memory, which appears as
an exact projection of the image and can be partitioned in the same way (see figure 1).

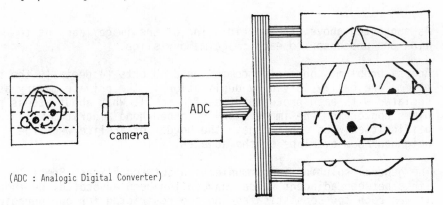

(ADC : Analogic Digital Converter)

Fig. 1 : From image partitioning to memory slicing

At this stage a very favourable feature of raster scan comes to

light. It is strictly a serial process and the best we can hope
to do is to capture one full image in the scanning delay (40 ms
in Europe), and this is done relatively easily.

The architectural problem boils then down to the single question :
"How to place the processors in relation to the memory" ?

The answer is obvious and does away instantly with the classical
(and absolete) multiprocessor organization (fig. 2).

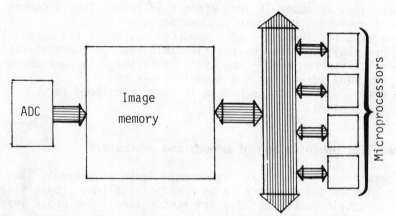

Fig. 2 : The (wrong) solution ...
We automatically escaped from

As suggested above, the partitioning of the image leads us to
link one processor to each image memory slice.

Access problems concerning common read objects (especially the pro-
gramm) are easily solved by duplication in the private memory associated with each processor (see figure 3). What about access problems concerning the image ? Any neighbourhood algorithm will raise
difficulties when applied near the border of a slice when the
neighbourhood overlaps on the adjacent slice.

The general solution is communication between slices. Special
links between adjacent slices have often been advocated. We thought
it was both too sophisticated and too restricted for our purposes
and means, and decided to implement a global processor able to address the whole image memory and allowing any transfer of any description between various locations. This is a very general measure,
as the expression "transfer" can be construed extensively : to
perform code placed anywhere in the memory space is also a kind
of transfer. Moreover we shall see that we need such a central
controlling device anyway.

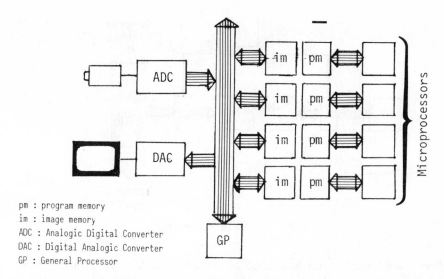

pm : program memory
im : image memory
ADC : Analogic Digital Converter
DAC : Digital Analogic Converter
GP : General Processor

Fig. 3 : The solution

Other simple solutions are possible : like slice overlapping, the problem being then passed over to the ADC.

2. Dynamic reconfiguration : dont displace data, dont talk about it, switch it !

The simple scheme of figure 3 could not work as shown, with free communication everywhere. We have to provide for gating of the data path, and decide which means we are going to resort to. As suggested by many good authors, a lot can be done (mostly by software) in the style of hand-shaking, semaphores, letter boxes, and any other variety of "polite exchanges". Our option is : "no niceties". Where sophisticated systems spend dozens of microseconds in soft exchanges, let us use a few nanoseconds to close or open a tristate bus driver. Then, the cost of gating being so low, you can use it freely...

Figure 4 shows a very simple example of that. Data contained in memory i can be used and/or modified by local processor i (G1 open, G2 closed) or by the external world (G1 closed, G2 open).The hardware and software necessary to generate REQuest was shown to be very simple in a previous article [4].

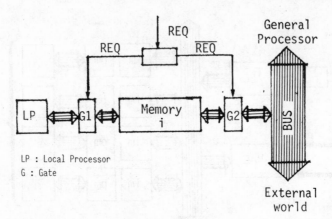

LP : Local Processor
G : Gate

Fig. 4 : Gate mecanism

A first implementation has been built (figure 5). By simply ope-
ning and closing tristate bus drivers we can create many different configurations of 5 different types :

1. Acquisition machine, ADC feeding the memory, slice after
   slice, from any TV type source (camera, video tape, etc.).
2. Display machine, the other way around.
3. A group of totally separated processors, each operating on
   its own memory. This concerns of course the 8 local proces-

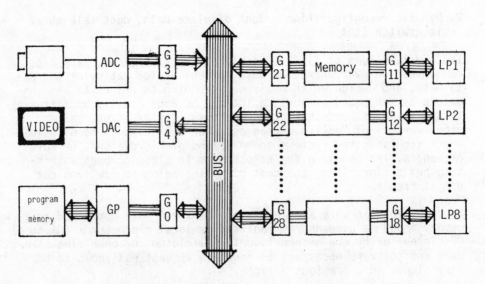

Fig. 5 : First implementation : the ROMUALD machine

sors, but also possibly the global processor, if needs be.
4. A memory intensive computer (monoprocessor). Power can be added to the global processor, by various means, in view of that.
5. Any combination of 3 and 4. Some local processor are at work independently, the global processor using the part of the memory left at its disposal by the idling local processors.

CONCLUSION

We are not ashamed to concede that large parts of our hardware will be idle most of the time, in interactive usage. Microprocessors, even very powerful ones, are really so cheap compared with the rest of the hardware (especially the image memory... but also cards, cages, and power supplies) that this is really of little consequence.

We are also not ashamed to admit that it is possible to built a faster machine. We just feel that the sort of architecture we plead for is well adequated to our needs : interactive experimentation of algorithms. We believe also that, with some upgrading such an architecture could also be implemented in production facilities.

REFERENCES AND BIBLIOGRAPHY

[1] M.J. FLYNN. Some computer organizations and their effectiveness. IEEE. Transactions on computers, C-21,9. Sept. 1972.

[2] G. MAZARE. Structure multimicroprocesseurs. Problèmes de parallélisme, définition et évaluation d'un système particulier. Thèse d'Etat, Grenoble, juin 1978.

[3] A.J. BERNSTEIN. Analysis of programs for parallel processing. IEEE Transactions on Electronic computers, EC 15,5, October 1966.

[4] B. BRETAGNOLLE, C. RUBAT DU MERAC, J. SEGUIN. Architecture of a multimicroprocessor for picture capture and processing : ROMUALD. 1st International Micro and Mini Conference,Houston, November 1979.

[5] P.E. DANIELSSON. Designs for parallelism in image processing. Linkoeping University. Internal Report, LITH-ISY-I-0388.

FINDING THE EDGE

S.Levialdi

Istituto di Cibernetica, CNR
Naples, Italy.

The problem of extracting an edge in real images is
briefly described. A number of methods grouped as local,
regional, global, sequential, heuristic, dynamic and
using relaxation are schematically illustrated with an
attempt to give some insight to the ideas which origi-
nated each approac-. Evaluation of results and critical
aspects of some methods, according to certain authors
are also included, as well as some hints on the latest
research trends in this area.

1. INTRODUCTION

In picture analysis, it is often required to extract subparts of the picture that have certain properties, either for recognition of the whole picture (scene analysis) or for the measurement of each single object which may be contained in those subparts. In this
very broad sense, a general purpose technique such as
segmentation may be also viewed, within a specialized
approach, as one means to obtain these subparts. However
such subparts require a definition in order to properly
develop a technique to be tested on real images.
As in many other instances of picture processing,
it is the starting hypothesys which sets the pace and
the elementary tools with which an algorithm is built
will finally establish its usefulness. Generally, the
goal is difficult to define in a clear cut way on one
hand and on the other the evaluation of a technique is a
function of the images on which it is applied. Let us
come closer to the subject of this work:namely a broad

105

J. C. Simon and R. M. Haralick (eds.), Digital Image Processing, 105–148.
Copyright © 1981 by D. Reidel Publishing Company.

discussion on the efforts that have been spent to automatically locate edges on a real digital image by means
of procedures which use, at the best possible level,
the information about these edges that the author has
been able to embed within his own method.

As stated above, one first point is the one regarding
the definition of an edge, or for some other authors, of
a contour, of a line segment, of an edge step, of a continuous line in the presence of noise, etc. The next point
to consider is connected to the basic ideas that form
the substratum of the algorithm and that, sometimes, are
deeply connected to the author's idea of an "edge".
These ideas are then translated into mathematical language and an algorithm results which also reflects the
author's personal tastes (determinism, parallelness,
heuristics, etc.).

Finally one should expect to have a nice classification of the algorithms according to their ideologies
and a comparative evaluation of these algorithms, hopefully on the same images (whether artificially generated or natural ones). The last part should deal with
the most recent ideas on this subject, which are the
new trends and how will they improve the present state
of the art of edge finding. The task is immense and I
am neither the only nor the first to have tackled this
subject. There are very good reviews of edge finding
algorithms both on specialized papers (1) and in books
(2), it seems an impossible task to improve them just
by being a little more updated. I will try instead to
consider the different author's point of view and to
quote only those papers which have indeed changed the
attitude towards the way in which an edge finder should
be designed and tested. A very subjective partition of
the different ways of thinking will also be given, just
to indicate the borders (sic) more than the contents of
the regions which contain similar algorithms for edge
finding.

Let us now discuss the motivations. In the early
days of Cybernetics, (1950s), a number of researchers
believed that an accurate study of the visual system
(in mammals) (3) could be of some use for designing machines that would act as living beings, at least on
some primitive level, i.e. detect objects in a scene,
evaluate contrast, distance from the observer, etc.
In those times a number of vision models were built and
the word "retina" was used to denote specific configurations
of electronic optical sensors which could, by means of a
special wired logic, extract the position of an object, its
contour, its connectedness degree, etc. Many analog machines
on the basis of the "lateral inhibition" (4) principle, exhibited

properties which were similar to the ones possessed by
natural living systems. But, as well known historically
the "copy" of living organisms by a machine is not necessarily the best way to perform the same task in an
economical way (see for instance the airplane and the
wheel as two good examples of convenient differences
with the way in which living organisms fly and walk).
The rapid evloution of the digital computers and the increased flexibility of programming such machines origi-
nated a completely new trend towards the implementation
of contour extractors (or edge detectors) which were not
machines any more but programs which reflected algorithms. In this way, many different ideas could be easily
tested and, although this is common practice today (1980)
we must not forget that the first programmable machine
is only 39 years old (ENIAC,1941).

The range of applications (in different fields and
for large quantities of image data) have also pushed
towards this direction and, nowadays, no real system for
image processing can avoid having some "subsystem" which
will, in general, perform pre-processing and very likely do some edge extraction (contour extraction,streak
detection,etc.) at an early stage of the global processing of the image.

One way of introducing the subject is by means of
defining the object of our research' namely, to find
edges in an image which is the digital version of an
analog picture, after it has been quantized both spatially (along the x and y coordinates) and in the intensity value of each local area, hereafter named pixel (or
picture element). This digital image is generally represented by a bounded non negative integer function which
may have a different value as a function of the coordinates of the pixel. If pixels share the property of having
similar values (grey levels), and if this similarity is
contained within a certain fixed small interval ε , we
may speak of regions, i.e. subsets of the image in which
the pixel values may vary between I and I+ ε .It may happen
that more than one region may be present in an image,
each one having a different average value, it is the
difference between these average values which originates
an edge. In other words, an edge may be viewed as a subset of pixels of the total image (it would be meaning-
less if this subset would coincide with the total number
of pixels in the image) for which, adjacent pixels share the property of having a significantly large grey
level difference. On the basis of which adjacent neighbours, what difference and for which width (pixels belong to an edge), a great number of variations have
been suggested in the literature.

But the diversity of approaches is not confined to these parameters only, local and regional methods (or more global), methods also differ in the extent of the neighborhood which must be inspected to decide about the existence of an edge pixel, parallel methods versus sequential ones' the first ones act on all the array at once substituting the values of each pixel on a given context (or weighted context) whilst the second ones will find a first pixel and then try to locate the remaining ones which are connected (in some way) to it and that constitute the edge within the image. Algorithms which do not use any a priori information about the the edge to detect in order to pick out the pixels belonging to the edge and, on the other hand, information dependent algorithms which use knowledge of the objects that have edges, that process parts of the picture and then decide using rules which embed such knowledge.

Tracking methods with forward direction only, that firstly find some edge pixels and then other new pixels in their neighborhood (as in the sequential methods briefly mentioned above) and tracking methods which use backtracking so as to confirm or discard new pixels on the basis of some local and past evidence about their membership to the edge class. Methods which firstly use an unexpensive line segment extractor (primary evidence about the existence of edges) and then continue with line builders so as to fill in existing gaps or remove pixels which do not really belong to the edge. More recent approaches include relaxation methods where some preliminary hypothesis is built about the properties of the pixels and of their immediate neighbors, some compatibility rules are given so as to impose a reinforcement of the probability of being "edge pixels" if their compatibility is high and viceversa if it is low, the process is iterated until at some step, there is a small variation in the values of pixels expressing their membership to the class of edge pixels. These methods are very fast if implemented on parallel processors and substantial work has been done in this area in these last years.

As in all feature extraction problems the main theme is' which are the best cues for extracting the feature? and the answer to this question is inevitably tied to the definition of the feature. If an edge involves large intensity variations between a region and another one, the obvious cue will be the intensity difference or a weighted difference of intensities within in a specified neighborhood and this is indeed an approach which has been followed, for many years, by a wide number of authors in the field. On the other

hand, if a different representation of the problem is
given, namely that the feature may be defined in terms
of a frequency spectrum, the cue will be obtained by
exploring the image in terms of the digital filtering
techniques which will provide a model of the image, of
its noise, and of the sets of pixels that may belong
to the edge class. On the other hand if we think that
the degree of detail in an image may constitute noise
from the point of view of an edge extractor, we can
firstly average locally the image and compact it into
a much smaller size, look for local intensity differences in this new image and use this knowledge (the
location of these differences) to perform a more refined analysis which is now based on some preliminary
analysis (planning approach). In some cases a set of images
(dynamic analysis) may play an important role, in others
the fact of using a varying threshold may also lead to
interesting results. As we have seen, there are many
ways of considering the principal cue for extracting
an edge and each one of these leads to a different me-
thodological approach and, therefore, to a different
algorithm. The literature is very rich in this area,
the yearly image processing literature surveys (5-12),
prove this fact and, perhaps, some similar approaches
still deserve publishing since they reflect each individual's experience when dealing with this problem in
his own specific application.

2. LOCAL METHODS

The first and most intuitive approach to extract the
pixels belonging to an edge is the one using the concept of a gradient transformation on the discrete (may
be traced back to 1962),(13) so as to detect local intensity variations. Within the realm of continuous two
dimensional functions, the gradient $\nabla f(x,y)$ is given

by $\nabla f(x,y) = (\frac{\partial f}{\partial x} + \frac{\partial f}{\partial y})$, its magnitude will be

$|\nabla f(x,y)| = \sqrt{(\frac{\partial f}{\partial x})^2 + (\frac{\partial f}{\partial y})^2}$ and the orientation of

the gradient vector $\alpha = \tan^{-1}(\frac{\partial f}{\partial x} / \frac{\partial f}{\partial y})$
But when operating on the discrete, like on digital
images, x,y and f(x,y) are non negative integer numbers so that the partial derivatives must be approximated with finite differences along the two orthogonal
directions x and y, so obtaining

$$\nabla_x f(x,y) = f(x,y) - f(x-1,y)$$

$$\nabla_y f(x,y) = f(x,y) - f(x,y-1)$$

x-1,y	x,y
x-1,y-1	x,y-1

and for any orientation α we will have

$$\nabla f(x,y) = f(x,y) \cos\alpha + f(x,y)\sin\alpha$$

and finally, the digital approximation to the gradient of $f(x,y)$ will be given by

$$|\nabla f(x,y)| = \sqrt{\nabla_x f(x,y)^2 + \nabla_y f(x,y)^2}$$

Since this expression may be cumbersome, it generally happens that the digital gradient is considered to be either the sum of the absolute values of the two directional increments or the maximum between these same two increments:

$$|\nabla_x f(x,y)| + |\nabla_y f(x,y)| \quad \text{or} \quad \max(|\nabla_x f(x,y)|, |\nabla_y f(x,y)|)$$

These approximations are dependent on orientation (14). In some cases other neighbors are also used (the eight neighbors) and we then have

$$f_{l,m}(x,y) = \max \{f(x,y) - f(l,m)\}$$

where l,m are the coordinates of the eight (or four) neighbors of the pixel located at x,y.

Another practical approximation is due to Roberts (15) and may be written as

$$f(x,y) = \max (|f(x,y) - f(x+1,y+1)|, |f(x+1,y) - f(x,y+1)|)$$

x,y+1	x+1,y+1
x,y	x+1,y

Since this approximation computes the finite differences about an ideal element located at $(x+1/2, y+1/2)$, it may be considered an approximation to the continuous gradient at that position.

If a three by three neighborhood is considered there are two well known approximations to the gradient, one due to Prewitt (16) and the other to Sobel (17), they are both given below with the corresponding labelled elements on the image.

x,y+2	x+1,y+2	x+2,y+2
x,y+1	x+1,y+1	x+2,y+1
ẍ,y	x+1,y	x+2,y

<u>Prewitt</u>

$$\{ \left| \sum_{y}^{y+2} f(x,y) - \sum_{y}^{y+2} f(x+2,y) \right|$$

$$- \left| \sum_{x}^{x+2} f(x,y) \right| - \left| \sum_{x}^{x+2} f(x,y+2) \right| \}$$

<u>Sobel</u> (introduces weights in the summation of the
values of the elements)

$$\{ | f(x,y)+2f(x,y+1)+f(x,y+2)-(f(x+2,y)+2f(x+2,y+1) +$$
$$+ f(x+2,y+2)) | - | f(x,y)+2f(x+1,y)+f(x+2,y)-(f(x,y+2)+$$
$$+2f(x+1,y+2)+f(x+2,y+2)) | \}$$

which have been written in carthesian notation for
uniformity reasons but will be re-written as masks,
i.e., as configurations of values which will be multiplied by the corresponding values of the pixels and
then added or subtracted as the sign of the coefficient in the mask suggests (see also (18)).

<u>Roberts</u>

$$H_1 = \begin{vmatrix} 0 & -1 \\ 1 & 0 \end{vmatrix} \qquad H_2 = \begin{vmatrix} -1 & 0 \\ 0 & 1 \end{vmatrix}$$

<u>Prewitt</u>

$$H_1 = \begin{vmatrix} 1 & 0 & -1 \\ 1 & 0 & -1 \\ 1 & 0 & -1 \end{vmatrix} \qquad H_2 = \begin{vmatrix} -1 & -1 & -1 \\ 0 & 0 & 0 \\ 1 & 1 & 1 \end{vmatrix}$$

<u>Sobel</u>

$$H_1 = \begin{vmatrix} 1 & 0 & -1 \\ 2 & 0 & -2 \\ 1 & 0 & -1 \end{vmatrix} \qquad H_2 = \begin{vmatrix} -1 & -2 & -1 \\ 0 & 0 & 0 \\ 1 & 2 & 1 \end{vmatrix}$$

This new form is easier to follow since we may observe
all the weights simultaneously on corresponding positions of the neighborhood. Since these last local approximations to gradients have more than two levels
(if compared to the initial digital gradient) they
are sometimes called 3-level gradient (for values
-1,0,1) and 5-level gradient (for values -2,-1,0,1,2).

Moreover, there is also another well known gradient operator, due to Kirsch (19), which uses stronger weights:

Kirsch

$$H_1 = \begin{vmatrix} 3 & 3 & -5 \\ 3 & 0 & -5 \\ 3 & 3 & -5 \end{vmatrix} \quad H_2 = \begin{vmatrix} 3 & -5 & -5 \\ 3 & 0 & -5 \\ 3 & 3 & 3 \end{vmatrix}, \text{ etc. and the}$$

other rotated versions of the masks which will be a total of 8 as with the 3 and 5-level masks. Finally, the compass gradient should also be mentioned, it is made of 8 masks, two of which are given below.

Compass gradient

$$H_1 = \begin{vmatrix} 1 & 1 & -1 \\ 1 & -2 & -1 \\ 1 & 1 & -1 \end{vmatrix} \text{ and } H_2 = \begin{vmatrix} 1 & -1 & -1 \\ 1 & -2 & -1 \\ 1 & 1 & 1 \end{vmatrix}$$

In all cases, the output of the digital gradient, $O_{x,y}$ one for each pixel of the image, will be given by the maximum value obtained after applying all the rotated versions of a mask:

i.e. $\quad O_{x,y} = \max_i \{ |G_i(x,y)| \}$

for i = 1,8

and the orientation (α) of the gradient will correspond to the $O_{x,y}$ which was found to be the maximum. An exhasperated version of this approach is the one using the digital version of the Laplacian, namely of

$$\nabla^2 f(x,y) = \frac{\partial^2 f}{\partial x^2} + \frac{\partial^2 f}{\partial y^2}$$

which may be approximated on the discrete, for $x,y,f(x,y)$ integer values, as

$$\nabla^2 f(x,y) = f(x+1,y) + f(x-1,y) + f(x,y-1) +$$

$$+ f(x,y+1) - 4f(x,y).$$

The graphic interpretation of the central pixel and its four adjacent neighbors helps in identifying the considered elements and, as before, the masks with the corresponding weights give the most direct way for following what the digital Laplacian really does:

$$DL_4(x,y) = \begin{vmatrix} 0 & 1 & 0 \\ 1 & -4 & 1 \\ 0 & 1 & 0 \end{vmatrix}$$

If the eight neighbors are considered we will have

$$DL_8 = \begin{vmatrix} 1 & 1 & 1 \\ 1 & -8 & 1 \\ 1 & 1 & 1 \end{vmatrix}$$

and if only positive values are requested, the absolute value of the summation may be taken.

The problem arising with this operator is that points will be detected (for the 4-neighbor case) 4 times as strongly as an edge, a line end will be 3 times as strongly and a line, 2 times as strongly since when the weighing of the neighbors is performed, its number will influence the obtained value.

All the methods which are gradient or Laplacian-based are intrinsically noise sensitive, since a small number of pixels having a different grey value from the background will be extracted independently from their organization, i.e. whether they lie on a line (edge) or whether they are a small dot-like cluster. These observations are certainly not new but; we may say that these consequences are inevitable because the edge's main cue is considered to be the grey level difference that may be measured between adjacent regions on the image.

It is difficult to compare the performance of all these local operators, whether they are gradients, laplacians or templates, in a way which is general enough and therefore does not depend on the particular images chosen for this test.

Different authors have approached this problem (the evaluation problem) and, for instance in (20) four local operators have been considered: Prewitt, Kirsch, 3-level and 5-level simple masks and applied them to three different images: a toy tank, a girl's face and a satellite image so as to have a diversified kind of edges to extract. As mentioned before, the mask which will provide the largest output is retained and thresholded, a binary edge map will be finally obtained. One general edge detection scheme (suggested by Robinson (21)) may be seen in the block diagram below.

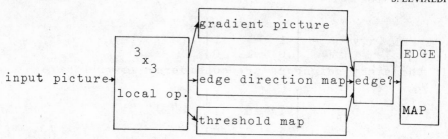

In this approach, the evidence of the gradient, its
magnitude, its direction and a local threshold are combined to build up an edge knowledge base after which a
binary decision will be taken so as to produce an edge
map.
If we now consider the 5-level simple masks their advantages may be summarized in the next five points (20).
1) they approximate the partial derivatives along both
   axis (x,y);
2) the 0-weights in the center help to suppress line
   jitter;
3) the use of only four masks (the ones having the 0s
   along the four principal directions) computes the
   analog of the gradient of the image, the sign of
   the output will establish either one direction (for
   instance positive sign, East direction) or the opposite one (negative sign, West direction);
4) a defocussing action may be simulated on the same

   3 by 3 neighborhood by using $M_o = 1/16 \begin{vmatrix} 1 & 2 & 1 \\ 2 & 4 & 2 \\ 1 & 2 & 1 \end{vmatrix}$ and

   convolving the image with it so as to compute a
   local threshold value (a weighted average);
5) the 5-level masks have a greater sensitivity to the
   diagonal directions than to the horizontal and vertical ones, so compensating the human visual response which has a smaller acuity for the diagonal directions.
One important aspect of building the edge knowledge
base is the use of local connectivity information which
might be simply stated as (21):
   if the direction of the central point is k (for
   k = 0, 1, ... 7 as in Freeman chain coding scheme)
   and
   if the directions of the preceeding and succeeding
   edge vectors are k, k±1 (modulo 8)
   then the edges are connected
In this way, we will require that the edge vectors exceed a threshold and are locally connected if they
must be considered as edges. In practice by adding the
local connectivity condition about 3/4 of the pixels

disappear and noise is greatly reduced; the details
are more prominent and spurious edges are cancelled.
The same happens for the Kirsch operator and a fixed
threshold, when the connectivity tests are applied
the results improve in a significant way.

Another parameter which may help in evaluating
the edge's strength is the edge activity index (EAI)
which is defined as the ratio of the maximum gradient
magnitude with respect to the average magnitude of the
gradients along the 8 compass directions,

$$EAI = Max\{|OP_k|, k=0,1\ldots,7\}/1/8\sum_{k=0}^{7} (OP_k^2)^{1/2} - 1$$

There might be no preferred orientation, in this case
EAI = 0. In this way the gradient will be accepted
only if EAI is high (above a certain threshold value),
otherwise a 0 is entered at the corresponding pixel
place.

Following this method, an analog gradient image
may be built where the grey level of a pixel represents
the gradient maximum found for that pixel or 0 if not
found (EAI under the threshold). There are many ways
to obtain a suitable threshold, one possibility (suggested in (21)) is to defocuss the image by applying
M_o and then a ratio is computed between the maximum
output at pixel x,y and the output of the mask M_o
applied to the same pixel and its environment. This
locally adaptive threshold (LAT) may therefore be defined by

$$LAT_{x,y} = \frac{Max\| OP_{x,y} \| (k=0,1,\ldots,7)}{OPM_{o_{x,y}}}$$

As may be seen by the figures reported in (21) the
results are much better than when a fixed threshold
is used. But the evaluation of the local techniques
for edge extraction is not only connected to the post
processing operations which are applied for improving
the results but should also consider some of the effects that directly descend from the structure of the
masks: weights and configuration. For instance, some
work has been done (14) on the influence of the orientation of an edge on its detection and a comparison
of the different masks on the basis of this feature
has been carried out, both when considering the magnitude and the square root of the maximum output for
each pixel with a given mask. If we take a look at
the reported curves in (14) we may see that the Sobel

and Roberts operators are worse than the template matching operators (Kirsch, Compass-Gradient, 3and 5-level
simple masks): the amplitude of the gradient increases
(from a normalized value of 1 to 1.5) with the variation of the angle α in radians (from 0 to 1) for the
Sobel operator and decreases (from 1 to .3) as the
angle increases in the same range as before. Another
interesting result relates the real and the detected
orientation for each mask; from the figure we may see
that the best accuracy is obtained from the Sobel operator whilst the template matching gives a step edge for
the transfer function between the orientations (real
and detected). We may then consider that the sensitivity may be described both as the amplitude response in
terms of the orientation of the edge and as the ratio
between real and detected orientations.

Another important feature of an edge detector is
its independence from the edge position (translation
independence). All the considered (20) operators
(Kirsch, Sobel, Compass Gradient and Prewitt) exhibit
a fall, as the edge is displaced, with a steep descent
around the normalized value .7 of the width of the
pixel until the displacement reaches 1.5 where the output of the operator is practically zero.
Other tests were performed on vertical and diagonal
edges on which white noise was added with signal to
noise ratios 1 and 10. All the detectors have been
tried (Sobel, Prewitt, Kirsch, 3-level and 5-level,
Compass Gradient) and it was shown, on a statistical
basis, that the best operators are Sobel and Prewitt.

A problem which has also been faced by the same
authors (20) is the one of assigning a figure of merit
for an edge extractor. In (22) a suggestion is made to
measure the detected correct and false edge probabilities, the figure of merit would be obtained by their
ratio. In (20) a model of an edge section going from
level $\underline{b}$ to level $\underline{b+h}$, in a width equal to $\underline{a}$ is considered. A mean square distance is used for evaluating the
quality of the edge detector. More precisely, a figure
of merit F is given by

$$F = \frac{1}{\max\{I_I, I_A\}} \cdot \sum_{i=1}^{I_A} \frac{1}{1 + \alpha d^2(i)}$$

where I_I and I_A are the number of ideal and real
points respectively, d(i) is the miss distance of the
i(th) edge detected and α is a weighting factor (1/9)
to penalize smeared and offset edges (see below).

ideal edge   fragment edge   offset edge   smeared edge

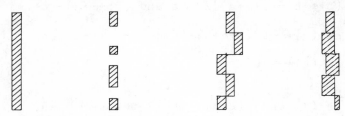

 An approximate graph representing the results of
measuring F (as %) on images having different signal
to noise ratio (from 1 to 100) for all the different
digital gradients described so far is given in (20).
It may be seen that a signal to noise ratio of 20 is
a standard value for obtaining a high F, the operators
may be ranked as follows:
Prewitt and Sobel provide a higher F than Roberts, the
Prewitt is better for vertical edges and approximately
equal to Sobel for diagonal edges. One possibility for
increasing F is to extend the mask size and therefore
to average out noise at low signal-to-noise ratios, but
pixels which are more distant should have a weight
which is inversely proportional to   its distance from
the center.
 Another way to consider a classification of local
edge detectors(23) is to separate them into linear ones
(in which the values resulting from the application of
a specific mask are directly applied and a threshold
is used to test the difference of local grey level
values) and non linear ones (in which a threshold is
used to establish both the difference across the regions and the difference along the regions). In this
approach, (23) the notion of semilinear detectors is
introduced, they look for sequences of points whose
average intensity is darker than the average intensity
of each of the adjacent sequences in the direction a-
cross the line.
In practice, for a 3 by 3 neighborhood like the one
labelled below
a   b   c   three kinds of detectors may be defined in an
d   e   f   Algol-like fashion
g   h   i
<u>Linear</u>
 If $[2(b+e+h)-(a+d+g)-(c+f+i)]/6 \geq t$
 then line (e)$\leftarrow[2(b+e+h)-(a+d+g)-(c+f+i)]/6$
 else line (e)$\leftarrow 0$

Semilinear
If [(b+e+h)-(a+d+g)]/3 ≥ t and [(b+ē+h)-(c+f+i)]/3 ≥ t
   then
      line (e)◄──[2(b+e+h)-(a+d+g)-(c+f+i)]/6
   else
      line (e)◄── 0

Nonlinear
   If (e-f) and (e-d) ≥ t If thin e > 0 and thin
                               b > 0 and thin h > 0
   then
      thin e◄─[(e-f)+(e-d)]/2   line (e)◄─[thin(e)+thin(b)
   else   thin(c)]/3
      thin e◄─ 0
   end;   else
                                           line (e) ◄─ 0
                                          end

The same will be for the other directions only that
the masks will be considered differently. The comparison between these three kinds of local edge detectors
shows that the linear detectors will output k/2 if the
grey level difference is k, they will also smear out
isolated noise points and for a single point which is
k darker in grey level, they will output 2k/6=k/3.
On the other hand, the semilinear detector will respond
to isolated noise points, since it is based on local
averaging, and might consider some noise regions as
edges (false positive response). Finally the nonlinear
detector will not respond to isolated noise points since it performs isolated tests. As we will see, there
are advantages and disadvantages when performing disjoint tests which must all be satisfied in order to
detect an edge. The experiments reported in (23) have
considered a synthetical vertical edge of grey level
1, 2 and 3 with a noisy environment of μ=32 (mean) and
σ=9 (deviation) from a normal distribution. A set of
images had one standard deviation above the mean, the
second had two standard deviations above the mean and
three standard deviations (the signal of the edge above
the noise). The general conclusions show that the low
signal to noise image is poorly processed with all
three detectors, that the threshold choice is very important: if too low, more noise is present, if too high
some gaps in the edge appear. The non linear detector
will clean the noise but breaks the edge, the linear
detector may detect wrong edges (false positives) and
semilinear detectors are not easily distracted by adjacent noise and bridge gaps better than nonlinear detec-

tors. A good approach might be to combine semilinear
with nonlinear detectors in a succession like: SNN.
This is in fact what is known as iterative local edge
detection and has been used by a number of authors.
In particular, in (24) a combination of local techniques defining an iterative procedure on an image has
been developed.

If $E_1(x,y) = G(x,y) - \overline{G}(x,y)^+$ (where $^+$ means positi-

ve value, else 0)

and $\overline{G}(x,y) = \dfrac{1}{(2r-1)(2s-1)} \displaystyle\sum_{m=x-r}^{m=x+r} \sum_{n=y-s}^{n=y+s} G(m,n)$ (local

average)

then, the second and subsequent iterations have $G(x,y)$
replaced by $E(x,y)$; for the kth iteration:

$$E_k(x,y) = (E_{k-1}(x,y)) - (\overline{E}_{k-1}(x,y))^{(+)}$$

After performing k times the averaging and subtracting
operations indicated in the above formulas, an edge
sharpening routine is iterated q times so as to enhance
straight lines along the four principal directions:
0°, 45°, 90°, 135° which may be analyzed by means of
the following 5 by 5 masks:

```
       bbbbb            .bb.a        b.a.c        a.cc.
       .....            bb.a.        b.a.c        .a.cc
0°=    aaaaa     45°=   b.a.c  90°=  b.a.c  135°=  b.a.c
       .....            .a.cc        b.a.c        bb.a.
       ccccc            a.cc.        b.a.c        .bb.a
```

The sharpening routine operates by summing three components in each mask

S_a, S_b, S_c and the sharpening effect is obtained

forming

$$S^\theta = 2S_a^\theta - S_b^\theta - S_c^\theta \quad \text{and choosing the maximum } S^\theta$$

The processed image will be described by $L_q E_k$ since
after k iterations of the edge extraction procedure then q iterations of the line sharpening procedure are applied. Typically, good results may be obtained with $L_2 E_2$. As before, the thresholding process is
the most critical one, namely the choice of a threshold for obtaining a binary image providing the edge
information. Generally, the grey level histogram of the

image is computed and a certain percentage of points
is established as those that must be retained after
thresholding, for instance 50%. As for the notation,
B_{50} will express this fact so that a totally processed
image may be represented by $B_{50}L_1E_2$ for a binarization
of 50% of the points, after one sharpening step
in turn after two edge extraction steps. A comparison
may be done with the 5 by 5 Laplacian where only the
positive values are preserved. The reasons for choosing
this operator are that a greater neighborhood for a
Laplacian should improve its performance if compared
to a three by three and, moreover, the described iterative technique uses a 5 by 5 window. As foreseen, the
noise levels increase due to the differential operator
and its output is not as good as the one from the technique described here. (See figures from (24)).

A more recent paper (25) reviews the past operators
and reframes them according to the $\{\theta,S,E,C\}$ parameters
that may vary in each local detection strategy since
we have already seen that the local operator is only
the first step in edge extraction. In this analysis, the
set of orientations is called $\{\theta\}$ and a corresponding
set of sampling configurations is $\{S\}$

$$\theta = \{\theta_1, \theta_2, \theta_3, \ldots, \theta_N\}$$

$$S = \{S_1, S_2, S_3, \ldots, S_N\}$$

for every point u,v and the digital image f, S_i determines three sets of values: a set $B_1 B_2 B_3$ with θ through
u,v and sets $A_1 A_2 A_3$ and $C_1 C_2 C_3$ parallel to the B line.
(see figure below for the positions of A, B and C)

$A_1\ B_1\ C_1$ $A_1\ B_1\ C_1$

   $A_2\ B_2\ C_2$ $A_2\ B_2\ C_2$ and other 12 configura-
                                   tions more on 5x5 windows
     $A_3\ B_3\ C_3$ $A_3\ B_3\ C_3$ which may be seen on fig.1.

An evaluation rule E is introduced which computes, from
9 values, the evidence of a line segment, for values
obtained from S_i rule E yields evidence E_i for a line
segment passing through u,v of orientation θ. Finally, a
comparison rule C (which generally is considered to be
the thresholding operation with a fixed or variable
threshold) will compute a vector from E_i whose orienta-

tion is the most adequate for a line segment and whose
magnitude measures the strength of this evidence.

This process may be characterized by $\{\theta, S, E, C\}$ and
analyzed by varying S, E and C independently. In the
first scheme, (see fig. 1) the As and Bs are adjacent
(as with the Cs) so that the width of the line segment
will alter the results considerably. Next, the Bs are
not always on a straight line, but 8 times out of 14
they bend at $\pi/4$. Moreover, $A_2B_2C_2$ should be perpendicular to $B_1B_2B_3$ but since $A_2B_2C_2$ is only horizontal or
vertical, it will cut $B_1B_2B_3$ at $90°$ only 2 out of 14
cases and this complicates the determination of diagonal line segments. To avoid the consequences of these
three flaws a greater neighborhood is suggested in (25),
a 9 by 9 neighborhood, having many empty entries. This
scheme will be called P1, blank spaces separate the Bs
from As and Cs which run parallel to the main line segment to detect and therefore different widths will be
catered for in this way. As a simple example, if the
input image is

```
0004400    the old configu-   0000000 and the 0004400
0004400    rations would give0000000 new ones0004400
0004400                       0000000         0004400
```

The new masks are given in fig. 2. These configurations are applied on the image and if all the six
differences $(B_1-A_1,\ B_1-C_1,\ B_3-C_3, B_3-A_3, B_2-A_2, B_2-C_2)$ are
positive, the maximum value of the average will indica-
te the gradient output and its direction the gradient
orientation.

The next important feature is the evaluation rule
(E) which should, in general, give a high value for the
edges and not be affected by noise. Unfortunately these
are two conflicting requisites. for noise immunity, the
linear rule is better than the semilinear one and, in
turn, better than the non-linear rule since an averaging process helps defeating the noise influence. These
results are analogous to those indicated in (23).

We are now left with the comparison ruele (C).
In previous work, the magnitude of the gradient was
given by the maximum E_k which is simple but perhaps
may be improved by considering the contributions of
all the E_ks as in the following formulas

$$X = \sum_k E_k \cos(2\theta_k)$$

$$\text{and } \bar{\theta} = 1/2 \tan^{-1}(Y/X)$$

$$Y = \sum E_k \sin(2\theta_k)$$

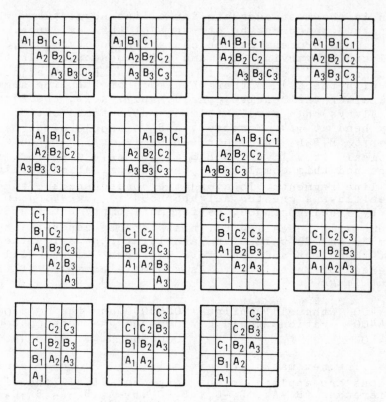

Fig. 1   5×5 Masks

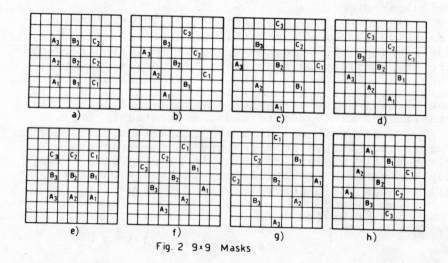

Fig. 2   9×9   Masks

and the circular variance defined as

$$Z = \sum_k E_k$$

$$\rho^2 = 1 - (X^2 + Y^2)/Z^2$$

The orientation will be found by

$$\theta = \text{nearest } \theta_k \text{ to } \bar{\theta}$$
$$\text{unanimity} = 1 - \rho^2$$

Another problem with the previous C rule was that the edge evidence had the same value whether it was a good fit to a dim edge or a bad fit to a strong edge with poor contrast. One way to overcome this contrast dependency, is to use a dimensionless approach and define an orientation. We have $\quad a = \sum_i A_i, \; b = \sum_i B_i$, $c = \sum_i C_i$ and we will assume $b > a$ and $b > c$

$$\mu = (a + 2b + c)/4$$

$$\sigma^2 = [(a - \mu)^2 + 2(b - \mu)^2 + (c - \mu)^2]/4$$

$$K = (2b - a - c)/4$$

$$\gamma = K/\sigma$$

then γ the "goodness of fit" of the pattern $\boxed{-1\,|\,1\,|\,1\,|\,-1}$ to the pattern $\boxed{a\,|\,b\,|\,b\,|\,c}$
γ is dimensionless (the ratio between a sort of average and the variance from the real mean) and we will use three functions of (u,v) (central point) for the C rule:

$$g_1(u,v) = \text{orientation}$$

$$g_2(u,v) = \gamma \text{ evidence}$$

and $g_3(u,v) = \text{unanimity (goodness of evidence)}$

The performance of sets of $\{S,E,C\}$ was evaluated on artificially generated images (with and without noise) for line segments and step edges of a diagonal orientation.

An ideal line segment of parameters $x_o y_o d_1 d_2 \omega \theta$
will be defined by the function p for which
$$p(x,y) = d_1 \quad \text{if } |(x-x_o)\sin\theta - (y-y_o)\cos\theta| \leq \omega$$

$$= d_2 \text{ otherwise;}$$

this line segment will pass near the origin,
$|x_o| \leq 1/2; \quad |y_o| \leq 1/2$

An ideal step edge will be characterized by

$$p(x,y) = d_1 \quad \text{if } (x-x_o)\sin\theta -(y-y_o)\cos\theta > 0$$
$$d_2 \quad \text{otherwise}$$

ω has no importance for the step edge. The digitized version of $p(x,y)$ is obtained by

$$f(u,v) = \left| \int_{u-1/2}^{u+1/2} \int_{v-1/2}^{v+1/2} p(x,y)dxdy \right|$$

The considered image and line segment extractor had parameters:

0, 0, 8, 0, 1, 3/4 π

$x_o \; y_o \; d_1 \; d_2 \; \omega \; \theta$

Some noise was added, normally distributed, and with a mean μ and a standard deviation σ.

$n(u,v) = \lfloor t \rfloor$ integer value of an independently extracted quantity from a Gaussian distribution, to be added to $f(x,y)$.
The test pictures examined by Paton may be specified by the following different test parameters:

σ = 0,30   type = line and step, $x_o = y_o = \mp$ 42, -.28, -.14, .18, .42
d_1 = 90, d_2 = 0,   ω = 0.5, 1.0, 1.5
θ = {kπ/32, k=1, 31, 2}

If now a threshold on g_2 and g_3 at 0,0 with two values γ_2 and γ_3 is introduced we will have a decision rule (C rule) such as
      if $g_2(0,0) > \gamma_2$ and $g_3(0,0) > \gamma_3$

then we have a line segment through 0,0 with orientation $g_1(0,0)$; else we do not have a line segment.
      We then have three possibilities: (as it is often the case in biomedical image analysis for tumor cells for instance) false positives, false negatives and fal-

se orientation. It may be easy, at a second stage, to
fill in gaps or to delete wrong line segments so that
in the first step we may tollerate high negatives but
should reduce high positives. To obtain few false positives we must have high γ_2 and γ_3 so that $g_2(0,0)$
and $g_3(0,0)$ must be very high to exceed them.
It is better to have a smaller error rate than a small
reject rate so that γ_3 should be high.

The different methods use the same set S, a comparison rule given by the values of $g_1(u,v)$, $g_2(u,v)$
and $g_3(u,v)$. The evaluation rules used in this study
are, a non linear one:

$$E = sum(B_1-A_1, B_2-A_2, B_3-A_3, B_1-C_1, B_2-C_2, B_3-C_3)$$

   if all the terms are positive, else

   = 0

and two variations of the semilinear rules:

   E = sum of (b-a, b-c) if both terms are positive,
        else
   = 0

and

   E = product of (b-a, b-c) if both terms are
        positive, else
   = 0

Now all the methods work comparably well with a
percentage of false positives which is under 15% and
an error rate which is under 2.5% and false negatives
under 40%.
In order to improve the performance of the line segment
detector, a "line builder" is suggested by Paton. He
embeds contextual information from a wider area in this
line builder as well as a priori knowledge about the
properties of the objects contained in the image.
This is considered in other methods which will be discussed later on.

In conclusion, the principal component of a line
segment detector is the set of configurations which
will be used on the image, both the evaluation and the
comparison rules are less important, especially if the
detection is followed by a constructive process which
makes up the line as more knowledge is gathered about
it.

3. REGIONAL METHODS

Another and, different, approach has been suggested by Hueckel (26) who considers what a theoretical edge should look like (within a circular diameter) and then finds, for each region, how distant from this configuration we are by using a best fit criterion such as the least squares one. But let us look at this suggestion more closely and at the different variations which have been proposed in order to reduce the computational cost of the algorithm without seriously impairing its performance.

As briefly mentioned above, a circular neighborhood is considered, having an area D in $F(x,y)$, two grey levels are assumed to exist in this area: b, and b+d which is higher and therefore represents a darker region. The analysis of this region involves looking for a two dimensional step function S such that

$$S(x,y,c,s,r,b,d) = \begin{array}{l} = \; b \quad \text{if } cx + sy > r \\ = \; b + d \quad \text{if } \quad cx + sy < r \\ = \; 0 \; \text{otherwise} \end{array}$$

the functional equation to be solved is

$$\int_D (F(x,y) - S(x,y,c,s,r,b,d))^2 \, dx \, dy = \min$$

The Fourier coefficients for F and S are

$$f_i = \int_D F(x,y) \, H_i(x,y) dx dy \quad S_i(c,s,r,d,b) =$$

$$= \int_D S(x,y,c,s,r,b,d) \, H_i(x,y) dx dy$$

$$\lambda (c,s,r,b,d) = (f_i - S_i(csrdb))^2 = \min$$

Hueckel's operator solves this functional equation and finds a set of functions (H_i functions) which approximate the series expansion only considering the first 8 base functions ($H_0 \ldots H_7$) which may be graphically represented by the figures shown below in which + and - are the signs of the functions and the lines denote the zero crossing in the D area. Note the resemblance with the receptor fields in the cat's cortex, (27).

$H_0 \qquad H_1 \qquad H_2 \qquad H_3 \qquad H_4 \qquad H_5 \qquad H_6 \qquad H_7$

The expansion of the grey values is done on an orthogonal basis and since only the low frequencies are considered, high frequency noise is reduced. The solution
is interesting because it gives a method which is orientation invariant and relatively noise immune but has
a heavy computational cost, for this reason (Mero-Vassy
(28)),have suggested a simpler minimization solution
although it is obviously an approximation to an approximation ...
They have proved that the equation may be rewritten
as a convolution and therefore a part of this equation
is equivalent to a template matching operation:

$$T(x,y,c,s,r,b,d) = \sum_{i=0}^{7} s_i(s,c,r,b,d). \quad H_i(x,y)$$

T will be an optimal template to the function $F(x,y)$
constructed as a linear combination from the set of
the original templates H_0, ..., H_7. In order to reduce
the parameters implied in the best fitting procedure
suggested by Hueckel, Mero Vassy have substituted
c,s,r,b,d by the direction cosines c and s so that only
two templates will be used to find the direction of
the zero crossing. Furthermore, to increase simplicity,
the functions of the template will be constant valued
and the shape of the region will be square instead of
circular. Now the procedure is as follows:

    1) determine the edge direction by constructing an
optimal template from two initial ones by linear combination;
    2) determine the exact location of the edge in the
analyzed region by its border points (one of them is
enough).

The two considered functions were H_2 and H_3 which may
be seen as the left and right semicircles for H_2 and
the top and bottom semicircles for H_3; their square
form will be approximated by the top right triangle
and bottom left one for f_1, and top left and bottom
right for f_2 as shown in the figure below:

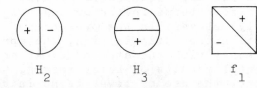

We may now express the approximation due to Mero-Vassy,
by means of which, the orientation of an existing line
and its position may be computed:

$$\frac{\int_J f \; f_2 \; dxdy}{\int_I f \; f_1 \; dxdy} = \tan \alpha$$

the integrals are fast to compute and noise free, the
region is a square of 8 by 8 pixels.

Once the square has been divided into four triangles $T_1 T_2 T_3 T_4$, the minimum T_i over the total number
of pixels, gives the minimum average value, the same
may be done for the maximum $\overline{T}_i$, in this way their difference is computed, $d = \overline{T}_{max} - \overline{T}_{min}$ and the approximated number of points with greater intensity (those
being b+d) will be calculated by

$$S' = \frac{S - 64b}{d}$$

On the other hand f_1 and f_2 are easily computed
and the angle (with respect to the subdiagonal) is obtained after inverting the tanα function. As with the
original Hueckel operator, the output also gives a measure of the estimate of having an edge in that position.
The total time to find an edge in a 256 by 256 image
with the Hueckel operator is about 40 seconds on a
PDP10, whilst the time taken when using this method for
a full 256 by 256 image on a minicomputer (8 by 8 mask)
is of a maximum of 800 microseconds (28).
Another simplified approach to Hueckel's operator
is due to (Nevatia, (29)) who generalizes the Mero-
Vassy procedure for a circular neighborhood, determining 3 parameters (orientation, position, step size)
for a step edge from three expansion coefficients.
The effect of ignoring some terms in the orthogonal expansion is particularly relevant in the presence of
noise: if the signal to noise ratio is high, the approximation fits perfectly with the step edge. After
showing some experimental results on noisy images, he
concludes by saying that the use of fewer terms in the
Fourier expansion produces an unacceptable performance
loss when the images have signal to noise ratios under

Nevertheless the Hueckel operator is widely used and the concept behind it is quite original.

4. GLOBAL METHODS

A completely different attitude has been taken by people who were experts in signal analysis and digital filtering and therefore wiewed the problem of edge extraction as one of filtering the image so that only the edge remains and all the rest is elliminated. For instance (Modestino Fries (30)) suggest an image model and then perform a linear shift invariant spatial filtering operation on the image in such a way that the mean square estimation error is minimized

$$I_e = E \{e^2(\underline{x})\} = E\{\ell(\underline{x}) - \hat{\ell}(\underline{x})\}^2$$

The parameters that enter in the system transfer function $H_o(r)$ are the following:

ρ correlation coefficient for the spatial evaluation of the random amplitude process- the degree of correlation across an edge-

ξ the signal to noise ratio of the edge structure with respect to the background noise

λ the number of events per sampled distance (of edges per unit distance)

r the neighborhood distance

If $\lambda_1 = \lambda_2 = \lambda$

then $H_o(r) = \dfrac{r^2 e^{-r^2/2}}{\dfrac{r^2 + 2(1-\rho)^2 \lambda^2 (r^2 + (1-\rho)^2 \lambda^2)^{1/2}}{8(1-\rho)\lambda\xi} + 1}$; $r \geq 0$

$\xi \simeq \dfrac{\sigma^2}{\sigma_n^2}$ S/N, ratio the model is completely defined in terms of ρ, λ and ξ.

Edge detectors designed on this basis prove quite efficient for a wide range of images. The digital approximation to the analog filter described above is $H_o(z_1 z_2)$ whose frequency response approximates $H_o(r)$. A window ($e^{-r^2/2}$) is chosen, empirically, for obtaining a frequency weighting of a Laplacian kind so as to avoid ringing effects on the edges. A two dimensional infinite impulse response filter (IIR) is then considered, of the kind

$$H_o(z_1 z_2) = \frac{\displaystyle\sum_{i=0}^{Mb} \sum_{J=0}^{Nb} b_{ij} z_1^{-i} z_2^{-j}}{1 + \displaystyle\sum_{\substack{i=0 \\ i=j}}^{Ma} \sum_{\substack{j=0 \\ i \neq j}}^{Na} a_{ij} z_1^{-i} z_2^{-j}} \quad ; \quad z_i = e^{j\omega i} \quad (i = 1,2)$$

this expression may be rewritten in terms of coefficients $a_{10} a_{11} b_{11}$ which may be found for each set of

values of ρ, λ and ξ.
For every point in a 3 by 3 neighborhood, pairs of grey
values in opposition are compared and if the grey level
difference exceeds a threshold than an edge is declared;
if there are many pairs exceeding this difference the
largest one is chosen . The final expression is

$$H_o(z_1 z_2) = A \left(\frac{1 - 1/2(b_{11} + 1)(z_1^{-1} + z_2^{-1}) + b_{11} z_1^{-1} z_2^{-1}}{1 + a_{10}(z_1^{-1} + z_2^{-1}) + a_{11} z_1^{-1} z_2^{-1}} \right);$$

where A is the gain.

The authors have applied this filter to an image
having different ρ and λ (events/sample dist.) and ξ
from ∞ to 10 and 3 db. A Laplacian was next applied
on a 4 by 4 image and then thresholded; if the noise
was high (low signal to noise ratio) the Laplacian did
not improve significantly the results. Nevertheless,
the filtered image does show the edges neatly as may
be seen from the figures in (30).

The examples contain an X ray photo of the chest
with a filter having $\lambda = 0.0125$ and $\rho = .5$ and
$\xi = 3db$, very good results were obtained for a wide class
of images.

Finally the authors conclude by saying that this
filter is an optimum compromise between simplicity, generality and noise immunity. As we have seen, this approach does not include any knowledge in the process
of the edge extraction, it is completely independent
from the kind of image that must be processed.

5.   SEQUENTIAL METHODS

Another possibility is to analyze the image so as
to find some elements which belong to an edge and then
proceed from these points onwards in a sequential way
(Rosenfeld-Kak (2)) in order to find the remaining

points belonging to the same edge. This reduces the
computational cost since some elements are never processed (in opposition to all the previous methods
which consider all the elements of the image) and on
this theme many variations can be envisaged. The first
way to observe an image is to scan it, rasterwise from
top to bottom and from left to right as in the classical television raster.

As an example, we may consider the one in (2) where
a thin dark continuous curve is to be detected in the
presence of noise. A typical condition is that this
curve will never turn abruptly, i.e. no angles will be
smaller than 90° between two successive elements. We
first detect an element whose grey level is above a
threshold, we then look around for its neighbors and
if there is one of them which is above some lower threshold than the above one (tollerance interval) we then
accept this new beighbor, and so on. If only one threshold would have been used, the number of detected points
would differ substantially and a wrong subset of elements would be obtained. More information can be included in the acceptance criterion: the number of accepted points may be fixed, the average grey level of
the region may influence the threshold value, the curvature may be also checked at each new added point
with respect to a number of previously found points,
the contrast (as obtained by some of the discussed methods of local gradient extraction) may also play a
role in the choice of a new point, (in fact the gradient also gives a direction which is normal to the
maximum variation direction and which also points to
where the next point should be).

Some of the problems that may appear when tracking
an edge are: an edge splits into more edges (and they
must all be followed, one at a time with a precedence
rule); more edges merge (and only edge will be followed); the edges run parallel to the tracking direction
(omnidirectional tracking should be considered); the
beginning of an edge is faint and difficult to find
(bidirectional tracking should be used: from the top
and from the bottom, although this doubles the computational cost).

But these variations include an apriori knowledge
of the kind of edge we are looking for and therefore
we are trying to refine the strategy of a search: i.e.
of a typical problem in artificial intelligence, namely the one of finding a best path (for some well defined cost function) in a graph which well represents
the problem space.

6. HEURISTIC METHODS

About nine years ago, (Kelly, (31)) suggested an
approach, called "planning" which borrowed some ideas
form the planning concept introduced by Minsky(1963).
Although Kirsch in 1957 had already suggested to defocuss the original image for later processing, this concept is used in Kelly's approach as the first step for
finding edge elements; the next stage uses the coordinates of the found pixels and PLANs future action on
this basis.

An algorithm, given below, is designed for the extraction of horizontal edges. The output from the first
step will be used for a search in a **tree** on the basis
of reasonable assumptions (no vertical edges may be
present in the middle of a horizontal edge, etc). A
head, the object used by Kelly, suggests that the top
and the sides should be firstly located and then an
attempt to connect them should be made, etc.

Algorithm for horizontal edge extraction:

```
                     if   b > h
   a b c             then
   d e f                 if [min(a,b,c)-max(g,h,i)] > 1
   g h i                            then edge
                         else       no  edge
                 else if  b < h  then
                     if [max(a,b,c)-min(g,h,i)] > 1
                                    then edge
                         else       no  edge
                 else no edge
```

Returning to the search, since the shoulders should be
under the head an attempt is made to locate outward
lines, finally the indentations that represent the ears
are searched.

The input to the plan follower is the original picture and the list structure containing the small head
outline. The output will be a new list structure with
the coordinates of the accurate outline of the head.
A narrow band is searched between the successive points
in the plan and then connected by the program. The band
is 16 pixels wide (the original image was reduced by 8),
several edges may be contained in this band, whenever
parallel edges are found, the external one (with respect to the center of the head) is chosen; a similar
edge detector will be used on the large image.

The quality of this program is difficult to measure
but it is certainly fast and easy to debug. It is far
better to work with some 600 pixels than with 40.000
(from a 256x256 input image). The time, on a PDP10 machine, for locating the head contour, is of about 6

seconds whilst without planning, the program requires
about 234 seconds. At this point one may suppose that if
more knowledge is accumulated in the list, better results will follow, but a heavier burden is layed on
the planner, more information to process and therefore
slower processing will result. According to Kelly his
program is a good compromise between high speed and
quantity of inserted knowledge in the program. A good
feature to add to the program would be a language for
expressing this knowledge so that a head might be described on a relatively high level, directly to the
planner.

An evolution of this approach is the one due to
(Martelli, (32)) who uses another heuristic scheme for
finding edges in noisy images. He firstly defines a
microedge (a pair of pixels having different grey values):

a = (i,j) and b = (i,j+1)

edge ab

An edge is a sequence of edge elements which starts
from the top row (Martelli's hypothesis), contains no
loops, ends in the bottom row and no element will have
an "upwards" direction. In this way an edge is a path
in the graph that must be found optimally between a
first node and a goal node. If MAX is the highest grey
level of the image, the cost of (expanding) a node
$n = a_r a_s$ will be given by $C(n) = (MAX - g(a_r) + g(a_s))$
where g is the grey level at a; C(n) is low if a_r has
a high grey level and a_s a low grey level; the cost
of getting to state n from some initial root node is
the sum of all the costs encountered along the path
from the sarting node to n; g(n) is low if the path to
n is of high contrast.

Davis (in his review paper, (1)) has noted that
the cost in terms of grey level differences only, is
unfortunate since it is difficult to evaluate the goodness of an edge on the basis of the contrast level only,
moreover the space representation (nodes associated to
micro edges) is not a very good one for a proper definition of the heuristic function. The criticsism to the
method, which has come from different sources, is concentrated on the choice of the two components of the
function (the evaluation function) which are supposed
to be additive. It is obviously a difficult problem
to choose a good heuristic function and, moreover, to
adequately weigh the global with the local information

for choosing the next node in the path. If the
contrast is low, a local gradient would still find the
edge whilst Martelli's method would not.

A still different approach is the one taken by
(Montanari, (33)), who uses dynamic programming for
extracting curves in a noisy environment. His method
defines a cost function in order to determine the best
path and therefore establish the choice of the elements
once they have all been computed. Figure of merit of
a path $z_1 \ldots z_{n-1}$ is $f(z_1 \ldots z_n) =$

$$= \sum_{i=1}^{n} g(z_i) - q \sum_{i=2}^{n-1} (d(z_{i+1}, z_i) - d(z_i, z_{i-1}) \bmod 8)$$

where $g(x)$ is the grey level of pixel x and $d(x,y)$
is the slope between pixels x and y; the minuend
(second term) is proportional to the curvature of the
line. The computation time is independent from the
noise level and the solution of this equation is constrained as follows:

$$\left. \begin{array}{l} \max(|x_{i+1} - x_i|, |y_{i+1} - y_i|) = 1 \\[2mm] (d(z_{i+1}, z_i) - d(z_i, z_{i-1}) \bmod 8) \le 1 \end{array} \right\} z_i = (x_i, y_i)$$

The image may be preprocessed with a local derivative operator so as to reduce the number of paths
and therefore decrease the computation efforts. It is
difficult to define a figure of merit of the method,
the global optimality requires large memory occupation
and high execution times.

Within the same kind of approach, namely, one in
which a heuristic methodology is employed, the work
of Ashkar-Modestino (34) is rather interesting.
They break up the problem of edge detection into the
usual two steps of preliminary edge extraction and
then, using some edge information and a priori information about the image, contour tracing generation—
which will output a contour.

They formalize the search space in the tree graph
where each node represents a pixel and its successors,
they may be 8 for eight neighbors, fanning out towards
the bottom of the tree. Each level is counted from
level 0 corresponding to the root, each node may only
have seven successors since the eighth neighbor is his
predecessor. After properly defining the root, an
optimum path is chosen for the successor (one out of
seven) by defining an associated metric $C_{n_{j-1}}(i), n_j(i)$.

This quantity expresses the "distance" associated to

jth branch along the ith path. This path (i) will move
from a node to the next $(n_0^{(i)} \, n_1^{(i)} \, n_2^{(i)} \, \ldots \, n_k^{(i)})$ up to
level k in the tree. The path having the minimum cumu-
lative cost will be preferred (from n_0 to a specified
set of goal nodes). In order that the algorithm may
perform efficiently, the branch metric must incorporate all the heuristic information available from the
problem, otherwise the method may turn erratic and unreliable or be tailored to one particular application
only.

After applying the IIR filter (developed in (30),
Modestino-Fries), they obtain a value for each pixel
which is either 0 or a number, and an orientation: normal to the direction of maximum grey level change. The
algorithm they employ for evaluating the minimum cost
and then choosing the nodes of the corresponding path,
is due to ZJ and will not be described here, (see (34).
The interesting part of this work is the one regarding
the composition of $\lambda^{(i)}$ in which the parameters may be
varied at will. Three separate components make $\lambda_{j+i}^{(i)}$:

$$\lambda_{j+1}^{(i)} = \alpha a_{j+1}^{(i)} + \beta b_{j+1}^{(i)} + \gamma c_{j+1}^{(i)}$$

α, β and γ are empirically adjusted in the interval
0-1, whilst a_{j+1}, b_{j+1} and c_{j+1} are adjusted to have

a maximum value of 100.
The factors (a,b,c) represent the local information obtained from the edge detector. If the contour passes
through the pixel then the tangent is specified by d_{kl};
$n_j^{(i)}$ is being extended towards the next pixel $n_{j+1}^{(i)}$ (next
node in the tree), the direction will be called $L_0^{(i)}$,

then an $a_{j+1}^{(i)}$ branch metric will be computed by

$$a_{j+1}^{(i)} = e_{kl} + H(g_{mn}) \qquad d_{mn} \neq 0$$
$$= e_{kl} \qquad d_{mn} = 0$$

where $g_{mn} = \text{diff}(L_0^{(i)}, d_{mn})$ represents the absolute
angular difference between the direction codes $L_0^{(i)}$
and d_{mn} measured in multiples of 45° and $H(g_{mn})$ is

obtained from a table. The main idea is to reward (high
a_{j+1} value) for high e_{kl} (edge value) and for parallel

directions between L_o and d_{mn} (g_{mn} = 0). Furthermore e_{kl} is scaled so as to exceed 75 with low probability (since the maximum value of $H(g_{mn})$ is 25 fo g_{mn} = 0).

Now the second term, b_{j+1}^{i} reflects knowledge concerning curvature of typical contours, supposed to be smooth, and will discriminate against irregular contours.

$$n_j^{(i)} \longrightarrow n_{j+1}^{(i)} \qquad\qquad L_p^{(i)} \qquad p = 0, \pm1, \pm2, \ldots$$

from $n_{j-p}^{(i)}$ to $n_{j-p+1}^{(i)}$ along the path (i). They consider the M direction codes associated with the moves of the last M nodes

$$L_o^{(i)}, L_1^{(i)} \ldots L_{M-1}^{(i)}$$

the horizontal displacement is Δx and the vertical one Δy , now the maximum $\{|\Delta x|, |\Delta y|\}$ = D and the curvature submetric will relate D with an f(D) which varies from -100 (for D=0) up to 0 for D=4 monotonically decreasing, with D.

so that $\qquad\qquad b_{j+1}^{(i)} = f(D)$

Finally, a priori information is contained in the third and last submetric component $c_{y+1}^{(i)}$. A nominal prototype is considered and the difference between the extracted line and such a prototype is computed (in a similar way to "planning").
If n_0, n_1, n_2 $\ldots$ n_n is a sequence of n+1 nodes and $n_j^{(i)}$ has coordinates $x_j^{(i)}$, $y_j^{(i)}$; $L_p^{(i)}$ (p=0, $\pm$1, $\pm$2, $\ldots$)

represents the direction code associated with the next move from n_{j-p} to n_{j+1-p} along the prototype outline, they define a distance

$$d = \min_{0 \le k \le J} \{|x_{j+1}^{(i)} - x_{j+k+1}| + |y_{j+1}^{(i)} - y_{j+k+1}|\}$$

which is the minimum distance between the proposed next node n_{j+1} and some node n_{j+k+1} $0 \le k \le J$ along the

prototype outline. J is empirically determined and allows some flexibility. In this case, $c_{j+1}^{(i)}$ is diminis-
shed in proportion to the distance and to the relative difference in orientation between the proposed outline and the prototype outline. If k_0 is the k that
minimizes d, they then define $G = \text{diff}(L_0^{(i)}, L_{k_0})$

where L_0 is the tentative direction code and L_{k_0} is

the code corresponding to the move $n_{j+k_0} \longrightarrow n_{j+k_0+1}$;
in order to have a similar orientation the difference
should be small. We now have

$$c_{j+1}^{(i)} = F(d) - 5G^2 \quad \text{where} \quad F(d) \quad \text{monotonically}$$

decrease with d and becomes -80 for d = 80 starting
from 0 for d=14.
      In conclusion the full branch metric may be subdi-
vided into three submetrics which account for the actual
information (a), the a-posteriori information (b) and
the a-priori information (c).
      The whole process may be summarized as follows:
a pseudoelement (root node, unknown) is considered, q
successors to it will be evaluated according to the
$\lambda_1^{(i)}$ branch metric and one of the seven successors of
this node will be chosen, then another branch metric
$\lambda_{j+1}^{(i)}$ will be computed and so on until a goal node is
reached.
      Many real images in the biomedical area (lung,
heart) are reported in this paper (34) and good results
are obtained from this approach even in the presence
of noise. Processed images may be seen in the figures
contained in this paper.

7. DYNAMIC METHODS

      Within a very comprehensive work in image analysis
of bone marrow cells (Lemkin, (35)) an interesting approach to the detection of a boundary seems useful to
report here.   The basic idea is to use a sequence of
thresholds (in proximity of an optimal one extracted
from considerations on the grey level histogram of
the cell image) and to consider the " stable regions".
of darker pixels which have small variations

whilst the image is thresholded at different values.
The **main** idea is to extract a boundary trace (hence
Boundary Trace Transform)which will then support the
evidence of an external boundary and of objects (gene-
rally darker than the boundary) within such a boundary.
The trace image T_{ti} is defined as follows

$$T_{ti}(x,y) = b \text{ if } (x,y) \text{ is a boundary pixel defined}$$

for the image G_i segmented at threshold t

$$= h \text{ if } (x,y) \text{ is a hole pixel defined for}$$
G_i such that $G_i(x,y) < t$ inside a

boundary
$$= 0 \text{ otherwise}$$

A trace image contains information about the boundary and about the holes which may be inside the boundary. The boundary trace transform builds up an image
BTT(T) such that after starting from a 0 value for all
its pixels (initialization) every element which has a
b value in its trace image (for a certain threshold
value) will be incremented by a constant value

$q = \dfrac{\text{max value}}{\text{threshold range}}$. After all the thresholds have

been tested, an image will result which has some pixels
with very high values (the so called stable pixels)
which correspond to the boundary. In practical cases,
the extraction of boundaries is severely complicated
both by the existence of artifacts and of pixels having
a high grey level and not belonging to the contour.

8. RELAXATION METHODS

A more recent approach (1976) to edge extraction
and to image analysis in general is the one employing
relaxation techniques. Its main advantages are that
within each iteration of the process the future value
of a pixel does not depend on any previous decision
taken in the same iteration and, at the same time, the
results obtained at the previous iterations are collected and gradually improve the reliability of the values
obtained for each pixel, for instance in edge extraction procedures. The main idea (Rosenfeld (36)) is to
use an initial (perhaps fuzzy) knowledge about the fact
that a pixel may belong to an edge, the same is done

for the probabilities at neighboring points, and if there is supporting evidence (which means high probabilities that the neighbors of this pixel have compatible assignments to the fact that the pixel belongs to the edge) then the probability of that pixel of belonging to the edge is increased. Conversely, if there is contradictory evidence (that the assignments of neighboring pixels are incompatible with the fact that the pixel belongs to the edge) then the probability that the pixel is an edge pixel is decreased. This process is done simultaneously for every pixel, and every probability corresponding to each pixel is varied, the process may be iterated a number of times and during the iteration the probabilities converge to a stable value. If the probability for a pixel is high, then this pixel will belong to the edge class, otherwise it will not, and this is the output of the relaxation process.

In particular (36), if there is a set of objects $A_1 \ldots A_n$ which must be probabilistically classified into $C_1 \ldots C_m$ classes which may obviously be m for n objects; in general $m \neq n$ and the objects may be curves, lines, regions, etc.

Each A_i has a well defined set of neighbors (A_j) and he then associates to each A_i a probability vector $(p_{i1} \ldots p_{im})$ which gives an estimate of the probability that A_i belongs to class $C_1 \ldots C_m$. Initially, the estimate of p_{ik} depends on a conventional analysis (gradient, for instance) for estimating the probability that a pixel belongs to an edge. He next assumes that for each pair of neighboring objects (A_i, A_j) and each pair of classes (C_h, C_k) a measure of compatibility between A_i belonging to class C_h and A_j belonging to class C_k may be defined.

As an example, $\boxed{\begin{array}{c} A_i \\ A_j \end{array}}$ are vertically adjacent pixels and C_h, C_k can be the classes of vertical edges having their dark sides on their right and on their left, respectively, See figure below.

We will have that $A_i \in C_h$ and $A_j \in C_h$ are highly compatible, the same will be true for

$$A_i \; \epsilon \; C_k \qquad \text{and} \qquad A_j \; \epsilon \; C_k \qquad \text{but for they}$$

$$A_i \; \epsilon \; C_h \qquad \text{and} \qquad A_j \; \epsilon \; C_k \qquad \text{will not}$$

be highly compatible but, just the opposite, highly incompatible.

A compatibility measurement r, has been introduced, ranging from -1 to 1 and having the value 0 for an irrelevanty (don't care) situation. The compatibility variable is $r(A_i, C_h, A_j, C_k)$ which has four arguments: the pair of pixels and the pair of classes. A numerical example from (36), illustrated below, is given to clarify matters; there are three objects which are neighbors of one object and their probability vectors are written for the three classes as well as their compatibilities.

$$A_2 \qquad (.9,.1,0)$$

$$r(A_1,C_1, A_2, C_1) = .9 \qquad\qquad r(A_1,C_2,A_2,C_1)=-.8$$
$$r(A_1,C_1, A_2, C_2) =-.6 \qquad\qquad r(A_1,C_2,A_2,C_2)= .7$$
$$r(A_1,C_1, A_2, C_3) =-.3 \qquad\qquad r(A_1,C_2,A_2,C_3)= .1$$

$$r(A_1\ C_3,A_2,C_1) \quad =-.1$$
$$r(A_1\ C_3,A_2,C_2) \quad =-.1 \qquad A_1(.4,.3,.3)$$
$$r(A_1\ C_3,A_2,C_3) \quad = .2$$

$$A_3(.2,.7,.1) \qquad\qquad\qquad A_4(0,.5,.5)$$

The iterations of the relaxation process will adjust the probabilities on the basis of the values of the probabilities of the neighboring pixels and their associated compatibility values. The rules may be explicity stated as follows:

1   if p_{jk} (probability of pixel j to belong to class k) is high and if $r(A_iC_hA_jC_k)$ is close to +1 then p_{ih} should be increased. (Let us remember that p_{ih} is the probability that pixel i belongs to class h),

2   if p_{jk} is low and $r(A_iC_hA_jC_k)$ is close to -1 then p_{ih} should be decreased,

3   if p_{jk} is low and $r(A_iC_hA_jC_k)$ is close to 0, then
p_{ih} should not change significantly.

A compact way of expressing these three rules is
by means of the product:

$p_{jk}r(A_iC_hA_jC_k)$ which will indicate, for all
A_j , C_k

whether the value of p_{ih} should be incremented or de-
cremented. The contribution of the neighbors (A_j) may
be expressed by

$\sum_j p_{jk}\cdot r(A_iC_hA_jC_k)$ which will range, in value,

between -1 and +1 since $\sum_j p_{jk} = 1$ and
$-1 \leq r \leq +1$

In order to have a positive value in all cases for the
contribution expression, weights may be introduced
which, when added, will sum 1 and then rewrite the net
contribution at each iteration as:

$p_{ih}(1 + \sum_j s_j \sum_k p_{jk}\cdot r(A_iC_hA_jC_k)$ since the interior

part ranges between -1 and +1 the full expression will
range between 0 and +2. In this way the value of p_{ih}
will always be multiplied by a non-negative number;
after all the p_{ih}'s have been updated, the expression
may be renormalized by dividing each p_{ih} by
$\sum_{h=1}^{m} p_{ih}$ so that their total sum will add to 1.

Let us now describe the process along the time
scale, where t is the time variable and $p_{ih}^{(t)}$ is the
value of p_{ih} at the tth iteration; the value of the
net increment will be given by:

$q_{ih}^{t} = \sum_j s_j \sum_k p_{jk}^{(t)}\cdot r(A_i,C_h,A_j,C_k)$

and the updated value of p_{jk} at time instant t+1 will
become:

$$p_{jk}^{(t+1)} = p_{ih}^{(t)} (1 + q_{ih}^{(t)} / \sum_h (1 + q_{ih}^{(t)})$$

The independence of the $(A_j C_k)$'s from the p_{ih}'s has been assumed and although this works well in some cases, a cooperative process in which neighboring probabilities affect the central p, may be introduced and lead to more refined models of the general relaxation procedure which is the basis of this approach.

The classes that interest us in this work are those concerning the fact that pixels belong to the edge or not (edge class, no-edge class). Therefore, the probability adjustment scheme is such that the edge probability at a pixel is incremented in accordance with the neighboring edge probabilities, if the orientations and the neighboring directions are such that the edges reinforce each other, the orientation might be adjusted in the direction of the strongest edge probability.

$A_1 \ldots A_n$ are image pixels and $\theta_1 \ldots \ldots \ldots \theta_{m-1}$ are the classes which correspond to edges having specific orientations, we will add two more classes: edge and no edge. The starting point is the assignment of a probability p_i at A_i according to the magnitude of the gradient at A_i (for instance) and estimate θ_i perpendicular to the gradient direction. The p_i's are adjusted in accordance with the probabilities p_j and orientations θ_j of edges at neighboring pixels A_j that extend or contradict the evidence of an edge along θ_i at A_i. If θ_{ij} is the direction from A_i, to A_j use an increment proportional to $|\cos(\theta_i - \theta_{ij})|$ and to $\cos(\theta_i - \theta_j)$;

the dark side will be assumed always on the clockway direction so as to distinguish edges with the same direction and opposite senses $(\cos \theta_i - \theta_j) = \cos \pi = -1$

and they will decrement one another's probabilities.

Finally, the edge reinforcement technique used to reduce or supress noise in the image is performed by assigning an initial edge probability to a pixel given by

$$p_e^{(0)} = \frac{\text{grad magn at pixel p}}{\max_{pixQ} \text{ grad magn at Q}}$$ the maximum value is taken over the whole image.

For any point (x,y) and neighboring pixel (uv) let α be the edge slope at x,y and β the edge slope at (u,v) and $D = \max(|x-u|,|y-v|)$, in(36) the compatibilities were defined as follows:
for the pairs of class assignments of pixels (x,y) and (u,v)

$$r(\text{edge, edge}) = \cos(\alpha - \gamma) \cos(\beta - \gamma)/2^D$$
$$r(\text{edge, no edge}) = \min[0, -\cos(2\alpha - 2\gamma/2^D]$$
$$r(\text{no edge, edge}) = (1-\cos(2\beta - 2\gamma))/2^{D+1}$$
$$r(\text{no edge, no edge}) = 1/2^D$$

these compatibilities have the following properties: parallel or perpendicular edges have no effect on one another, collinear edges reinforce on one another (inversely proportional to the distance), non edges collinear with edges weaken them, non edges alongside edges have no effect on them, edges alongside nonedges strengthen them; edges collinear with non edges have no effect and nonedges reinforce one another·
     The convergence of the relaxation process has been proved (although the proof holds for sufficient conditions rather than for necessary conditions), see Zucker (37) and the results of a large number of experiments using relaxation may be found in Davis-Rosenfeld (38), or more recently in Eklundh, Yamamoto, Rosenfeld (39).
     In conclusion, the relaxation method is based on an initial estimate of the class probabilities for each object and then, by iteration, in updating such probabilities on the basis of the probabilities of the neighboring objects and their mutual compatibilities.
     A recent report (40) discusses the difficulties that arise when using relaxation techniques, they stem from the problem of adequate knowledge representation which can be expressed as "... how do we represent what we know about our problem in terms of an interactive local process?" (Davis Rosenfeld, (40)). The compatibility coefficients, the neighborhood model and the formalization of the interactivity between the properties of the pixels, are the three main points that should be adequately taken care of in any relaxation process. Finally, the control of the process, how many times should it be iterated and how can the partial results be evaluated still remains as an open question.
Both convergence and unambiguity have been analyzed and recently, the consistency of both the initial labelling and the compatibilities (model assumed) has been fully investigated (Peleg (41)).

9. CONCLUSION

A number of methods for edge extraction have been very schematically described and there is no claim to be exhaustive in the approaches that have been considered here. There is a great amount of work in this area, especially because low level vision must precede any "intelligent" system for scene analysis and edges play such an important part in the description of shape, whether they are boundaries, contours, streaks, line segments, step edges, etc. We may recollect the methods discussed so far by listing the headings used to group similar approaches; they recall the idea which is to be found at the back of the algorithm:

Local methods which use a gradient or which use masks, these may either consider arrangements of pixel values or weigh them. The repetition of the operations in local environments leads to iterative methods. The next step to regional methods in which a best **fit is** searched with an ideal edge in a larger context and variations on this theme produce different approximations to the best fit whilst reducing the computational cost of implementing the algorithm.

The global and sequential methods use digital filtering and a series of actions to pick up edge pixels if certain conditions are satisfied, respectively. Another line of thought is the heuristic approach which uses planning, or methods from the general problem solver described in the artificial intelligence literature, the key point is the goodness of the evaluation function and its matching with the real image situation. Finally, the dynamic approach considers the variation of the pixel values as the threshold is changed and relaxation combines some of the features contained in other methods attempting to solve the general labelling problem of an image on a starting hypothesis (probabilistic in nature), on a model of neighborhood and of compatibility relations between the properties of each pixel and its neighbors.

10. ACKNOWLEDGEMENTS

Thanks are due to Mr. U. Cascini for the figures and to Mrs. M. Izzo for the careful typing of the manuscript.

11. REFERENCES

1. L.S. Davis, "A Survey of Edge Detection Techniques" CGIP, vol 4, 3, 1976, pp. 248-270.

2. A. Rosenfeld, A. Kak, Digital Picture Processing, Academic Press, 1976.

3. N.S. Sutherland, "The methods and findings in experiments on the visual discrimination of shape by animals", Exp. Psychol. Soc. Monograph 1,Heffer's Cambridge, England, 1961.

4. D. Varjú,NATO ASI on Cybernetics of Neural Processes, Quaderni della Ricerca Scientifica,CNR, 31,1965, pp. 291-316.

5. A. Rosenfeld, "Picture Processing by Computer", Computing Surveys, vol 1, 3, pp. 147-176, 1969.

6. A. Rosenfeld, "Progress in Picture Processing:1969-71, Computing Surveys, vol 5, 2, pp. 81-108, 1973.

7. A. Rosenfeld, "Picture Processing:1972", CGIP, vol 1, 4, pp. 394-416, 1973.

8. A. Rosenfeld, "Picture Processing:1974", CGIP, vol 4, 2, pp. 133-155, 1975.

9. A. Rosenfeld, "Picture Processing:1975", CGIP, vol 5, 2, pp. 215-237, 1976.

10. A. Rosenfeld, "Picture Processing:1976", CGIP, vol 6, 2, pp. 157-183, 1977.

11. A. Rosenfeld, "Picture Processing:1977", CGIP, vol 7, 2, pp. 211-242.

12. A. Rosenfeld, "Pictnre Processing:1978", CGIP, vol 9, 4, pp. 354-393.

13. W.S. Holmes, "Design of a photointerpretation automaton", Proc. Fall Joint Comp. Conf., pp. 27-35, 1962.

14. E.S. Deutsch, J.R. Fram,"A quantitative study of the orientation bias of some edge detector schemes",IEEE Transactions on Comp., vol. C-27, 3, pp. 2o5-213, 1978.

15. L.G. Roberts, "Machine perception of three dimensio-
    nal solids", Optical and Electrooptical Information
    processing, edits. J. Tippett, D. Berkowitz, L.Clapp,
    C. Koester, A. Vanderburgh, MIT Press, pp. 159-197,
    1965.

16. J.M.S. Prewitt, "Object enhancement and extraction"
    in Picture Processing and Psychopictorics, edits.
    A. Rosenfeld, B. Lipkin, Academic Press, pp. 75-149,
    1970.

17. J.M. Tennenbaum, A.C. Kay, T. Binford, G. Falk,
    J. Feldman, G. Grape, R. Paul, K. Pingle, I. Sobel,
    "The Stanford hand-eye project", Proc. IJCAI, edits.
    D.A. Walker, L.M. Norton, pp. 521-526, 1969.

18. F.C.A. Groen, Course on Pattern Recognition and
    Image Processing, Delft University of Technology,
    Department of Applied Physics, pp. 130-144,1978.

19. R.A.Kirsch, L. Cahn, C. Ray, G.H. Urban,"Experimen-
    ts in processing pictorial information with a digi-
    tal computer", Proc. Eastern Joint Comp. Conf.,
    pp. 221-229, 1957.

20. I.E. Abdou, W.K. Pratt, "Quantitative design and
    evaluation of enhancement/thresholding edge detect-
    ors", Proc. IEEE, vol 67, 5, pp.753-763, 1979.

21. G.S. Robinson, "Edge detection by compass gradient
    masks", CGIP, vol 6, 5, pp. 492-501, 1977.

22. J.R. Fram, E.S. Deutsch, "On the quantitative eva-
    luation of edge detection schemes and their compa-
    rison with human performance",IEEE Trans. on Comp.
    vol. C-24, 6, pp. 616-628, 1975.

23. G.J. VanderBrug, "Semilinear line detectors", CGIP,
    vol 4, 3, pp. 287-293, 1975.

24. T. Kasvand, "Iterative edge detection", CGIP, vol 4,
    3, pp. 279-286, 1975.

25. K. Paton, "Line detection by local methods", CGIP,
    vol 9, 4, pp. 316-332, 1979.

26. M.H. Hueckel, "An operator which locates edges in
    digital pictures", Journal ACM, 18, pp. 113-125,
    1971.

27. D.H. Hubel, T.N. Wiesel, "Receptive fields of single neurons in the cat's striate cortex", J. Physiol., vol 148, pp. 574-591, 1959.

28. L. Mero, Z. Vassy, "A simplified and fast version of the Hueckel operator for finding optimal edges in pictures", Proc. 4th IJCAI, Tbilisi, USSR, vol 2, pp. 650-655, 1975.

29. R. Nevatia,"Evaluation of a simplified Hueckel edge-line detector", CGIP, vol 6, 6, pp. 582-588, 1977.

30. J.W. Modestino, R.W. Fries,"Edge detection in noisy images using recursive digital filtering", CGIP, vol 6, 5, pp. 409-433, 1977.

31. M.D. Kelly, "Edge detection in pictures by computer using planning", in Machine Intelligence, edits. B. Meltzer, D. Michie, Edinbrugh UP, pp. 397-409, 1971.

32. A. Martelli, "Edge detection using heursitic search methods", Proc. 1st IJCPR, Washington, pp. 375-388, 1973.

33. U. Montanari, "On the optimal detection of curves in noisy pictures", Com. ACM, 14, pp. 335-345, 1971.

34. G.P. Ashkar, J.W. Modestino, "The contour extraction problem with biomedical applications", CGIP, vol 7, 3, pp. 331-355, 1978.

35. P. Lemkin, "The boundary trace transform", CGIP, vol 9, 2, pp. 150-165, 1979.

36. A. Rosenfeld, "Iterative methods in image analysis", Pattern Recognition, vol 10, pp. 181-187.

37. S.W. Zucker, "Scene labeling by relaxation opera-tions", IEEE Trans. on Syst. Man and Cybernetics, 8, pp. 41-48, 1976.

38. L. S. Davis, A. Rosenfeld, "Application of relaxa-tion labeling, 2: spring-loaded template matching", Proc. 3rd IJCPR, Coronado, 76CH1140-3C, pp. 591-597, 1976.

39. J. O. Eklund, H. Yamamoto, A. Rosenfeld,"Relaxation methods in multispectral pixel classification", IEEE Trans. PAMI, (in press).

40. L.S. Davis, A. Rosenfeld, "Cooperating processes
    for low-level vision:a survey". Technical Rep. TR-
    123, University of Texas, Dep. of Computer Sciences,
    1980.

41. S. Peleg, "Monitoring relaxation algorithms using
    labeling", Technical Rep. TR-842, University of
    Maryland, Computer Science Centre, College Park,
    1979.

CONTOUR TRACING AND ENCODINGS OF BINARY PATTERNS

Michael T. Y. Lai and C. Y. Suen

Department of Computer Science
Concordia University
Montreal, Canada

ABSTRACT

    This paper presents an algorithm for tracing the boundary of
binary patterns. A 3x3 window operator is employed and the tracing
scheme can be applied to detect both outer and inner boundaries.
Direction codes, concave and convex features extracted from a
"coding" process, are used to describe the shape of the patterns.
This algorithm can be applied in various areas of pattern recognition, shape analysis, and computer perception of two-dimensional
pictures.

1. INTRODUCTION

    Researchers of various disciplines have studied shape analy-
sis in many different ways. Their approaches usually begin by
extracting the boundaries of the object which is followed by a
description of the object to establish information about its size
and shape. In this paper, we shall present an algorithm which
traces the entire outer and inner boundaries of binary (black and
white) patterns. A new coding process will be introduced to describe the digitized patterns.

2. CONTOUR TRACING OF PATTERNS

    A 3x3 window operator is used throughout the tracing process.
Fig. 1 shows the operator with a center point X_{ij} and its eight
neighbors, where i and j are the ith row and j^{th} column of
the input pattern matrix respectively. The process can be divided

149

J. C. Simon and R. M. Haralick (eds.), Digital Image Processing, 149–153.
Copyright © 1981 by D. Reidel Publishing Company.

into three steps described below.

A	B	C
D	X_{ij}	E
F	G	H

Fig. 1

Step 1. Extraction of boundary points

All points in the pattern are scanned by moving X_{ij} across it. A black point in the center will be eliminated (i.e. assign it as a white point) if
a) all its eight neighbors are white, or
b) points B, D, E, G, are black.
As a result, all black points between boundaries are eliminated, and the remaining image is then processed in the following steps.

Step 2. Outer contour tracing

The window center is moved point by point, starting from the leftmost column to the rightmost column, and from the top row to the bottom row of the pattern until a black point is encountered. After locating this black point, the eight surrounding neighbors are scanned in a clockwise sequence EHGFDABC, i.e. the scanning starts from the center's right neighbor E in a predefined sequence, and X_{ij} will shift to the first black point it encounters. The sequence of scanning the eight beighbors depends on the position of the black point (now the 'new' center point) in the previous window. For example, if the center point was in position 'E' of the previous window, the scanning of its eight neighboring points will be in sequence 'ABCEHGFD' until a black point is encountered, and this black point becomes the center element of the next window. Table 1 shows the scanning sequence for each position point.

Table 1.   Scanning sequence for each position point.

Previous window position of the center point	Scanning sequence
'A'	GFDABCEH
'B'	FDABCEHG
'C'	DABCEHGF
'D'	HGFDABCE
'E'	ABCEHGFD
'F'	EHGFDABC
'G'	CEHGFDAB
'H'	BCEHGFDA

The tracing process continues by repeatedly scanning and shifting the window center to new positions, and it terminates when the operator completes its clockwise trace of the entire pattern and the window center returns to the starting position.
We have to note that the location (x-y coordinates) of all

scanned black points are recorded during the tracing process.

Step 3. Inner contour tracing

The tracing process is the same as the one described in the above step. After locating the first black point, which has not been scanned in Step 2, the eight neighboring points are scanned in the sequence 'GFDABCEH' until a black point is encountered. The process continues by repeating the scanning and shifting operations, and it terminates when the operator completes an anti-clockwise trace of the inner boundary and the window center returns to the starting point. We have to note that this step works well for detection of holes in non-skeletonized patterns.

Fig. 2 shows the results obtained from these three steps. '+' represents a boundary point of the pattern, '*' and '.' represent the first and second inner boundary points respectively.

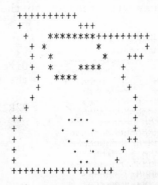

Fig. 2. Results from tracing algorithm.

3. ENCODINGS OF PATTERN BOUNDARIES

During the outer contour tracing, the locations of all the scanned black points are recorded by examining the changes in the x-y coordinates of each successive pair of points, the vertices in the boundary can be located. By making use of the vertices, boundary can be expressed as a sequence $[d_i]$ where d_i is a code for the direction from vertex (x_i, y_i) to vertex (x_{i+1}, y_{i+1}). Eight principal direction codes are generated from two successive vertices and can be defined as follows:

1. 'T' - direction pointing upward
2. 'B' - direction pointing downward
3. 'L' - direction pointing to the left
4. 'R' - direction pointing to the right
5. '1' - direction pointing between 'T' and 'R'
6. '2' - direction pointing between 'R' and 'B'
7. '3' - direction pointing between 'B' and 'L'
8. '4' - direction pointing between 'L' and 'T'

Fig. 3 illustrates a pattern with the generated direction codes to describe its shape, '*' denotes a vertex of the pattern. Several features for describing the patterns can also be derived from the codes, e.g.

i) Direction code features - This is the percentage occurrences of each direction code with respect to the total number of generated direction codes in the pattern.

ii) Direction length features - This measures the distance between two successive vertices (x_i, y_i) and (x_{i+1}, y_{i+1}) for each generated direction code. The distance, f_i, can be computed from the following equation:

$$f_i = \sqrt{(x_{i+1} - x_i)^2 + (y_{i+1} - y_i)^2} \, ,$$

and each direction length feature can be represented by the percentage occurrences with respect to the perimeter of the pattern boundary.

```
     *MMMMMMMM*
     *MMMMMMMMM*M*
     MMMMMMMMMMMM*MMMMMMMM*
      *MM              MMMMMMMM*       Direction codes generated from
      MMMM           MMMM*M*           the pattern:
      *MMM       MMMMMMMM*
      *MMMMMMMMMMMMMM*                 R2R2RB3L3B2BL3LTR1T4T4T
      *MMMMMMMMMMMMMMMM
      MMMMMMMMMMMMMMMMMM*
      MMMMMMMMMMMMMMMMMMMM
     **MMMMMMMMMMMMMMMMMMMM
      MMMMMMMMM     MMMMMMMM
      MMMMMMMMMM   MMMMMMMM**
      MMMMMMMMMMM    MMMMMM
      MMMMMMMMMMMMMMMMMMMM
      *MMMMMMMMMMMMMMMM*            Fig. 3. Direction codes.
```

iii) Curvature features – **In** the outer contour tracing, the boundary is scanned in a clockwise fashion, and by considering each direction code and its successor in the direction chain (the sequence d_i), two main types of curvature information can be generated. They are

1) Concave features: which are generated from the outside angle between two successive direction codes in the direction chain. This approximate angle should be greature than 0° but less than 180°, and

2) Convex features: which are also generated from the outside angle between two successive direction codes, and the approximate angle should be greater than 180°.

Each type of curvature information is further subdivided into two groups which can be represented by two sets of successive direction codes as follows:

Concave features:	(a)	T3, TL, T4	(b)	1L, 1T, 14
		B1, BR, B2		2T, 21, 2R
		L2, LB, L3		3R, 32, 3B
		R4, RT, R1		4L, 43, 4B

Convex features:	(a)	T1, TR, T2	(b)	1R, 12, 1B
		B4, BL, B3		2B, 2L, 23
		L4, LT, L1		3L, 34, 3T
		RB, R3, R2		4T, 41, 4R

4. SUMMARY

A contour tracing algorithm and a "coding" scheme for processing binary patterns have been described. The new approach has been applied to a character recognition system with excellent results (1). It is believed that these methods could be applied to a wide variety of pattern recognition problems, shape analysis and visual perception in describing geometric patterns.

ACKNOWLEDGEMENTS

This research was supported by the Department of Education of Quebec and the Natural Science and Engineering Council of Canada. The continued support from the Computer Center of Concordia University is much appreciated.

REFERENCE

(1) Lai, M. T. Y. and Suen, C. Y.: "Automatic recognition of characters by Fourier descriptors and boundary line encodings," Conference on Pattern Recognition, Oxford, Jan. 1980. Also in press, Pattern Recognition.

OPTIMAL EDGE DETECTION IN CELLULAR TEXTURES

Amar Mitiche and Larry Davis

Department of Computer Sciences
University of Texas at Austin
Austin, Texas   78712

ABSTRACT

Because edges represent a powerful description of images,
detecting edges is an important preliminary process in many
image analysis systems. This paper proposes a general procedure
for detecting edges in cellular textures. This procedure is
based on a class of one-dimensional edge detectors and involves
peak selection among above threshold values of the edge detector.
This paper also investigates the effect of different parameters
on the performance of the edge detection process.

INTRODUCTION

Detecting edges is an important first step in the solution
of many image analysis tasks. Edges are used primarily to aid
in the segmentation of an image into meaningful regions, but are
also extensively used to compute relatively local measures of
textural variation (which, of course, can subsequently be used
for segmentation purposes). Although there has been a considerable amount of research concerning quantitative models for edge
detection (e.g., Nahi [1], Modestino and Fries [2], Shanmugan
et al [3], Cooper and Elliot [4]), very little work has been
devoted to developing such models for images described by
texture models. This paper addresses the problem of detecting
edges in what are called macro-textures, i.e., cellular textures
where the cells, or texture elements, are relatively large (at
least several pixels in diameter).

Once edges are detected in texture regions, they can be

155

J. C. Simon and R. M. Haralick (eds.), Digital Image Processing, 155–176.
Copyright © 1981 by D. Reidel Publishing Company.

used to define texture descriptors in a variety of ways. For
example, one can compute "edge per unit area" (Rosenfeld [5]).
More generally, one can compute first-order statistics of edge
properties [6,7], such as orientation, contrast, fuzziness, etc.,
or higher-order statistics which can measure the spatial arrangement of edges in the texture. Such statistics can be computed
from generalized cooccurrence matrices (Davis et al [8,9] which
count the number of times that specific pairs of edges occur in
specific relative spatial positions. Clearly, the utility of
such tools depends on the reliability with which edges can be
detected in textures.

This paper is organized as follows: Section 2 contains a
description of the image texture models which will be considered.
These models are one-dimensional, since the edge-detection procedure, described in Section 3, is one-dimensional. The edge
detector is a 2 step process involving minimum error thresholding
and non-maxima suppression procedures. Section 4 contains
derivations of the expected value and variance of the edge
operator and an analysis of the thresholding step. Sections 5
and 6 consider the problem of non-maxima suppression. Finally,
Section 7 contains conclusions.

TEXTURE MODELS

We will consider a simple class of one-dimensional models.
A texture model is an ordered pair <P,C> where

1) P is a cell width model, which successively drops
intervals along a line, and

2) C is a coloring model, consisting of coloring processes,
$c_1, \ldots, c_m$, and probabilities, $p_1, \ldots, p_m$. As P produces cells, C
colors the cells.

If we let w be the random variable corresponding to cell
width, then the following are examples of cell width models:

1) Constant cell width model

$$P_c(w) = \begin{cases} 1 & w=b \\ 0 & w \neq b \end{cases}$$

2) Uniform cell width model

$$P_u(w) = \begin{cases} 1/b & 0 \leq w \leq b \\ 0 & w > b \end{cases}$$

3) Exponential cell width model

$$P_e(w) = \lambda \exp[-\lambda w].$$

To simplify the analysis, we assume that there are only two coloring processes, c_1 and c_2. Therefore, there is only one relevant probability for choosing cell colors, which will be denoted by p. Each coloring process colors a cell by choosing the intensity of each point in that cell independently from a normal distribution of intensities. The distributions are denoted $N(m_i, v_i)$ i=1,2, with mean m_i and variance v_i.

Notice that given a one-dimensional cell structure model, <P,C>, one can derive a one-dimensional component structure model, where a component is a contiguous set of identically colored cells. For example, for the cell model <P_c,C> the corresponding component model has components whose lengths are distributed geometrically. Of course, in a component model, the various colors alternate, since by definition two adjacent components must have different colors. The component model is required to determine the prior probabilities of various types of pixels (see Section 4). The analysis described in the subsequent sections could be done by synthesizing the image with one-dimensional component models directly. Instead, an ordered pair <P,C> as described above has been used to allow generality for possible extension of the analysis to two-dimensional models.

A ONE-DIMENSIONAL EDGE-DETECTOR

In this section we will describe a simple, one-dimensional edge detection procedure. There are a variety of reasons for considering one-dimensional edge detectors including computational efficiency, suitability for implementation on special purpose image processing hardware and mathematical tractability.

The class of edge operators which we have considered is based on differences of averages between adjacent, symmetric one-dimensional image neighborhoods. Specifically, let f be a one-dimensional image function. Then the edge operator is:

$$e_k(i) = (1/k) \sum_{j=1}^{k} (f(i-j)-f(i+j))$$

$$= (1/k) (LS(i)-RS(i))$$

where

$$LS(i) = \sum_{j=1}^{k} f(i-j) \qquad \text{and}$$

$$RS(i) = \sum_{j=1}^{k} f(i+j)$$

By noticing that

$$e_k(i+1) = (1/k) \ [LS(i)-f(i-k)+f(i)-RS(i)+f(i+1)-f(i+k+1)]$$

we see that e_k can be computed in a constant number of operations per picture point, independent of k, on a conventional sequential computer.

The operator e_k is used to detect edges by the following three step process.

1) Compute $e_k(i)$ for all points i.

2) Discard all i with $|e_k(i)| < t$. This <u>thresholding</u> step is intended to discriminate between points which are edges of texture elements and points which are in the interior of texture elements, but far from edges.

3) Discard all i with $|e_k(i)| < |e_k(i+j)|$, $|j| < d$. This <u>non-maxima suppression</u> step is intended to discriminate between points which are edges and points which interior to texture elements, but are close to edges.

Step 3 is crucial since e_k gives high response not just at edges, but also near edges, so that thresholding alone would result in a cluster of detections about each true edge point. The above procedure involves three classes of texture pixels:

1) <u>edge pixels</u>, which are pixels located directly at the edges between texture elements;

2) <u>near-edge pixels</u>, which are located within distance d of an edge pixel and are discarded by the non-maxima suppression step; and

3)   interior pixels, which are located distances greater
than d from the nearest edge pixel, and are ordinarily eliminated
by the thresholding step (but may be eliminated due to proximity
to above threshold, near-edge pixels).

Optimizing the above edge detection procedure involves
choosing k, t and d in order to minimize the probability of
error--i.e., minimizing the frequency of discarding edge points,
and not discarding near-edge and interior pixels.

ANALYSIS OF e_k.

In this section we will derive the expected value and
variance of e_k at edges and at interior points. We will regard
an interior point as a point whose distance from the nearest
edge is greater than k. Expressions for the prior probabilities
of edge and interior points are developed. Finally, by assuming
that e_k is normally distributed at edge/interior points, a
minimum error thresholding procedure for distinguishing between
edge points and interior points is developed.

The Expected Value of e_k, $E[e_k]$

The definition of e_k was originally given for a discrete
function f. If f is continuous, then we can redefine e_k as

$$e_k(i) = 1/k \left[\int_{-k}^{0} f(i+j)dj - \int_{0}^{k} f(i+j)dj \right]$$

Then, the expected value of e_k is

$$E[e_k(i)] = 1/k \left[\int_{-k}^{0} E[f(i+j)]dj \right.$$

$$\left. - \int_{0}^{k} E[f(i+j)]dj \right]$$

If i is an interior point, then all points $f(i+j)$, $-k \leq j \leq k$
are colored by the same process. Therefore, the expected values
are all the same, and thus $E[e_k(i)|i$ is an interior point$] = 0$.

Now suppose that i is an edge point, and assume, without
loss of generality, that the cell to the left of i, C_ℓ, is

colored by process c_1, and that the cell to the right of i, C_r, is colored by process c_2, and that $m_1 > m_2$. Let w_ℓ be the width of C_ℓ and w_r be the width of C_r (see Figure 1). We will also make the simplifying assumption that the points to the left of C_ℓ (or the right of C_r) are individually colored by processes c_1 and c_2 with probabilities p and (1-p). For k much greater than w_ℓ or w_r, this assumption is not unreasonable. As w_ℓ or w_r approaches k, it is more likely that only one cell will be found to the left of C_ℓ or the right of C_r. However, large cells are ordinarily less likely than small cells. Letting $a = pm_1 + (1-p)m_2$, we can then write

$$E[e_k(i)] = 1/k \left[\int_0^k (m_1 w_\ell + a(k-w_\ell)) P(w_\ell) dw_\ell \right.$$

$$+ \int_k^\infty m_1 k P(w_\ell) \, dw_\ell$$

$$- \int_0^k (m_2 w_r + a(k-w_r)) P(w_r) dw_r$$

$$\left. + \int_k^\infty m_2 k P(w_r) dw_r \right]$$

Since w_ℓ and w_r are drawn from the same distribution, terms can be grouped to obtain

$$E[e_k(i)] = \frac{(m_1 - m_2)}{k} \left[w_0 - \int_k^\infty (w-k) P(w) dw \right]$$

where

$$w_0 = \int_0^\infty w P(w) dw.$$

The Variance of e_k.

From the definition of e_k, we have

$$\text{Var } [e_k(i)] = \text{Var } \left[1/k \int_0^k [f(i-j)-f(i+j)]dj \right]$$

Suppose i is an interior point and is in a cell colored by c_1. Then

$$\text{Var}[e_k(i)] = 2v_1/k$$

If is is an interior point and is in a cell colored by c_2, then

$$\text{Var}[e_k(i)] = 2v_1/k.$$

Next we will consider the case of i, an edge point.

$\text{Var}[e_k(i)$ i is an edge$]$

$$1/k^2 \left[k(v_1+v_2) \int_k^\infty P(w)dw \right.$$

$$+ (v_1+v_2) \int_0^k wP(w)dw$$

$$+ 2(pv_1+(1-p)v_2) \int_0^k (k-w)P(w)dw$$

$$+ 2p(1-p)(m_1^2+m_2^2) \int_0^k (k-w)P(w)dw$$

Simple expressions can be obtained for the constant, uniform and exponential models. For more details, see [10].

In order to derive a minimum error threshold for discriminating between edge and interior points using e_k, it is necessary to determine the prior probabilities of edge and interior points and specify a form for the distribution of e_k at edges and at interior points.

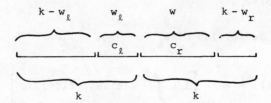

Figure 1.   Neighborhood of an edge point, i.

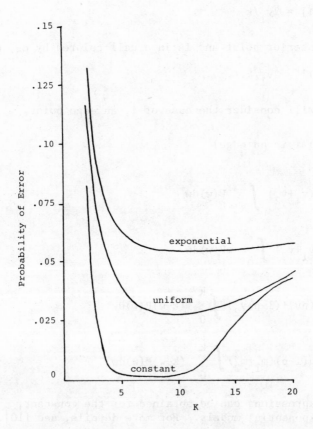

Figure 2.   Plots of er(k) as a function of k for the three cell width
models. All parameters are the same for all three models:
$m_1 = 10$, $m_2 = 5$, $v_1 = v_2 = 2$, $p = .5$ and $w_0 = 10$.

To compute the prior probabilities of edge and interior points, we must derive a <u>component model</u> from the cell width model. A component is a set of connected, identically colored cells. Let a c_1-component (c_2-component) be a component whose cells are colored by process c_1 (c_2). Then, the length of a component is defined to be the number of cells that compose it, and the width of a component is its actual measure.

In the following we will show how a component model can be derived from a cell model. We will consider only c_1-components and simply refer to them as components. The same analysis will hold for c_2-components.

If n is the random variable that represents component length and $P_\ell(n)$ is the probability mass function for it, then

$$P_\ell(n) = (1-p)p^{n-1}$$

Now, if w and r are random variables that describe the cell width and the component width respectively, then r is the sum of a random number n of independent identically distributed experimental values of w.

$$r = \sum_{i=1}^{n} w_i$$

Expressions for the expected value and variance of r can be obtained in terms of the expected value and variance of n and w:

$$E[r] = E[n] \, E[w]$$

$$Var[r] = E[n] \, Var[w] + (E[w])^2 \, Var[n]$$

If w is continuous and if f and g are the probability density function for w and r, then

$$g_\ell^T(s) = p^T(f^T(s))$$

where f^T and g^T are the exponential transforms (s-transforms) of f and g respectively, and p_ℓ^T is the z-transform (discrete transform) of P_ℓ.

If w is discrete, then the above expression holds with f
and g being the probability mass functions for w and r, and
f^T and g^T being their respective z-transforms. By taking the
inverst transforms one can obtain the distribution of component
widths.

In order to perform minimum error thresholding on a given
image texture, it is necessary to know the prior probabilities
of an edge point, p_e, and of an interior point, p_i. An edge
point is defined to be a point on the image line which is no
more than a fixed distance ε away from a true edge. We choose
$\varepsilon > 0$ so that $p_i \neq 0$. An interior point, as before, is a point
that is at least distance k away from a true edge.

Given that a point is picked at random, let h be the
probability density function for the distance y_1 to the next
edge (the forward edge or edge on the right). From general
results on random incidence into a renewal process we have [11].

$$h(y_1) = (1-prob[r \leq y_1])/E[r]$$

The function h is also the probability density function for the
distance y_2 to the backward edge (the nearest edge on the left
of the point). Then we have

$$h_j(y_1,y_2) = h(y_1)h_c(y_2/y_1)$$

where h_c is the conditional probability function for y_2 and h_j
is the joint probability function for y_1 and y_2. If y_1 and y_2
are independent, then

$$h_j(y_1,y_2) = h(y_1)h(y_2)$$

From the above we now can derive expressions for the priors

$$p_i = \int_k^\infty \int_k^\infty h_j(y_1,y_2)dy_1dy_2$$

$$p_e = 1-\int_\varepsilon^\infty \int_\varepsilon^\infty h_j(y_1,y_2)dy_1dy_2$$

For the discrete case we have

$$p_i = \sum_{y_1=k}^{\infty} \sum_{y_2=k}^{\infty} h(y_1,y_2)$$

$$p_e = 1- \sum_{y_1=\varepsilon+1}^{\infty} \sum_{y_2=\varepsilon+1}^{\infty} h(y_1,y_2)$$

The set of interior points and the set of edge points are not complementary in the sense that they do not form a partition of the set of all points of the image line. Thus it is necessary that we normalize p_e and p_i.

$$p_e \leftarrow p_e/(p_e+p_i)$$

$$p_i \leftarrow p_i/(p_e+p_i)$$

In order to use p_e, p_i and the expected values obtained above for minimum error edge detection, we will make the assumption that e_k is normally distributed at edge points as well as interior points. More precisely, we assume that e_k is $N(0,2v/k)$ at interior points and $N(E[e_k],Var[e_k])$ at edges. See [10] for a discussion of this assumption.

Given the normality assumption, the following two step process can be used to compute an optimal k and t for a minimum error edge detector.

1) For a range of k, find the minimum error threshold for discriminating between edges and interior points. Since both e_k at edges and e_k at interior points are modeled by normal distributions with known parameters and priors, this is straightforward. Let er(k) be the probability of error for the minimum error threshold for e_k and let t(k) be the threshold.

2) Choose k such that $er(k) \leq er(k')$, for all k' considered. Then (k,t(k)) defines the minimum error edge detector.

Figure 2 shows plots of er(k) as a function of k for the three cell width models presented in Section 2. Figures 3a-e

a) original texture

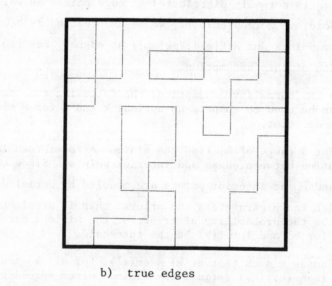

b) true edges

Figure 3. Application of e_k to a checkerboard texture.

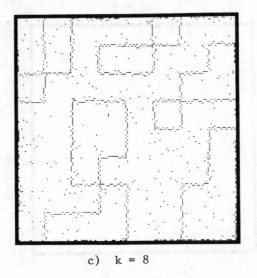

c) k = 8

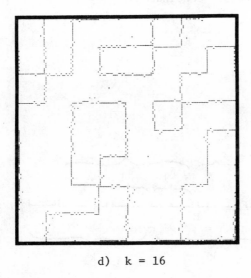

d) k = 16

Figure 3. Continued

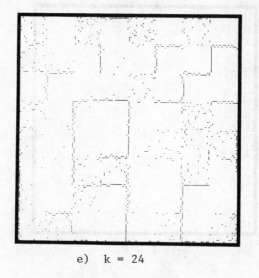

e)   k = 24

Figure 3.   Continued

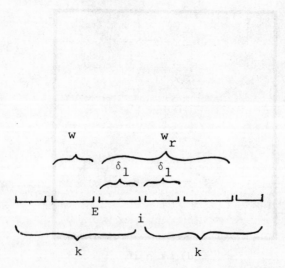

Figure 4.   Composition of the neighborhood of point i.

show an example of the effect of k on the performance of e_k.

Figure 3a contains a checkerboard texture. Figure 3b shows the true edges, while Figures 3c-e show the results of applying e_k, k=8, 16, and 24, thresholding at the minimum error threshold for the appropriate k, and then performing non-maxima suppression across 8 pixels. Note that the results for the optimal value of k (16) are significantly better than choosing k too small (8) or too large (24).

NON-MAXIMA SUPPRESSION

To get a better insight into the problem of non-maxima suppression we must be able to answer the following key question - what is the probability that the value of $|e_k|$ at a point which is not an edge point is a local maximum? The answer to this question will give the probability of detecting an edge distance δ away from a true edge.

Let D be the interval of width $r = 2d + 1$ centered at a point of interest i, where non-maxima suppression is to be performed. We will restrict our attention to values of r such that $d < k$.

If $|e_k(i)|$ is a local maximum then $|e_k(i)| > |e_k(j)|$ for each j in D. To evaluate Prob[$|e_k(i)|$ is a local maximum] we cannot simply evaluate Prob[$|e_k(i)| > |e_k(j)|$] individually for every point j in interval D. This is because the expressions for e_k at points i and j may involve evaluating the picture function at common points. This means that for points j_1 and j_2 in interval D Prob[$|e_k(i)| > |e_k(j_1)|$, $|e_k(i)| > |e_k(j_2)|$] is not equal to Prob[$|e_k(i)| > |e_k(j_1)|$] Prob[$|e_k(i)| > |e_k(j_2)|$].

To determine Prob[$|e_k(i)|$ is a local maximum] in interval D we will instead model the pattern of responses as a multivariate normal distribution of 2d random variables, each one of which corresponds to a particular point $j \neq i$ in D. More specifically we will consider

$$\text{Prob}[|e_k(i)| - |e_k(j_1)| , \ldots ,$$
$$|e_k(i)| - |e_k(j_{2d})|] = N(m,c)$$

where m is the mean vector and c is the covariance matrix

$$m = (m_1,\ldots,m_{2d}) \quad m_j = E[|e_k(i)| - |e_k(j)|] \text{ and } j \neq i$$

$$c(j_1,j_2) = Cov(|e_k(i)| - |e_k(j_1)|, |e_k(i)| - |e_k(j_2)|)$$

and $j_1,\ldots,j_{2d}$ are the points in D other than point i.

The Mean Vector

Let i and j be two points in D, $i \neq j$ and let i be the distance δ_1 away from the nearest edge E. Notice that $\delta_1 < k$. For the purpose of analysis we will assume without loss of generality that this edge is located to the left of point i as illustrated in Figure 4. Also let μ_1 and μ_2 be the means of coloring processes that colored the cells c_ℓ and c_r on the left and right of E respectively and P_ℓ and P_r be the probability density functions that describe the width w_ℓ of c_ℓ and the width w_r and c_r respectively. Finally let a be the average $a = p\mu_1 + (1-p)\mu_2$ where p is the probability of choosing the coloring process with mean μ_1.

We will derive expressions for the expected value of RS(i). Expressions for the expected value of LS(i), RS(j) and LS(j) can be found in [12]. There are two cases to consider:

a) $w_r > k + \delta_1$

b) $2\delta_1 \leq w_r \leq k + \delta_1$

If $w_r > k + \delta_1$ then E[RS(i)] is $\delta_1\mu_2 + (k-\delta_1)\mu_2$ or $k\mu_2$. Next suppose that $2\delta_1 \leq w_r \leq k + \delta_1$. If we assume that the $k + \delta_1 - w_r$ pixels not in c_r are colored independently by processes C_1 and C_2 with probability p and 1-p, then the expected value of RS(i) is $\delta_1\mu_2 + x\mu_2 + (k-\delta_1-x)a$ where x is $w_r - 2\delta_1$.

Remembering that w_r is at least $2\delta_1$ since i is distance δ_1 away from the nearest edge, we combine the two cases a) and b) above and obtain

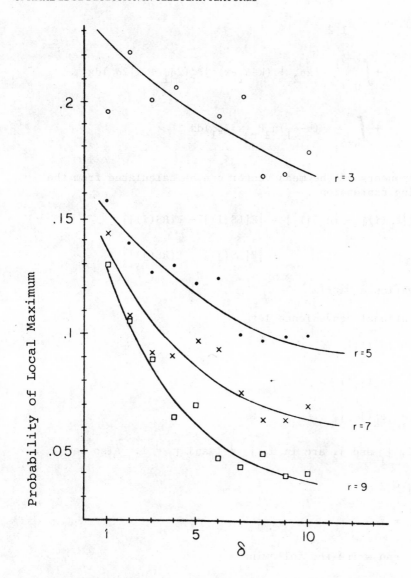

Figure 5.   Probability of a point being a local
             maximum as a function of its distance
             from the nearest edge and for different
             values of the non maxima suppression
             interval width, r.

$$E[RS(i) = \delta_1 \mu_2$$

$$+ \int_0^{k-\delta_1} [x\mu_2 + (k-\delta_1-x)a]P_r(2\delta_1 + x|2\delta_1)dx$$

$$+ \int_{k+\delta_1}^{\infty} (k-\delta_1)\mu_2 P_r(x|2\delta_1)dx$$

The components of the mean vector can be calculated from the following expression

$$E[|e_k(i)| - |e_k(j)|] = |E[LS(i)] - E[RS(i)]|$$

$$- |E[LS(j)] - E[RS(j)]|$$

The Covariance Matrix

For notational convenience let

$$X = |e_k(i)|$$

$$Y = |e_k(j_1)|$$

$$Z = |e_k(j_2)|$$

where i, j_1 and j_2 are in D, $j_1 \neq i$ and $j_2 \neq i$. Also let

$$d_1 = X - Y$$

$$d_2 = X - Z$$

Then we can write the following

$$Cov(d_1,d_2) = E[d_1 d_2] - E[d_1]E[d_2]$$

$$E[d_1 d_2] = E[(X - Y)(X - Z)]$$

$$= E[X^2] - E[XZ] - E[XY] + E[YZ]$$

If m_x, m_y, and m_z are the expected values of $\overline{X}$, Y and Z respectively, then,

$$E[d_1] = m_x - m_y$$

$$E[d_2] = m_x - m_z$$

$$Cov(d_1,d_2) = E[X^2] - E[XZ] - E[XY] + E[YZ]$$

$$- m_x^2 + m_x m_z + m_x m_y - m_y m_z$$

$$= Var[X] - Cov[X,Z] - Cov[X,Y] + Cov[Y,Z]$$

Assuming equal variance v for the regions on either side of an edge we can write

$$Var[X] = 2kv/k^2 = 2v/k$$

We make the additional mild assumption that $e_k(i)$, $e_k(j_1)$ and $e_k(j_2)$ are all of the same sign. This assumption is reasonable since we are assuming no interference between edges. Then,

$$Cov[X,Z] = (n_1 - n_2)v/k^2$$

where n_1 is the number of terms with the same sign and n_2 the number of terms with opposite sign common to the expressions of $e_k(i)$ and $e_k(j_2)$

If $|i - j_2| = \Delta$ then

$$n_1 = 2(k - \Delta)$$

$$n_2 = \begin{cases} \Delta-1 & \text{if } \Delta > 0 \\ 0 & \text{if } \Delta = 0 \end{cases}$$

Similar simple expressions can be derived for $Cov(X,Y)$ and $Cov(Y,Z)$. Finally, we have the following expressions for the covariance terms, with two cases being distinguished.

a) $j_1 = j_2$

In this case $d_1 = d_2 = d$ and

$$\text{Cov}[d_1, d_2] = \text{Var}[d]$$

$$= 6v|i - j_1|/k^2$$

b) $j_1 \neq j_2$

$$\text{Cov}(d_1, d_2) = v(3|i - j_1| + 3|i - j_2| - 3|j_1 - j_2| - 1)/k^2$$

And, of course, the covariance matrix elements are specified by

$$c(j_1, j_2) = \text{Cov}(d_1, d_2)$$

INTERVAL WIDTH SELECTION

The above analysis has been applied to image models with exponentially distributed cell widths. The curves in Figure 5 show $\text{Prob}[|e_k(i)|$ is a local maximum] as a function of the distance δ from point i to the nearest edge and for several values of the width r of the interval D in which non-maxima suppression is performed. These curves have been computed using an IMSL (International Mathematical and Statistical Libraries) routine for generating samples from a multivariate normal distribution. For each value of r, a total of 1000 (r-1)-dimensional points are sampled in order to compute $\text{Prob}[e_k(i)$ is a local maximum] for each value of the distance δ from i to the nearest edge.

$\text{Prob}[|e_k(i)|$ is a local maximum] is highest at the edge and decreases when we go farther from the edge, assuming there is no interference from other edges. Also, it can be noticed that the curves for different values of r are not only similar but close for small values of δ. This means that, at least for points not too far from edges, no sensible increase in the performance of the non-maxima suppression step should be expected when the width of the interval in which it is performed is increased. In other words, the error rate around edges will remain approximately the same over a range of values of r. Since non-maxima suppression is a process intended to discriminate between edge points and near edge points (which are points with small δ values) this suggests that non-maxima suppression need only be performed in small intervals. This also enhances the computational efficiency of non-maxima suppression.

CONCLUSIONS

Our ultimate goal for designing edge detectors for textures is to produce as reliable a "cartoon" of a texture as we can using computationally simple techniques. Various first- and second-order statistics of the distributions of edges in the texture can then be used to discriminate between different classes of textures, as in [8,9]. The effectiveness of such an approach clearly depends on the reliability with which edges can be detected in textures.

The analysis presented in this paper prescribes how the parameters of a minimum error edge detection procedure can be chosen given parametric information about the distributions of size and colors of the texture elements. Given an unknown texture, one must estimate the latter parameters; or, more generally, given an image which contains many different textures, the image must be segmented into regions where these parameters are constant. The design of procedures to estimate these parameters as well as their application to a variety of natural textures will be discussed in a subsequent paper.

REFERENCES

1. N. Nahi and S. Lopez-Mora, "Estimation detection of object boundaries in noisy images", IEEET-Automatic Control, 23, 1978, 834-846.
2. J. Modestino and R. Fries, "Edge detection in noisy images using recursive digital filtering", Computer Graphics and Image Processing, 6, 1977, 409-433.
3. K. Shanmugan, F. Dickey and R. Dubes, "An optimal frequency domain filter for edge detection in digital pictures", IEEE-Pattern Analysis and Machine Intelligence, 1, 1979, 37-49.
4. D. Cooper and H. Elliot, "A maximum likelihood framework for boundary estimation in noisy images", Proc. IEEE Computer Society Conf. on Pattern Recognition and Image Processing, Chicago, IL, May 31-June 2, 1978, 25-31.
5. A. Rosenfeld and A. Kak, Digital Picture Processing, Academic Press, New York, 1976.
6. J. Weszka, C. Dyer and A. Rosenfeld, "A comparative study of texture features for terrain classification", IEEET-Systems, Man and Cybernetics, 6, 1976, 269-285.
7. D. Marr, "Early Processing of Visual Information", Phil. Trans. Royal Society, B, 275, 1976, 483-524.

8. L. Davis, S. Johns and J. K. Aggarwal, "Texture analysis
    using generalized cooccurrence matrices", <u>IEEET-Pattern
    Analysis and Machine Intelligence, 1</u>, 1979, 251-258.
9. L. Davis, M. Clearman and J. K. Aggarwal, "A comparative
    texture classification study based on generalized cooccur-
    rence matrices", to appear in IEEE Conf. on Decision and
    Control, Miami, FL, Dec. 12-14, 1979.
10. L. Davis and A. Mitiche, "Edge detection in textures", to
    appear in <u>Comp. Graphics and Image Processing</u>, 1980.
11. A. Drake, <u>Fundamentals of Applied Probability</u> Theory,
    McGraw-Hill, New York, 1967.
12. L. Davis and A. Mitiche, "Edge detection in textures -
    maxima selection", TR-133, Computer Sciences Department,
    University of Texas at Austin.

RESTORATION OF NUCLEAR IMAGES BY A CROSS-SPECTRUM EQUALIZER FILTER.

V. Cantoni, I. De Lotto

Istituto di Informatica e Sistemistica -
Pavia University and I.A.N. - C.N.R.

ABSTRACT
Images gathered by scintillation camera are often degraded and consequently, for medical diagnosis, methods to restore the information are needed.
An estimation of the ideal image which minimizes the integral mean square error leads to an optimum linear shift invariant filter, the restored image obtained by this filter presents oscillations around peaks which can mask small peaks or generate spurious peaks close to a large one. Then, the discrimination between true anomalies and artifacts due to such oscillations can become a serious problem. A different solution that improves the resolution without introducing significant artifacts is here proposed. This filter gives an average power spectrum of the estimated image equal to the expected cross-spectrum between the recorded and the ideal images. The comparison with results of the optimum filter shows that oscillations around peaks are considerably reduced with a slight increase of the total mean square error.

This work has been supported by C.N.R.Grant n.79.00704.02, Special Program on Biomedical Engineering Grant n.8000315. 86 and by HUSPI Project.

J. C. Simon and R. M. Haralick (eds.), Digital Image Processing, 177–187.

1. INTRODUCTION

Nuclear medicine images are often of poor quality,
owing to the low resolution of the point spread fun-
tion of the scintillation camera, and to the radioiso-
tope dose to be injected into the human body that must
be as small as possible. Therefore, the gathered image
is also degraded because of the emission statistical
fluctuations.

It has been stated by many authors that a γ-camera
can be considered as a linear system with a shift invariant point spread function with a component of noise
added to the output to take into account the poissonian fluctuations of the nuclear emission and the noise
due to the spatial quantization.

So

$$g(x,y) = f(x,y) * h(x,y) + n(x,y) \qquad (1a)$$

or in frequency domain

$$G(u,v) = F(u,v) \cdot H(u,v) + N(u,v) \qquad (1b)$$

Where $g(\cdot)$ is the recorded image, $f(\cdot)$ the ideal one,
$h(\cdot)$ the point spread function and $n(\cdot)$ the additive
noise. An optimum estimate of the ideal image f which
minimizes the total mean square error E:

$$E = \left\langle \iint_{xy} (f(x,y) - f(x,y))^2 \right\rangle dxdy \qquad (2)$$

is given by /1/ :

$$F(u,v) = G(u,v) \cdot L(u,v) = \frac{G(u,v)}{H(u,v)} \cdot \frac{1}{1 + \dfrac{M}{\Phi_f(u,v) \cdot |H(u,v)|^2}}$$

$$= G(u,v) \cdot L_i(u,v) \cdot L_s(u,v) \qquad (3)$$

where M is the total number of counts of the image and
$\Phi_f(u,v)$ is the power spectrum of the ideal image (usual
ly unknown but it can be approximated as shown in/2/)

$L_i(u,v)$ is the inverse filter $(L_i(u,v)=\dfrac{1}{H(u,v)})$ and
L_s is a smoothing filter.
By substituting eq.1 in eq.3 it results:

(4) $F(u,v)=F(u,v)\cdot L_s(u,v)+N(u,v)\cdot L_i(u,v)\cdot L_s(u,v)$

so the smoothing filter determines the degree to which
the original image is reproduced in the output.
The second term of eq.4 expresses the noise component
included in the output.
It is well known that total mean square error measures
do not correspond closely to subjective quality estimates; in fact results obtained by filter of eq.3, show
a notable increase in the resolution (the full width
at half maximum (FWHM) of the composite point spread
function [1] (in the following called cpsf) can be even
a half of that of the original point spread function)
but a lot of ringings can spring out around large peaks.
They are due mainly to the pass-band characteristic of
the transfer function $L(u,v)$ with relative high gains
due to $L_i(u,v)$ and to the sharp cut-off frequently shown
by the smoothing filter $L_s(u,v)$./3,6/.
These oscillations can mask small peaks or generate
spurious peaks. The trouble is that sometimes true anomalies and artifacts so introduced by the image processing technique in scintigraphic images cannot be discri-
minated. To avoid these spurious oscillations an alternative suboptimum solution is here presented in which
the resolution is improved less than with the optimum
filter given in eq.3 but with an appreciable decrease
of ringings, in practice without introducing artifacts.

2. THE CROSS-SPECTRUM EQUALIZER FILTER.

A suitable restoration of biomedical images can be performed by using a figure of merit which takes into account the artifacts introduced during the processing.
The cpsf defined above can be used as an index to measure the oscillations due to the filtering technique/4/.
The total mean square error is added to a measure of
the energy concentrated in the spurious oscillations of

the cpsf. It must be noted that, for an impulse source
image and a poissonian noise for a gaussian approximation of the point spread function of a scintillation
camera, cpsf becomes:

$$cpsf=H(\cdot)L(\cdot)=L_s(\cdot) = \frac{1}{1+\dfrac{e^{\sigma^2(u^2+v^2)}}{M}}$$

where the last term has been written for the Wiener
optimum filter. It can be approximated with a unit function :

$$L'_s(\cdot)=1 \qquad (u^2+v^2)\leqslant \frac{1}{\sigma^2}\ \ln M=\omega_o^2$$

$$L'_s(\cdot)=0 \qquad \text{otherwise} \tag{5}$$

where ω_o^2 is the value of (u^2+v^2) at which $L_s(u,v)=1/2$.
The transform of eq.5 is a Bessel function $\omega_o J_1(\omega_o(x^2+y^2)^{1/2})/(x^2+y^2)^{1/2}$ which shows rings of a maximum relative amplitude of $\alpha_o=0.132$ and an extension proportio-
nal to $1/\omega_o$.
A measure of the energy of the spurious ringings can
then be defined extending the approximation quoted above
to cpsf different from the Wiener one, evaluating the relative amplitude and the extension of oscillations.
In the following we limit an analysis to cpsf which can
be described by:

$$cpsf=H^n(\cdot)\cdot L_s^m(\cdot)$$

where $L_s(.)$ is the cpsf of the Wiener filter and $H(\cdot)$
is the point spread function; n,m two real positive numbers less than 1.
In such a case , the relative maximum amplitude of oscillations is:

$$\alpha = H^n(u,v)\Big|_{u^2+v^2=\omega_{om}^2}\cdot\alpha_o$$

where $u^2+v^2=\omega_{om}^2$ is the value for which:

$$L^m_{s\omega}(u,v)=1/2$$

Consequently an index of the spurious ringings can be written as proportional to:

$$E^2_{sr} = \alpha^2 M^2 \omega^2_o$$

The figure of merit to minimize in order to take care both of bias and noise components, but particularly of the spurious ringings which are judged dangerous, is suggested to be:

$$E^2 = E^2_b + E^2_n + kE^2_{rs}$$

where k is a weighing factor to control the importance of the spurious ringing effects.

As a special case, consider again the cpsf of a impulse input function of amplitude M, a poissonian noise and a gaussian point spread function. The figure of merit becomes:

$$E^2 = E^2_b + E^2_n + k\ E^2_{sr} = M^2 \iint\limits_{uv} \left\{ 1 - e^{-\frac{\sigma^2(u^2+v^2)n}{2}} \right. \cdot$$

(7)
$$\left[\frac{1}{1+e^{\frac{\sigma^2(u^2+v^2)}{M}}} \right]^m \left. \right\}^2 du\ dv + M^2 \iint\limits_{uv} \left\{ e^{-\frac{\sigma^2(u^2+v^2)}{2}(n-1)} \right. \cdot$$

$$\left[\frac{1}{1+\frac{e^{\sigma^2(u^2+v^2)}}{M}} \right]^m \left. \right\}^2 du\ dv + k\ \alpha^2_o \frac{M^{2-n}}{(2^{1/m}-1)^n} \frac{\ln M(2^{1/m}-1)}{2}$$

where $\omega^2_{om} = 1/\sigma\ \ln M(2^{1/m}-1)$ and $H^{2n}(u,v)\bigg|_{u^2+v^2=\omega^2_{om}} = \frac{M^{-n}}{(2^{1/m}-1)^n}$

Note that for the Wiener filter n=0, m=1 and consequently:

$$E^2_{sr} = k\ \alpha^2_o\ M^2\ \frac{\ln M}{\sigma^2}$$

For a wide range of the total number of counts minimization of figure of merit of eq.7 in the monodimensional
case /5/ gives obtimum values of m and n around the point
n=m=0.5.This filter is the geometric mean between the
obtimum smoothing of eq.3 $L_s(.)$ and the inverse filter
$L_i(.)$. In this case the restored image is given by:

$$F(u,v) = \frac{F(u,v) \cdot |H(u,v)|^{1/2}}{(1+\frac{M}{\Phi_f(u,v)\,|H(u,v)|^2})^{1/2}} \cdot \left\{ \frac{1}{1+\frac{M}{\Phi_f(u,v)\,|H(u,v)|^2}} \right\}^{1/2} + \frac{N(u,v)}{|H(u,v)|^{1/2}} \quad . \tag{8}$$

and the cpsf is:

$$\text{cpsf} = |H(u,v)|^{1/2} \left\{ \frac{1}{1+\frac{M}{\Phi_f(u,v)\,|H(u,v)|^2}} \right\}^{1/2} \tag{9}$$

that is,cpsf is composed of two low-pass filters: the
former with usually the lowest cut-off frequency and an
easy slope; the latter which has the aim of reducing the
noise component,with a higher and sharper cut-off.
Let us point out some characteristics of this filter:
-the resolution gain.By comparing the first term of
eqs.1b,4 and 8 it can be shown that the resolution of
the proposed filter is usually lower than that of the
optimum one of eq.3, but it is still sensible. For example in the case of a point spread function of gaussian
shape, $h(x,y)=e^{-(x^2+y^2)/2\sigma^2}$, the resolution gain of
the optimum filter (computed with the assumption of
eq.5) is $(\ln M)^{1/2}/1.78$ with respect to the original
resolution, instead, from eq.9 (disregarding the second
term) it can be seen that the gain with this filter is
$\sqrt{2}$.
Fig.1 shows for the monodimensional case a normalized
impulse restoration performed by several techniques:
(A) the filter of eq.9, (B) the Wiener filter of eq.3,
(C) the homomorphic filter /7/ and (D) is the original
pulse.

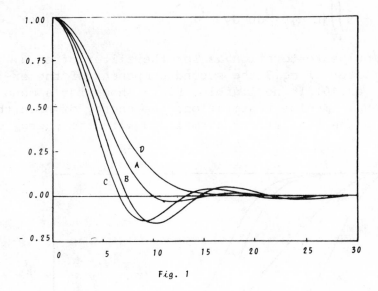

Fig. 1

-the amplitude of oscillations. With reference again to
the case of a gaussian point spread function the former
term of eq.9 alone does not generate ringings, which
are due to the truncation effect produced by the latter
term. While the former one depends only on the σ of the
acquisition system, the smoothing has a cut-off frequen
cy depending on M: the higher is M the higher is the
cut-off frequency ($\omega_o=(\ln(3M))^{1/2} / \sigma$) and the smaller
is the truncation effect on the first term and then
the amplitudes of oscillations produced. For the usual
values of M in nuclear medicine oscillations of cpsf
are negligible.
In Fig.2 for the monodimensional case the cpsf of the
filter of eq.9 (A) of the Wiener filter of eq.3 (B),
of the homomorphic filter (C) and the original point
spread function of the system (D) are given. Note that
the shape of the filter of eq.9 approximates a low pass
"minimum ringing filter " $(\sin x/x)^2$ (dashed line).
- the mean square error. The mean square error can be
computed by the addition of two contributions: the
bias error and the component due to the noise:

$$<E^2> = <E_b^2> + <E_n^2> = \iint_{uv} \Phi_f(u,v) \cdot |1-H(u,v) \cdot L(u,v)|^2 du\ dv$$

$$+ \; M \iint_{uv} |L(u,v)|^2 du \; dv \tag{10}$$

In the impulse restoration, as for the filter of eq.3, in the filter of eq.9 the second component of the error (see eq.10) is negligible. Fig.3 shows for a mono-dimensional impulse restoration, the energy of both the total and the bias errors normalized with the energy of

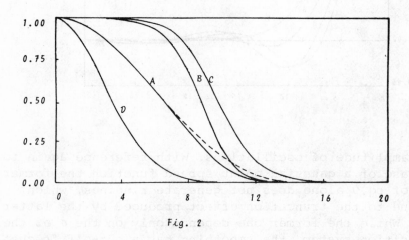

Fig. 2

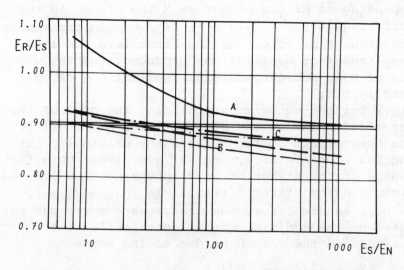

Fig. 3

the ideal pulse versus several signal to noise ratios:
(A) are the original errors(continous line); (B) and
(C) are the resulting errors obtained respectively by
the filters of eq.3 (dashed line) and eq.9 (dotted line). Note that the total error reduction of the filter
of eq.9 is always a large portion of that of eq.3.
- the power spectrum of the restored image. It is well
known that the expected power spectrum of the restored
image using the homomorphic filter is equal to that of
the ideal image, and in the case of the Wiener filter
of eq.3 it is given by $\Phi_f(\cdot)$ $L_s(\cdot)$; it can be easily
seen that the power spectrum of the restored image by
the filter of eq.9 is given by $\Phi_f(\cdot)$ $H(\cdot)$; that is,it
is equal to the expected cross spectrum between the
ideal image and the original degraded one.

3. CONCLUSION

Some experimental results obtained by applying to scintigraphic images the filter of eq.9 show an increase
of resolution with a great reduction of the noise component. The computational cost is increased with respect to the case of the Wiener filter, due to the square root operator, but with an appreciable decrease of
ringings.
Fig.4 shows a restoration of two kidneys: a) is the
original image and b) the restored one. Note the lack
of ringings around the high value peak of the kidney
on the left.

NOTES

[1] This term is used for the convolution between the original point spread function h(·) and the restoration
filter and gives the final resolution in f(·). In the
case of the optimum linear filter of eq.3(which is an
extension of the Wiener filter of the stationary case,
and therefore sometimes is here called Wiener filter)
and when the restoration filter is composed of the inverse filter and a smoothing filter, the cpsf is given

by the impulse response of the smoothing filter itself.

[2]In this approximation we suppose $L_s(\cdot)$ is an ideal low pass filter. A more precise approximation of $L_s(\cdot)$ can be:

$$L_s''(\varrho)=1 \qquad \varrho \leq \omega_o - \frac{\Delta\omega}{2}$$

$$L_s''(\varrho)=1-\frac{\varrho-\omega_o+\Delta\omega/2}{\Delta\omega} \qquad \omega_o - \frac{\Delta\omega}{2} < \varrho < \omega_o + \frac{\Delta\omega}{2} \qquad (11)$$

$$L_s''(\varrho)=0 \qquad \varrho \geq \omega_o + \frac{\Delta\omega}{2}$$

where $\varrho = (u^2+v^2)^{1/2}$; $\omega_o = \ln(M)/\sigma^2$ is the value of ϱ at which $L_s(\varrho)=1/2$ and $1/\Delta\omega$ is the angular coefficient of a linear approximation of the cut-off shape. In the case $\Delta\omega/\omega_o$ approaches to 0 eq.5 closely approximates eq.11; otherwise as $\Delta\omega/\omega_o$ is drawing near 2 eq.11 becomes a triangular pulse in such a case the transform of this function is $k/(\omega_o\varrho^3)\int_o^\varrho J_o(t)\ dt - k/(\omega_o\varrho^2)J_o(\varrho)$ which shows rings of a maximum relative amplitude $a_o=0.02$. Intermediate values of $\Delta\omega/\omega_o$ can be analyzed considering $L_s'' = (\omega_o/\Delta\omega+0.5)\cdot\Lambda\ (\omega_o+\Delta\omega/2) - (\omega_o/\Delta\omega-0.5)\Lambda(\omega_o--\Delta\omega/2)$ where Λ is the triangular pulse function. The relative amplitude a_o of the ringings showed in such

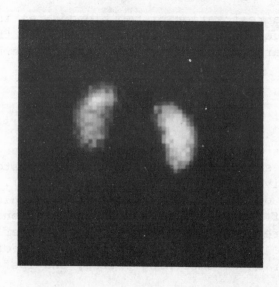

Fig.4a

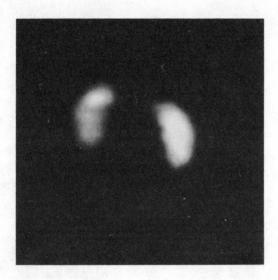

Fig.4b

cases is of course $0.02 \leqslant \alpha_0 \leqslant 0.132$. In nuclear medicine images high values of $\Delta\omega/\omega_0$ are unusual so eq.5 can be use to determine the upper bound of the ringings.

REFERENCES

1. Cantoni,V., De Lotto,I., Valenziano,F. and Favino, A.:1977,Proceedings Conf. Digital Image Processing, Monaco, pp.161-177.
2. Cantoni, V., De Lotto,I. and Dotti,D.: 1975, Alta Frequenza 45, pp.761-764.
3. Cantoni,V., De Lotto,I. and Valenziano,F.: 1979, Nucl. Instr. & Methods 166, pp.551-554.
4. Chu,N.Y., McGillem,C.D.:1979, IEEE Trans. ASSP, pp.457-465.
5. Huang,T.S.:1972, IEEE Trans. Audio and Electroacust. 20, pp.88-89.
6. Cantoni,V., De Lotto,I., Ghirardi,G.:1978,Nucl. Instr. & Methods 153,pp.199-205.
7. Openheim,A.V.,Schafer,R.W., Stockahm,T.G.:1968, Proc. IEEE 56,pp.1264-1291.

IMAGE TEXTURE ANALYSIS TECHNIQUES - A Survey

Larry S. Davis

Computer Sciences Department
University of Texas at Austin
Austin, Texas   78712

ABSTRACT

This paper contains a survey of image texture analysis
techniques. Three broad classes of methods are discussed:
pixel-based, local-feature based and region-based. The pixel-
based models include grey level cooccurrence matrices, difference
histograms and energy-measures. The local feature-based models
mostly rely on edges as local features and include Marr's primal
sketch model and a generalization of cooccurrence matrices.
Region-based models include a region-growing model and a topo-
graphic model which treats the texture image as a digital
terrain model.

INTRODUCTION

This paper presents an overview of techniques for image
texture analysis, emphasizing methods for describing image
textures for the purpose of image classification.

A textured area in an image is characterized by a non-
uniform spatial distribution of image intensities. Although
color images also contain textures, we will limit our attention
to grey scale images. Very little research has been devoted
to computational models of color textures which make essential
use of color information.

The variation in intensity which characterizes a texture
ordinarily reflects some physical variation in the underlying
scene. Although it is possible, in principle, to account for

189

J. C. Simon and R. M. Haralick (eds.), Digital Image Processing, 189–201.
Copyright © 1981 by D. Reidel Publishing Company.

the texture by modeling this physical variation, in practice
this is quite difficult to do. Horn (1) discusses such image
models. Rather, the approaches which we will discuss will treat
texture as a two-dimensional pattern of intensities, and will
not consider the physical basis of the texture. In fact, we
will adopt an intuitive model for image textures in which a
texture is composed of pieces: the size, shape, shades, and
spatial arrangement of the cells are the critical factors in discriminating between different textures. Notice that the cells
might actually form a partition of the image (i.e., form a
mosaic) or might be scattered on a homogeneous background (as
"bombs" dropped on a field). Although it is possible to develop
formal mathematical image models for such patterns (see, e.g.,
Ahuja (2), Schachter et al (3) or Zucker (4)), this paper will
not consider the development of texture analysis procedures
based on such models. Rather, the texture description models
which we will discuss are more heuristically motivated. They
have been applied to a wide variety of practical problems,
including, e.g., texture analysis of many biomedical images and
satellite images.

Texture description models can be broadly classified into
three main classes:

1) pixel-based models, where the texture is described by
statistics of the distribution of grey levels, or intensities,
in the texture,

2) local feature-based models, where the statistics are
computed with respect to the distribution of local features,
such as edges or lines, in the texture, and

3) region-based models, where the texture is first
segmented into regions, and then statistics on the shape and
spatial arrangement of the regions are used to characterize
the texture.

Section 2 of this paper reviews pixel-based models;
Section 3 deals with local feature-based models, and Section 4
discusses region-based models.

PIXEL-BASED TEXTURE MODELS

Perhaps the most widely used pixel-based texture model
is the grey level cooccurrence matrix, or GLCM. GLCM's were
first introduced by Haralick et al (5), and are defined as
follows. Let f be a digital image and let $D = \{(dx_i, dy_i)\}$ be
a set of image displacement vectors. Then the GLCM of f with

respect to D, C_D, is:

$$C_D(g_1,g_2) = \#\{((x,y),(x',y')): f(x,y) = g_1, f(x',y') = g_2,$$

and for some i,

$$x = x' + dx_i, \text{ and } y = y' + dy_i\}$$

where #S is the size of set S.

Thus, $C_D(g_1,g_2)$ is a count of the number of pairs of points in f which have grey levels g_1 and g_2, respectively, and are separated by one of the displacement vectors in D.

We can intuitively relate the structure of the cooccurrence matrix to the structure of the texture on which it is computed. Suppose, for example, that the set D includes all displacement vectors of length less than 2 (i.e., we consider a point and its eight neighbors in computing the cooccurrence matrix), and that the texture itself is very "busy" – i.e., contains very many small pieces. Then many pairs of adjacent points will have different grey levels, so that the cooccurrence matrix will have large values at positions far from its main diagonal – i.e., where $|g_1-g_2|$ is high. Suppose, on the other hand, that the texture is composed of a relatively few, large pieces. Then most pairs of adjacent pixels will have similar grey levels (they will either both be in one of the pieces or in the background) and therefore the cooccurrence matrix will have high values only on those elements which lie on or near the main diagonal. Any statistic computed from the cooccurrence matrix which is sensitive to the spread of values away from the main diagonal should be helpful in discriminating between such textures. One such statistic is the CONTRAST statistic defined as:

$$\text{CONTRAST} = \sum_{i,j} (i-j)^2 \cdot C_D(i,j)$$

Haralick et al (5) contains a much more extensive list of such statistics.

A second important aspect of texture which can be captured in cooccurrence matrices is directionality – i.e., the differences in grey level correlation as a function of direction. To compute directionally sensitive texture statistics, one can design the set D to include only displacement vectors of a fixed direction, and then compute cooccurrence matrices for a variety of

directions, and compare the statistics for each direction. One
tool which has been developed to accomplish this is the
polarogram which is a polar plot of a directionally sensitive
texture statistic (such as the contrast statistic of a directionally specific cooccurrence matrix) as a function of
direction (6). Statistics which measure the shape and size of
the polarogram are quite sensitive to texture directionality,
while at the same time being invariant to the orientation of
the texture in the field of view. Figure 1 contains two texture
samples, while Figure 2 contains polarograms for the two textures.
Notice that they have quite different shapes and sizes.

The utility of statistics computed from GLCM's depends on
the choice of the displacement vector set, D. Zucker (7)
describes a procedure for choosing good displacement vectors
based on measures of statistical independence of the rows and
columns of the cooccurrence matrices. Applying this technique
to the LANDSAT data set used by Weszka et al (8) in their
experimental study described below enabled him to achieve comparable classification rates to those reported in (8) without
exhaustively classifying a large training set using many sets
of cooccurrence statistics.

A second pixel-based texture analysis tool which is closely
related to the cooccurrence matrix is the difference histogram.
A difference histogram is a frequency count of the number of
pairs of pixels whose grey levels differ by a fixed amount.
More specifically, let D be a set of image displacement vectors
as defined above. Then the difference histogram of a texture
with respect to D, H_D, is defined by:

$$H_D(v) = \#\{((x,y),(x',y')): \left| f(x,y) - f(x'y') \right| = v,$$

$$\text{and for some } i, \ x = x' + dx_i, \ y = y' + dy_i\}$$

Notice that the grey level difference histogram for a set
of displacements D can be directly obtained from the GLCM for
D by summing along diagonals of the GLCM which are parallel to
the main diagonal. Pairs of points which contribute to the
main diagonal of the GLCM, e.g., have difference of grey levels
0. Thus, the grey level difference histogram contains strictly
less information than the GLCM. If, however, it turns out in
practice that texture classifications based on statistics
derived from the difference histograms are as high as those
based on statistics derived from GLCM's, then the difference
histogram would be the preferable tool, since its storage
requirements are lower than GLCMs, and it is computationally
less costly to compute statistics from difference histograms

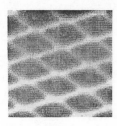

Figure 1a
A grating texture

Figure 1b
Scrap metal texture

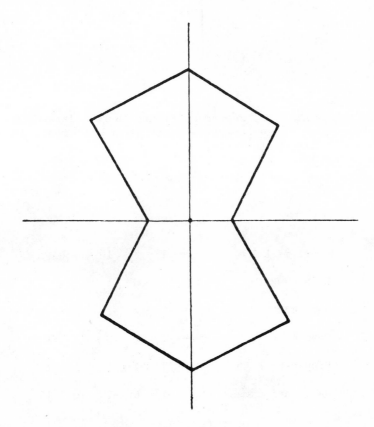

Figure 2a- Polarogram of Figure 1a for distance 3
         cooccurrence matrices and the contrast
         statistic.

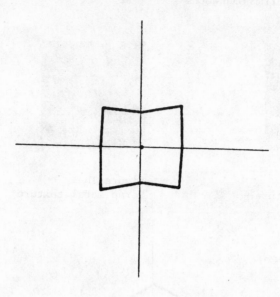

Figure 2b – Polarogram of Figure 1b for distance 5
cooccurrence matrices and the contrast
statistic.

Level mask L3	= 1	2	1			
Edge mask E3	= -1	0	1			
Spot mask S3	= -1	2	-1			
Wave mask W5	= -1	2	0	-2	1	
Ripple mask R5	= 1	-4	6	-4	1	
Oscillation mask O7	= -1	6	-15	20	-15	6	-1

Figure 3 – One-dimensional convolution masks.

than it is to compute them from GLCMs.

The structure of difference histograms can also be intuitively related to texture structure. For example, suppose that the set D contains all image displacements vectors of length less than 2, as we did for GLCMs. Then, for a busy texture, since most pairs of points will have different grey levels, we would suppose that the difference histogram will have relatively high values for large differences. For coarse textures, on the other hand, since most pairs of adjacent points will either both be in the same texture element, or both be in the background, we would expect that the difference histogram would have relatively high values for small differences. One statistic defined for difference histograms which can capture such distinctions is the MEAN statistic defined as:

$$\text{MEAN} = \sum_{i} i * H_D(i)$$

Weszka et al (8) describe a comparative experimental study where classification results based on statistics derived from GLCMs and difference histograms were compared for several texture discrimination problems. They found that the difference histograms performed just as well on those problems as the GLCMs. That study also investigated statistics derived from the texture's power spectrum, but they resulted in lower overall classification rates.

A texture model which is computationally very similar to difference histograms has recently been introduced by Faugeras and Pratt (9). They propose a texture synthesis model whereby a stochastic texture array is produced by applying some spatial operator to an array of independent identically distributed (i.i.d.) random variables. They suggest, therefore, that to analyze a texture one should attempt to decorrelate the grey levels in the texture to compute an approximation to the i.i.d. field; first-order statistics of the decorrelated texture should, then, be valuable texture statistics. This decorrelation can be achieved by a whitening transformation, but due to the computational cost of applying the whitening transformation, they suggest as an alternative that a gradient operator, such as a Sobel operator (10) or a Laplacian (11) be applied to the texture.

One last pixel-based model deserving attention is a recent set of "texture energy" transforms introduced by Laws (12). These transforms are fast since they can be computed by one-dimensional convolutions and simple moving-average techniques. They can also be made invariant to changes in luminance,

contrast and orientation without any image preprocessing.
Figure 3 contains one set of the one-dimensional convolution
masks which Laws describes. The one letter names are mnemonics
for Level, Edge, Spot, Wave, Ripple, Undulation and Oscil-
lation. The 1x3 vectors form a basis for the remainder of the
vector set. For example, each 1x5 vector can be generated by
convolving two 1x3 vectors.

Two-dimensional masks can be produced by convolving a
vertical 3-vector with a horizontal 3-vector. Texture information can then be extracted from an image by convolving the
3x3 masks with the texture. The fact that the two-dimensional
masks are separable means that they can be computed very fast.
A 5x5 convolution, for example, can be achieved by performing
two 3x3 convolutions.

Laws described the results of a comparative classification
study where statistics derived from his texture energy transforms were compared to grey level cooccurrence statistics. The
textures used were digitized version of the textures in
Brodatz (13). The texture energy measures gave 87% classification accuracy versus an accuracy rate of only about 70% for
the cooccurrence statistics.

LOCAL FEATURE-BASED TEXTURE MODELS

Local feature-based models describe image textures using
statistics based on the distribution of local image features,
such as edges, in the texture. They are most useful for what
are called macro-textures, i.e., textures where the cells, or
texture elements, are relatively large, say several pixels in
diameter. For such textures, pixel-based statistics depend
more on the transitions between grey levels in the texture
elements than they do on the size, shape and spatial arrangement of the texture elements.

Perhaps the most salient local feature in textures are
edges. Davis (14) contains a survey of edge detection techniques
for general image analysis. When we restrict our attention to
macro-textures, then it is possible to construct mathematical
models which describe the size and spatial arrangement of the
texture elements and to use such models to guide the design of
optimal edge detection procedures. Davis and Mitiche (15-16)
present a minimal error edge detection procedure based on such
mathematical models for macro-textures.

The simplest texture model based on edges describes textures
using first-order statistics of the distribution of edges. For
example, the average contrast of edges, or the variability in

			L		
	H	L	R		
	V			R	
	V				V
		R	H		V
	V				H

a)   edge map H(-), V(1), L(/), R(\)

	H	V	L	R
H	0	0	1	1
V	2	4	0	0
L	0	1	2	0
R	0	2	0	2

b)   GCM of (a)

Figure 4 - A simple example of a GCM.

their orientations are first-order edge statistics. However,
such first-order statistics do not seem to result in very
reliable texture discriminations.

Marr (17) suggested that instead of computing first-order
statistics of a raw edge map of a texture, one should first
construct relatively large segments of edges in those maps by
applying certain similarity grouping operators to the edge map.
He claims that first-order statistics of such extended edges can
be used to account for most texture discriminations which humans
can perform. This conjecture is consistent with recent psychophysical results reported in Julesz (18).

In practice, however, the computational cost of Marr's
similarity grouping operators might prohibit their application
to a texture analysis problem. Davis et al (19) suggested as an
alternative that second-order statistics of the raw edge map
might be a computationally cost-effective alternative. Such
second-order statistics can be collected from cooccurrence
matrices computed from the edge map. In order to compute useful
cooccurrence matrices, it is necessary to generalize the notion
of a cooccurrence matrix. This generalization involves replacing
the set of displacement vectors, D, with a spatial predicate, P.
The spatial predicate has at its arguments a pair of edges and,
based on properties of the edges such as their position,
orientation and contrast, returns a value of either TRUE or
FALSE. Those pairs of edges which satisfy the spatial predicate
(i.e., result in a TRUE value) are used to construct the edge
cooccurrence matrix. Figure 4 contains a simple example.
Figure 4a contains an edge map where the edges are labeled H
(horizontal), V (vertical), L (left diagonal), and r (right
diagonal). Figure 4b contains the edge cooccurrence matrix.
The spatial predicate assigns value TRUE to the pair of edges
(e_1, e_2) if:

1) e_1 and e_2 are neighbors, and

2) e_2 smoothly continues e_1 - i.e., if e_1 is a vertical

edge, then e_2 must be a vertical neighbor of e_1.

Davis et al (20) contains a comparative classification
study of edge cooccurrence with grey level cooccurrence for a
database of natural textures. The classification rates achieved
using statistics computed from edge cooccurrence matrices were
20% higher than classification rates achieved using statistics
computed from grey level cooccurrence matrices.

One can use a variety of other local features to describe

textures. For some textures it might be more appropriate to
detect linear features than edges. Or, as another alternative,
one can regard the intensity image as an elevation map and
describe the spatial distribution of peaks and valleys. Ehrich
and Foith (21) develop a hierarchical representation of such
peaks and valleys called a relational tree which they have
found useful for texture description. Such a structure is
invariant to linear transformations of the image grey scale.

REGION-BASED TEXTURE MODELS

All of the previously described texture models attempt to
describe the structure of an image texture in relatively indirect ways. Inasmuch as our intuitive texture model describes
textures in terms of the size, shape and spatial arrangement of
texture elements, a reasonable approach is to model texture by
directly computing such factors. Such approaches can be generi-
cally referred to as region-based texture models. Ideally, the
regions should coincide with the cells which compose the texture.

Although in principle such an approach might be preferable
to either pixel-based or local feature-based models, it is
important to keep in mind that computing the texture cells on
which the description of the texture will be based is an
instance of the image segmentation problem (hopefully for non-
textured segments). Thus, it might be very difficult to reliably compute the texture cells in practice.

The first region-based texture model was proposed by
Maleson et al (22) who suggested that standard region-growing
segmentation techniques be applied to a texture to extract the
texture cells. Zucker (23) contains a survey of region growing
procedures. Each cell is then described by an enclosing
ellipse. Properties of the ellipse (such as its eccentricity
or size) as well as cooccurrence statistics between ellipses
(e.g., how many have parallel axes) can be used to describe
texture.

Haralick (24), in an extensive survey paper, suggests as
texture regions the reachability sets of local grey level ex-
trema. The reachability set of a local maximum is the set of all
pixels reachable by that local maximum by a monotonically decreasing path, and not reachable from any other local maximum
by such a path. For the ellipses that Maleson employed elongation was related to eccentricity. For reachability sets,
elongation can be more generally defined as the ratio of the
larger to smaller eigenvalue of the 2x2 second moment matrix
obtained from the coordinates of the boundary pixels of the
set (25). It should be pointed out that there has not been

a substantial amount of research devoted to applying region-
based texture models to the discrimination of different textures.

CONCLUSIONS

This paper has attempted to provide a non-comprehensive
survey of techniques for describing image textures. It has
concentrated on techniques which have either proved to be of
value or which are potentially of value in solving texture
discrimination problems. The techniques discussed were classified into three categories - pixel-based, local feature-based
and region-based. Pixel-based models have been extensively
applied to real texture analysis problems; neither local feature-
based nor region-based models have received the extensive testing
which will be required to gauge their actual utility in image
analysis.

This research was supported in part by funds derived from the
Air Force Office of Scientific Research under Contract
F49620-79-C-0043.

REFERENCES

1. B.K.P. Horn, "Understanding Image Intensities", Artificial
    Intelligence, 8, pp. 201-231.
2. N. Ahuja and A. Rosenfeld, "Mosaic Models for Textures", in
    Proc. IEEE Conf. on Decision and Control, Miami, FL., 1979,
    pp. 66-70.
3. B. Schachter, A. Rosenfeld and L. Davis, "Random Mosaic
    Models for Texture", IEEE Trans. Systems, Man and Cyber-
    netics, 9, pp. 694-702, 1979.
4. S. Zucker, "Towards a Model of Texture", Computer Graphics
    and Image Processing, 5, pp. 190-202,1976.
5. R. Haralick, B. Shanmugam and I. Dinstein, "Texture Features
    for Image Classification", IEEE Trans. on Systems, Man and
    Cybernetics, 3, pp. 610-622.
6. L. Davis, "Polarograms: A New Tool for Image Texture
    Analysis", University of Texas Computer Sciences TR-116,
    November, 1979.
7. S. Zucker and D. Tersopoulos, "Finding Structure in Co-
    occurrence Matrices for Texture Analysis", to appear in
    Computer Graphics and Image Processing, 1980
8. J. Weszka, C. Dyer and A. Rosenfeld, "A Comparative Study
    of Texture Measures for Terain Classification", IEEE Trans.
    on Systems, Man and Cybernetics, 4, pp. 269-285, 1976.

9. O. Faugeras and W. Pratt, "Decorrelation Methods of Texture Feature Extraction", to appear in IEEE Trans. Pattern Analysis and Machine Intelligence, 1980.

10. W. Pratt, Digital Image Processing, Wiley-Interscience, New York, 1978.

11. A. Rosenfeld and A. Kak, Digital Picture Processing, Academic Press, New York, 1976.

12. K. Laws, "Texture Energy Measures", University of Southern California Image Processing Institute semiannual report, September, 1979.

13. P. Brodatz, Textures: A Photograph Album for Artists and Designers, Dover, New York, 1956.

14. L. Davis, "A Survey of Edge Detection Techniques", Computer Graphics and Image Processing, 4, pp. 248-270, 1975.

15. L. Davis and A. Mitiche, "Edge Detection in Textures", to appear in Computer Graphics and Image Processing, 1980.

16. L. Davis and A. Mitiche, "Edge Detection in Textures-- Maxima Selection", University of Texas Computer Sciences TR-133, Feb. 1980.

17. D. Marr, "Early Processing of Visual Information", Phil. Trans. Royal Society, B, 275, pp. 483-524, 1976.

18. B. Julesz, E. Gilbert and J. Victor, "Visual Discrimination of Textures with Identical Third Order Statistics", Biological Cybernetics, 31, pp. 137-140, 1978.

19. L. Davis, S. Johns and J.K. Aggarwal, "Texture Analysis Using Generalized Cooccurrence Matrices", IEEE Trans. Pattern Analysis and Machine Intelligence, 1, pp. 251-258, 1979.

20. L. Davis, M. Clearman and J.K. Aggarwal, "A Comparative Texture Classification Study Based on Generalized Co- occurrence", University of Texas Computer Sciences TR-110, August 1979 (also in IEEE Conf. on Decision and Control, Miami, FL., December 1979).

21. R. Ehrich and J. Foith, "Topology and Semantics of Intensity Arrays", in Computer Vision, Eds. E. Riseman and A. Hanson, Academic Press, NY., 1978.

22. J. Maleson, C. Brown and J. Feldman, "Understanding Natural Texture", University of Rochester Computer Science Tech. Report, Sept., 1977.

23. S. Zucker, "Region Growing: Childhood and Adolescence", Computer Graphics and Image Processing, 5, pp. 382-399, 1976.

24. R. Haralick, "Statistical and Structural Approaches to Texture", Proc. IEEE, 67, pp. 786-804, 1979.

25. Y. Frolov, "Measuring the Shape of Geographical Phenomena: A History of the Issue", Sov. Geog.: Rev. Transla. 16, pp. 676-687, 1975.

TEXTURE FEATURES IN REMOTE SENSING IMAGERY

DESACHY, JACKY

Laboratoire CERFIA (Université P. SABATIER)
II8 , route de Narbonne
3I077 Toulouse-cédex FRANCE

The remote-sensing main automatic approach is nowadays based
on pixel radiometry and so necessarily incomplete as it takes no
account of its environment. Our aim is to improve the real
efficiency of spatial information (texture) , the principle being
that information derived from one pixel should be regarded as a
"spectral-spatial" information pair. It seems indeed obvious
that texture is going to assume an ever increasing importance as
spatial resolution of satellite imagery becomes more and more
precise.

In this paper we shall expose a texture feature extraction
method trying to take into account as much of spatial information
as we can. To prove the interest of such a method we shall apply
it first within a supervised classification method and then to
texture based image partitionning.

1. TEXTURE FEATURE EXTRACTION

Remote sensing imagery texture appears to have a large variety of
appearances, so it needs an enough general texture feature
extraction method . Moreover , texture primitives are often
pixels or small groups of pixels (micro-textures) .
For these reasons we chose to introduce Harralick 's grey level
cooccurence matrices (1) . Let 's remember their formulation :

- let $I(x,y)$ be an image function on a domain D with NG possible
  grey tones.
- let R be a spatial relation between two pixels (for example
  "distance=1")

J. C. Simon and R. M. Haralick (eds.), Digital Image Processing, 203–210.
Copyright © 1981 by D. Reidel Publishing Company.

The associated grey level cooccurence matrix is a NG X NG symetric matrix whose general term is the relative frequency of cooccurences of pixel pairs verifying the spatial relation R , the first pixel grey tone being i-1 , the second one j-1 .

$$m(i,j) = \frac{\#\{((x,y),(x',y')) \in D, \text{verifying } R \,/\, I(x,y)=i\text{-}1 \text{ and } I(x',y')=j\text{-}1\}}{\#\{((x,y),(x',y')) \in D, \text{verifying } R\}}.$$

Provided domain D is not too small and with a spatial relation R involving a relatively small distance between the pixels , it is easy to draw grey tone histogram and to estimate cumulated autocorrelation function only once having cooccurence matrix relative to relation R . These properties prove the wealth of such an approach especially for texture measure. But it suffers from heavy cost in memory space.

16x16 image with	⟺	256x256 real symetric
256 grey tones		cooccurence matrix
⬇		⬇
256 bytes		131 584 bytes

Then feature extraction based on matrix element statistics (Harralick defines fourteen of them (2)) , such as

$$f_1 = \underset{ij}{\Sigma\Sigma}\, m(i,j)^2 \quad , \quad f_2 = \overset{NG-1}{\underset{0}{\Sigma}}\, n^2\, (\, \underset{ij}{\Sigma\Sigma}\, m(i,j)\,)\ ,\ \ldots$$

which permits to isolate special texture properties looses a part of the texture information contained within the cooccurence matrix itself .

So we developped the idea of a compromise for a simple but general remote sensing imagery texture feature measure :

- to use cooccurence matrices.
- by the way of artificially reducing grey tone number , to reduce memory space and CPU time needed by such matrices .
- to keep the cooccurence matrix itself as texture signature .

2. OUR TEXTURE FEATURE EXTRACTION METHOD

It begins with preprocessing the image , by reducing the channel number in multispectral imagery and properly reducing the grey tone number .

2.1. Grey tone number reducing method

There are necessary constraints for this method :
- must keep unchanged the initial energy mean in a small neighborhood of each pixel .

- must keep unchanged the texture .
In this sense , among all binary transforms the best one was
"ordered dither method" exposed by JF Jarvis (3) :

First step consists in building a threshold matrix $D^n(i,j)$ nxn
with n= 2^k k= 1,2, ...

for ex. $D^2 = \begin{bmatrix} 0 & 2 \\ 3 & 1 \end{bmatrix}$ then if $U^n = \begin{bmatrix} 1 & \cdots & 1 \\ \vdots & & \vdots \\ 1 & \cdots & 1 \end{bmatrix}$

$$D^n = \begin{bmatrix} 4 D^{n/2} + D^2(0,0) U^{n/2} & 4 D^{n/2} + D^2(0,1) U^{n/2} \\ 4 D^{n/2} + D^2(1,0) U^{n/2} & 4 D^{n/2} + D^2(1,1) U^{n/2} \end{bmatrix}$$

Suppose we have chosen n and let 's call T the binary transform .
Then the transform T has the following steps :

$$i = x \quad \text{modulo n}$$
$$j = y \quad \text{modulo n}$$

if $I(x,y) > D^n(i,j)$ then T $\left[I(x,y)\right]$ =1 else =0

The transform T works as if we had superposed on the initial
image a threshold image composed of contiguous threshold matrices.
n=8 gives almost the same result as n=16 .
A parasitic texture is superposed, but regularly on the whole
image (high spatial frequency regular noise), what we could
call a "texture translation" on the image.An interesting extension
of this transform is to obtain a transformed image with four or
eight grey tones , as we did. For ex. , for 4 grey tones we have
three threshold matrices $D^{n,1}$ $D^{n,2}$ $D^{n,3}$ with

$$D^{n,1}(i,j) = \frac{D^n(i,j)}{255}^2 \qquad D^{n,2}(i,j) = D^n(i,j)$$
$$D^{n,3}(i,j) = 2 \times D^n(i,j) - D^{n,1}(i,j)$$

remark :
Flattenning the grey tone histogram can be done at the same time
by properly choosing the thresholds , which is usually done on
images before all calculations about texture

2.2. Texture signature

We shall discuss it in the case of a four grey level transform.
Let D be a texture homogeneous domain , then we distinguish two
cases :

* <u>monospectral image</u> with the following steps :

$$M = \left\{ m(i,j) \right\}_{4 \times 4}$$

M : normalized cooccurence matrix for spatial relation R
    "distance = 1"

As this matrix is symetric, we define the texture signature of
domain D as being :

$$[X_1 , X_2 , X_3 , X_4 , X_5 , X_6 , X_7 , X_8 , X_9 , X_{10}] =$$
$$[m(0,0),m(0,1),m(0,2),m(0,3),m(1,1),m(1,2),m(1,3), m(2,2),m(2,3),$$
$$m(3,3)]$$

The ten parameters are quite correlated , but this correlation is
a regular one , introduced by the grey tone reducing transform T.

* <u>multispectral image</u>

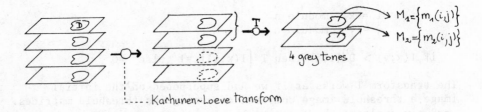

..... Karhunen-Loeve transform

Then we obtain two vectors , one for each KL component (1°,2°)
$[X_1, X_2 \ldots X_{10}]$ and $[XX_1 , XX_2 \ldots XX_{10}]$ which are uncorrelated.

3. APPLICATIONS

We applied our texture signature within a supervised classification
method and then to texture based segmentation.

3.1. supervised classifier

Each class is defined by samples on which we calculate m vectors
$[X_1 \ldots X_{10}]$ for a monospectral image and m additional vectors
$[XX_1 \ldots XX_{10}]$ for a multispectral image (these vectors being
calculated on m 16x16 subimages extracted from the samples
(16x16 for LANDSAT imagery)).
Class A will then be defined by $[\mu_1 , \mu_2 \ldots \mu_{10}]$
                              and $[\sigma_1 , \sigma_2 \ldots \sigma_{10}]$

$$\mu_i = \frac{1}{m} \sum_{j=1}^{m} X_i(j) \qquad \sigma_i^2 = \frac{1}{m} \sum_{j=1}^{m} (X_i(j) - \mu_i)^2$$

Let then be S whatever subimage with texture signature
$[X_1, \ldots X_{10}]$, then similarity between S and class A will be

defined by :

$$d(S, \text{class } A) = \sum_{i=1}^{10} \frac{|X_i - \mu_i|}{\sigma_i}$$

In the MSS case :
$$d(S, \text{class } A) = \max \left\{ \sum \frac{|X_i - \mu_i|}{\sigma_i} , \sum \frac{|XX_i - \mu\mu_i|}{\sigma\sigma_i} \right\}$$

Now processing the whole image : we divide it in non overlapping contiguous images (8x8 on LANDSAT)
of " K-type " surrounded by four
16x16 subimages : ①, ②, ③, ④.

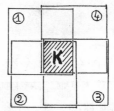

Then ,

$$d(K, \text{class } A) = \min_i \left\{ d(\text{①}, \text{class } A) \right\}$$

These four subimages permit region
edge smoothing because K is agregated
with the most similar textured region.

Then among all apriori defined classes , we look for the class B
verifying :
$$d(K, \text{class } B) \leqslant d(K, \text{class } A) \quad \forall A$$

$K \in$ class B iff $d(K, \text{class } B) \leqslant S_B$, S_B being defined in the

hypothesis that class B is gaussian , or being defined in an
interactive process (See fig. 1 for an example of supervised
classification of a LANDSAT image of Palni hills , South India)

3.2. Texture based image partitionning

The first step is to build the texture feature images (simply ,
the initial image may be divided in non overlapping contiguous
subimages and then their texture signatures are calculated)

In the case of a monospectral image , we obtain 10 images
corresponding to the ten parameters $X_1 \ldots X_{10}$. We calculate
the features histograms and look for valleys , which permits
the feature values clustering. First partition is obtained from
X_1-histogram , then the obtained partition is "deepened" with
X_2-histogram ...etc...At end the found classes number was not
apriori chosen (See fig. 2 for an example of an aerial image,
(a binary transform being used , we have only three parameters :
X_1 , X_2 , X_3)) .

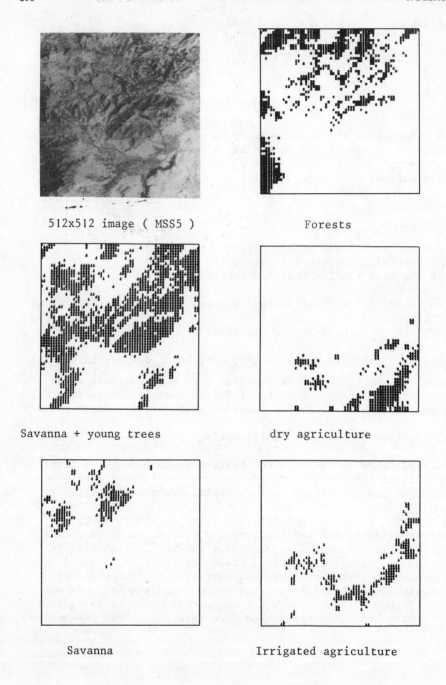

512x512 image (MSS5)   Forests

Savanna + young trees   dry agriculture

Savanna   Irrigated agriculture

Figure 1 . Supervised classification

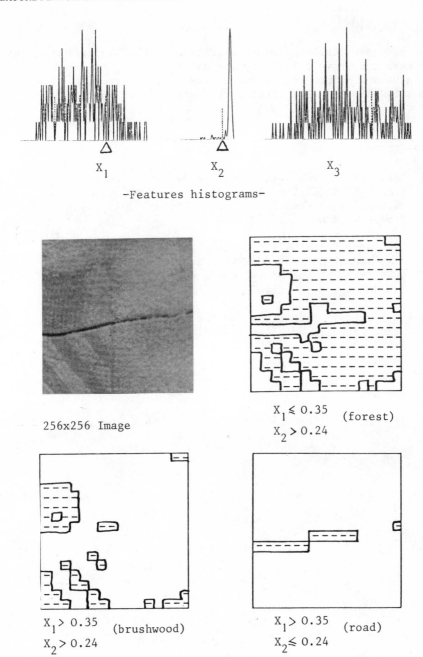

X_1

X_2

X_3

−Features histograms−

256x256 Image

$X_1 \leqslant 0.35$ (forest)
$X_2 > 0.24$

$X_1 > 0.35$ (brushwood)
$X_2 > 0.24$

$X_1 > 0.35$ (road)
$X_2 \leqslant 0.24$

Figure 2   . Texture based segmentation .

4. CONCLUSION

The proposed texture feature extraction method is very easy to
process and above all a very speedy one needing few memory space.
We have now in project to deepen the realised spectral-spatial
relationship and to introduce our method in an interactive
process.

REFERENCES

(1) "Statistical and structural approaches to texture"
    R.M. HARRALICK Department of computer science
    University of Kansas Lawrence , Kansas 66045
    Proceedings of 4° IJCPR 1978

(2) "Textural features for image classification"
    R.M. HARRALICK , K. SHANMUGAM and ITS'HAK DINSTEIN
    IEEE Transactions on systems , man , and cybernetics
    Vol SMC-3 , n. 6 nov 73

(3) "A survey of techniques for the display of continuous
     tone pictures on bilevel displays"
    J.F. JARVIS , C.N. JUDICE and W.H. NINKE
    Computer graphics and image processing 5 pp 13-40 (1976)

TWO DIMENSIONAL TIME SERIES FOR TEXTURES

J. P. Gambotto*

ETCA, Fort de Montrouge, Arcueil

ABSTRACT

This paper presents an approach to synthesis, analysis and recognition of textures based on two dimensional time series. In this approach, two dimensional autoregressive models are used to generate textures having given statistical properties. Some new estimation techniques are proposed, and used to fit an autoregressive model to a reference region in an image. The interest of a recognition procedure based on two dimensional inverse filtering method is then emphasized. This procedure is applied to natural pictures for classification and segmentation using the second order statistical properties of the textures only

I)   INTRODUCTION

In recent years, much work has been done in the field of texture recognition. However, no simple and general model seems to be available for image modeling. A straightforward generalization of the one dimensional (1D) case is the two dimensional (2D) autoregressive moving average model. This simple recursive model has not been much used in the field of image processing. The main reason seems to be the complexity of the stability tests. Generally, the models used are not really two dimensional. For example, such a model is obtained when the 2D filter is separable [4] ; in this case the 2D signal can be decomposed into two 1D processes.

*Formerly with ENST, Paris

J. C. Simon and R. M. Haralick (eds.), Digital Image Processing, 211–229.

In this paper, the advantages of general 2D models over separable ones are emphasized. Then, some of the well known techniques and algorithms currently used for 1D signal processing [6] are developped in the 2D case. Several recursive algorithms can be implemented at the estimation stage : some are closely related to the vector version of Levinson algorithm [7] . A recognition procedure based on 2D inverse filtering method is described, and is used for partitioning the image into regions with uniform textural characteristics.

II) 2D SIGNAL MODELING

1) 2D ARMA Models :

A simple representation model for a stationary and zero mean 2D signal $y_{m, n}$, is the autoregressive moving average model :

$$\sum_{i=0}^{p} \sum_{j=0}^{r} a_{i,j} \, y_{m-i, \, n-j} = \sum_{i=0}^{h} \sum_{j=0}^{l} b_{i,j} \, u_{m-i,n-j} \qquad (1)$$

where $\{a_{i,j} b_{i,j}\}$ are parameters to be chosen, and without loss of generality we can set $a_{0,0} = 1$; $u_{m,n}$, is a white noise with covariance d.

This model can also be represented by its transfer function :

$$H(z,w) = \frac{B(z,w)}{A(z,w)} \qquad (2)$$

where $A(z,w)$ and $B(z,w)$ are polynomials in the 2 variables z and w :

$$A(z,w) = \sum_{i=0}^{p} \sum_{j=0}^{r} a_{i,j} z^i w^j \qquad B(z,w) = \sum_{i=0}^{h} \sum_{j=0}^{l} b_{i,j} z^i w^j$$

In the general case polynomials $A(z,w)$ and $B(z,w)$ are not separable - i.e. they do not factorize into two one-variable polynomials - ; several authors [8] [9] have shown that serious difficulties may occur with these models ; in particular, we note that checking stability represent a considerable computational task. Very often, however, as in the 1D case, model (1) with the autoregressive part only, is sufficient for signal representation.

2)   2D Autoregressive Models :

The 2D autoregressive model is defined from the transfer function (2) by letting $B(z,w) = 1$:

$$\sum_{i=0}^{p} \sum_{j=0}^{r} a_{i,j} \; y_{m-i,n-j} = u_{m,n} \tag{3}$$

A more general model in which the estimate $\hat{y}_{m,n}$ utilizes all observations $y_{m-i,\,n-j}$ in the asymmetric upper half plane (i.e. $\{i > 0\}$ and $\{i = 0$ for $j > 0\}$) is defined by :

$$y_{m,n-h} = -\sum_{(i,j) \,\in\, E_{p,r}^{h}} \sum a_{i,j} \; y_{m-i,\; n-j} + u_{m,n-h} \tag{4}$$

where $a_{0,h} = 1$ and $E_{p,r}^{h}$ is the set given by :

$$E_{p,r}^{h} = \{(i,j) \; / \; 0 \leqslant i \leqslant p, \; 0 \leqslant j \leqslant r \text{ and } j > h \text{ for } i = 0\}$$

and the covariance sequence of the signal $y_{m,n}$,

$$r_{s,t} = E\left[y_{m,n} \; y_{m-s,n-t} \right] \tag{5}$$

has the property :

$$r_{s,t} = r_{-s,-t}$$

It can be shown that for a separable autoregressive model,

$$r_{s,t} = \frac{r_{s,0} \; r_{0,t}}{r_{0,0}} \tag{6}$$

Model (4) is known as the half plane asymmetric filter ; in the following, it will be refered to using the inverse of its transfer function $A_{p,r}^{h}(z,w)$.

Ekstrom and Woods [10] have shown that there exist, 7 other asymmetric half planes where recursions similar to (4) can be defined ; we also remark that the causal model (3) is obtained from (4) when $h = 0$.

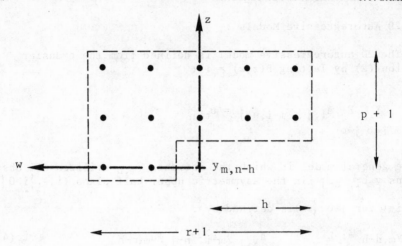

Figure 1 : Half plane asymmetric filter $A^h_{p,r}$ (z,w)

A stability theorem was first given by Shanks [11] who stated that (3) is a stable filter if and only if :

$$A (z,w) \neq o \text{ for } \{(z,w) / |z| \leqslant 1 , |w| \leqslant 1\} \qquad (7)$$

This test requires, in theory, an infinite amount of computational time ; Huang [12] simplified it by showing that (7) is equivalent to the following conditions :

$$A(z,o) \neq o \qquad |z| \leqslant 1 \qquad (8)$$

$$A(z,w) \neq o \qquad |z| = 1 \quad |w| \leqslant 1$$

These stability conditions still hold for the half plane asymmetric filter ; however, it was noted [9] that w and z are not interchangeable as it is the case for filter (3).

III) TEXTURE SYNTHESIS

The synthesis of natural looking textures is of great importance in image processing : first for visual check of the quality of the analysis, second, as a needed complement of analysis in coding and storage of digitized pictures.

Many authors suggested methods for generating textures [2] , [5] [3] .

In particular, one dimensional seasonal time series models were first used by Mc Cormick and Jayaramamurthy [13] for texture synthesis purposes.

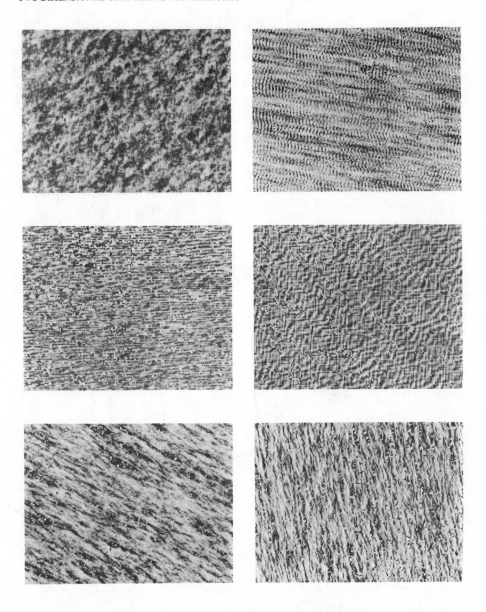

Figure 2 : Synthetic textures generated with 2D autoregressive
        model (9).

Figure 2 (cont.)

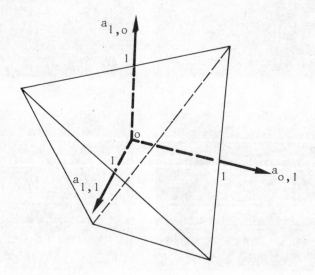

Figure 3 : Stability domain for the 2D filter (9).

For textures with dominant statistical properties, the linear dependence between neighbouring pixels can be represented using only the autocorrelation function ; in this case, one can use 2D autoregressive models, instead of 1D ones, to have a better representation of 2D signals.

The 2D linear model is in fact the simplest one to generate textures with given statistical properties ; we assume here that the output of the 2D autoregressive filter $y_{m,n}$ is the grey level value of a texture :

$$\xrightarrow{\quad u_{m,n} \quad} \boxed{1/A(z,w)} \xrightarrow{\quad y_{m,n} \quad}$$

white noise   Texture

Textures are generated with low-order models, such as :

$$A(z,w) = 1 + a_{o,1}w + a_{n,o}z^n + a_{n,1}z^n w \qquad (9)$$

$n = 1, 2, 3 \ldots$

For this simple model the stability is easily checked using the efficient algorithm proposed by Anderson and Jury [14] ; the stability domain of this filter in the parameter space is shown in figure 3 ; It was found that textures having many different aspects can be generated by model (9) depending on which point is chosen in the stability domain (see figure 2).

IV)   TEXTURE ANALYSIS

The correlation coefficients $\{r_{i,j}\}$ are estimated on a given area of the image, where texture is assumed to be homogeneous, or on a known texture from a data set. Parameter estimates of the autoregressive filter are then computed from these correlation coefficients and characterize the texture of this area. Figure 4 shows an example of resynthesized texture after analysis.

We emphasize here, that to have a good representation of 2D signal, with a low-order autoregressive model, it is necessary to consider general nonseparable models. In fact, relationship (6) generally does not hold, and the parameter estimates being uniquely determined from the correlation cofficients, the corresponding autoregressive model is, in the general case nonseparable.

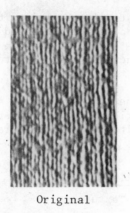

Original   Synthesis

Figure 4 : automatic analysis and synthesis of a natural texture.

Let $\hat{y}_{m,\ n-h}$ be the prediction of $y_{m,\ n-h}$, then the prediction error is given by :

$$u_{m,n-h} = y_{m,n-h} - \hat{y}_{m,n-h} \tag{10}$$

The least mean square criterion consists in minimizing the variance of the prediction error :

$$E\left[u_{m,n-h} \cdot u_{m,n-h}\right] \tag{11}$$

For the autoregressive model, the value of $u_{m,\ n-h}$ is obtained from equation (4) and minimization of (11) with respect to the parameters a_{ij} then leads to the generalized Yule-Walker (or

Normal)equations :

$$\sum_{(i,j)\ \varepsilon\ E^h_{-p,r}} \sum a_{i,j}\ r_{s-i,t-j} = \begin{cases} o & (s,t)\ \varepsilon\ E^h_{p,r} \\ d^h & (s,t) = (o,h) \end{cases} \tag{12}$$

where $E^h_{-p,r} = E^h_{p,r} + (o,h)$ and d^h is the covariance of input noise $u_{m,n}$.

This set of linear equations can be rewritten in matrix form as :

$$\left[\underline{a}'_o,\ \underline{a}_1,\ \cdots,\ \underline{a}_p\right]\ \underline{R}^h = \left[d^h, o,\ \cdots,\ o\right] \tag{13}$$

where $\qquad \underline{a_i} = \begin{bmatrix} a_{i,0} & \cdots & a_{i,r} \end{bmatrix} \quad i = 0, \ldots, p$

and $\underline{a'}_0 = \begin{bmatrix} a_{0,h} & \cdots & a_{0,r} \end{bmatrix}$.

R^h is a fxf symetric and positive definite matrix with
$f = (p + 1).(r + 1) - h$

The matrix corresponding to the higher order $\underline{R}^0$ is block Toeplitz :

$$
\underline{R}^0 =
\begin{vmatrix}
R_0 & R_1 & \cdot & \cdot & \cdot & R_p \\
R^T_1 & & & & & \\
\cdot & & & & & \\
\cdot & & & & & R_1 \\
\cdot & & & & & \\
R^T_p & & & R^T_1 & R_0 &
\end{vmatrix}
\tag{14}
$$

and its $(r + 1).(r + 1)$ blocks R_i are Toeplitz matrices with entries
$(R_i)\ k,l = r_{i,\ k-1}$.

For $h = 0, \ldots, r - 1$ the matrix $\underline{R}^{h+1}$ is a submatrix of $\underline{R}^h$, and is obtained by removing first row and first column from $\underline{R}^h$.

By taking into account the special structure of matrix R^h, it is possible to develop an algorithm for solving system (13) without performing a matrix inversion.

In the 1D case the Levinson algorithm has been widely utilized to solve the Yule Walker equations ; this fast and efficient algorithm recursively computes the estimated models from order $n = 1$ to $n = p$; this attracting property of the Levinson algorithm makes it possible to find the optimal order of the model - with Akaike FPE criterion [15] for example - within the recursion itself. Unfortunately, no extension of this criterion to 2D data seems to be available.

A Two Dimensional version of the Levinson algorithm was first derived by Justice [16] , who used orthogonal properties of bivariate Szegö polynomials. However, in two dimensions there is no unique way, in which models can be ordered ; this fact can be easily understood by noting that the order of our general model $A^h_{p,r}$ (z,w) is the

3-tuple (p,r,h) ; consequently, other recursions may be found for solving system (13).

A different approach based on the relation between 2D models
(5) and autoregressive vector models has been developped by the
author [18] : let us consider the (r+1)-vector Ym built up with r+1
successive pixels on line m. We define the autoregressive vector
model by :

$$A_0 Y_m = -\sum_{i=o}^{p} A_i Y_{m-i} + U_m \tag{15}$$

where the $(r + 1)$-vector U_m is a white noise and matrices A_i are
the parameters of the vector model. The following normalization
is used : A_0 is upper triangular with "1s" along the main diagonal.

We point out that R^o is the correlation matrix for the autoregressive vector model. By combining equation (13) for h=o,. . .,r
the following result was established [19] :

Theorem 1 : The estimated 2D models $A^h_{p, r}$ (z,w) h = o,. . ., r

and the estimated vector model $A_p(z)$ are related by :

$$A_p(z) \underline{w} = \left[A^o_{p,r}(z,w) \ . \ . \ . \ A^r_{p,r}(z,w) \right]^T \tag{16}$$

where $\underline{w}^T = \left[1, w, \ . \ . \ ., w^r \right]$ and $A_p(z) = \sum_{i=o}^{p} A_i z^i$

One consequence of relationship (16) is that 2D models can also
be identified using a straighforward vector version of the Levinson
algorithm [18] .

V)   RECOGNITION AND CLASSIFICATION

The recognition procedure can use a distance measure between
the unknown 2D signal (or its model) and a learned reference ;
this reference can be taken from a data base (dictionnary of textures), or computed on a known area of the image itself.

However, due to the nonstationarities in most pictures, the
recognition problem is strongly related to a segmentation problem.
Following our approach classifying texture will lead to a corresponding segmentation of the image.

1)   Recognition Criteria :

A first criterion is deduced from the value of the mean
square prediction error (11) ; for the 2D autoregressive model,
equation (13) yields :

$$d^h(k) = \underline{a}'(k)\ \underline{R}^h\ \underline{a}'(k)^T \tag{17}$$

where $\underline{a}'(k)$ is the parameter vector of the kth reference model :

$$\underline{a}'(k) = \left[\underline{a}'_o, \underline{a}_1, \ldots, \underline{a}_p\right](k) \tag{18}$$

and $\underline{R}^h$ is the correlation matrix of the 2D signal under test. (17) is then compared with the minimum value :

$$d^h = \underline{a}'\ \underline{R}^h\ \underline{a}'^T \tag{19}$$

It is noteworthy, that this criterion is related to the inverse filtering of the 2D signal under test by the reference models $\underline{a}'(1)$:

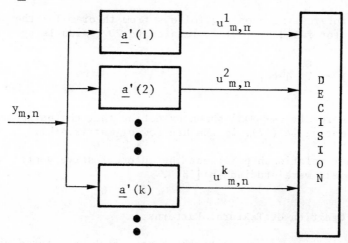

The recognition is made by observing the statistical properties of the output signal $u^1_{m,n}$; in particular, if the signal $y_{m,n}$ has

been generated by model $\underline{a}'(k)$, the variance of $u^1_{m,n}$ will be minimum for $1=k$. In this case, the 2D inverse filtering method is identical to the former criterion.

When the order of the 2D autoregressive model is increased, the prediction error tends to the innovation sequence of the signal which is uncorrelated ; therefore, another good test of fit should be the whiteness of the residue ; however this criterion is more difficult to evaluate in practice, and is not uniquely defined.

For the autoregressive vector model (15), similar recognition criteria can be derived. The value of the mean square prediction error is here given by :

$$D(k) = \text{Trace } A(k) \, \underline{R}^o \, A(k)^T \tag{20}$$

A(k) being the $(p+1)(r+1) \times (r+1)$ parameter matrix of the kth reference vector model :

$$A(k) = \left[A_o, A_1, \ldots, A_p \right] (k) \tag{21}$$

(20) is then compared with the minimum value :

$$D = \text{Trace } A\underline{R}^o \, A^T \tag{22}$$

An interesting property follows from theorem 1 : the value of the vector residual (20) is related to 2D residuals by

$$D(k) = \sum_{h=o}^{r} d^h(k) \tag{23}$$

This result is easily shown by noting that the parameter vector $\underline{a}'(k)$ appearing in (17), is the hth row of matrix A(k).

Other relationship between the autoregressive vector model and 2D models were studied in [18] .

2)   Recognition of Textural Patterns :

In digitized pictures textures are often associated with the high or medium spatial frequency components of the signal and are considered as homogeneous areas of various sizes. When these areas are not too small, a reference model can be learned, and one of the preceding criteria be used to recognize all the areas of the picture having the same texture.

The problem of partitioning the image into regions having uniform textural characteristics is, in fact, not trivial. A simple method, we developped consist in dividing the image in small areas where the 2D signal is assumed to be second order stationary, and then the inverse filter is used for all areas along with a low order model. Other techniques also using the inverse filtering method were reported in [20] .

For very homogeneous textures, it was shown that the inverse filtering method give good results [17] . The same technique was used with the 18 different textures shown in figure 5a, each 60 x 70 pixels. For each texture a reference model was computed with all the pixels, and 4 samples of size 30 x 30 were chosen for the classification (see fig. 5b). The following models were used :

$A_{2,2}^2$ (z,w), $A_{2,2}^1$ (z,w), $A_{2,2}^0$ (z,w) and also the associated vector model A_2 (z).

Figure 6 shows that a better classification is achieved with the vector criterion than with 2D ones : other vector criteria were also used [18] giving similar results (not shown on figure 6). We remark that sample [2,4] which was rejected by class (1,2), was classified in (2,1) by all the criteria. This result seems in accordance with the recognition done by the human visual system. Likewise, sample [5,7] which belongs to class (3,4) was classified in (1,2), and we remark that textures (1,2) and (3,4) are very similar.

The procedure has been also applied on natural pictures such as the bell-tower image (Fig. 7a). Figure 7b shows the value (in grey levels) of the vector criterion when computed on small areas of 10 x 10 pixels, and with a reference model taken on the clouds. Figure 7c is the same, except the reference model is taken on the stones of the steeple ; we note that the clouds have about the same texture as the sky, except for the transition areas between the clouds and the sky. For this experiment the vector model used was A_2 (z).

3)   Segmentation

Preceding results show the importance of second order statistics in images ; in particular Figures 7b and 7c strongly suggest that the inverse filtering procedure can be used to segmentate pictures.

A very simple scheme is used : the image is first divided into small squares of size N x N. Considering one of these squares, the criterion is evaluated twice : once with the reference model on the upward neighbour, and once with the reference model on the lefthand neighbour ; in each case this measure is compared to a threshold, and if greater an edge is detected. With this technique, horizontal and vertical edge elements of lenght N can be detected.

The boundary can be refined using overlapping areas, so that edge elements of lengh N' < N are detected. Such a technique was used with a part of the bell tower image ; resulting segmentation is shown on figure 7c.

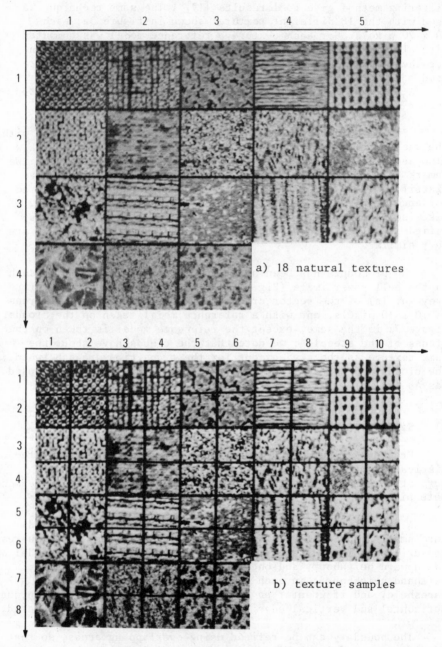

a) 18 natural textures

b) texture samples

Figure 5 : The 18 textures used in the recognition experiment

Classes	Rejected samples belonging to the class	Number of misclassified samples with the following models :			
		$A^2_{2,2}$	$A^1_{2,2}$	$A^{\circ}_{2,2}$	Vector model
(1,1)	0	0	0	0	0
(1,2)	1 [2,4]	3	7 [5,7]	0	1 [5,7]
(1,3)	0	1	8	0	0
(1,4)	0	0	0	0	0
(1,5)	0	0	10	0	0
(2,1)	0	1 [2,4]	1 [2,4]	1 [2,4]	1 [2,4]
(2,2)	0	1	0	0	0
(2,3)	0	0	1	0	0
(2,4)	0	0	2	6	0
(2,5)	0	0	1	0	1
(3,1)	0	3	0	0	0
(3,2)	0	0	0	0	0
(3,3)	0	5	1	0	1
(3,4)	0	2	0	0	0
(3,5)	0	0	1	1	0
(4,1)	0	0	0	0	0
(4,2)	0	0	0	0	0
(4,3)	0	0	0	0	0

Figure 6 : Result of the classification for the 72 texture samples of fig.5b, with 3 2D models and related vector model.

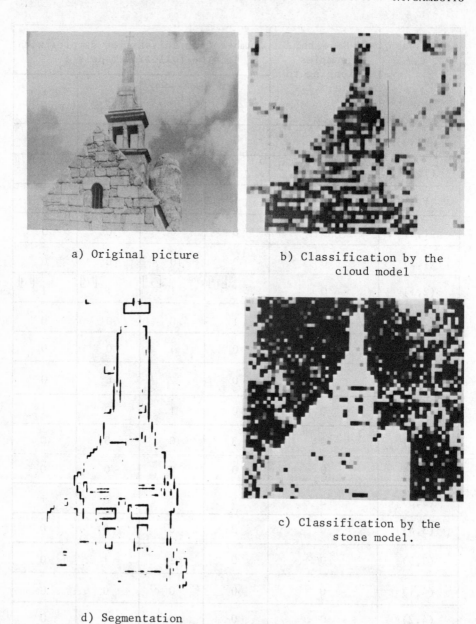

a) Original picture

b) Classification by the
   cloud model

c) Classification by the
   stone model.

d) Segmentation

Figure 7 : Recognition and segmentation using textural information only.

VI)   CONCLUSION

An approach to the problem of synthesis, analysis, and recognition of textures has been presented. It was found that different
textures can be generated with low order nonseparable recursive
filters. These models were then fitted to image samples using 2D
estimation techniques.

A recognition procedure based on inverse filtering methods has
been developped and used on real images. Two important properties
of this procedure were outlined : (1) the use of small samples
along with low order models, and (2) the possibility to have overlapping windows. The first one enabled us to characterize small
homogeneous areas in the image, as well as textures with slowly
varying characteristics ; the second was used in order to find
the boundary between homogeneous areas more precisely.

Further researchs should be directed towards the implementation
of a more adaptive technique ; such a technique would combine a
change in the model order as well as a change in the window size.
We also suggest to use a more general model in order to include
deterministic components of the texture, such as local mean, slopes,
and periodicities.

REFERENCES

1.   Pratt W., Faugeras O., Gagalowitz A. : "Visual discrimination
of Stochastic texture fields". IEEE Trans. on Systems, Man and
Cybermetics, Vol SMC - 8, n°11, Nov. 1978.

2.   Monne J. Schmitt F. : "Synthèse de Textures par une méthode
probabiliste bidimensionnelle". 2ème congrès AFCET-IRIA Reconnaissance des formes et intelligence artificielle - Toulouse Sep. 1979.

3.   Haralick R.M. : "Statistical and Structural approaches to
Texture", Proc of the IEEE, Vol. 67, n°5, May 1979.

4.   Too J.T, Chang Y.S. : "An approach to texture pattern analysis
and recognition" in Proc. IEEE conf. on décision and control, 1976.

5.   Davis L.S. : "Image texture analysis techniques - A survey"
"NATO ASI on Image Processing - Bonas - France - 1980".

6.   Gueguen C. : "Apport de la modélisation au traitement du signal", 7e colloque GRETSI, Nice 1979.

7.   Levinson N. : "The Wiener rms error criterion in filter design
and prediction". J. Math. Phys., Vol 25, pp 261-278, Jan 1947.

8.   Goodman D. : "Some stability properties of two-dimensional
shifts invariant digital filters". IEEE Trans. on Circuits Syst.,
Vol. CAS-24 pp 201-208, Apr. 1977.

9.   Jury E.I. : "Stability of multidimensional Scalar and matrix
polynomials", Proc. of the IEEE, vol. 66, n°9, Sep. 1978.

10.   Ekstrom M.P., Woods J.W. : "Two-dimensional spectral facto-
rization with applications in recursive digital filtering", IEEE
Trans. on ASSP, Vol ASSP-24 n°2, April 1976.

11.   Shanks J.L., Treitel S. Justice J.H. : "Stability and synthesis
of two dimensional recursive filters". IEEE Trans. Audio Electro-
acoust., vol AU-20, p 115, June 1972.

12.   Huang T.S. : "Stability of two dimensional recursive filters".
IEEE Trans. on Audio and Electro. Vol AU-20, n°2, June 1972.

13.   Jayaramamurthy : "Computer methods for analysis and synthesis
of visual texture", Ph.D. Thesis, Sep. 1973, Univ. of Illinois,
Urbana Illinois.

14.   Anderson B.D.O.., Jury E.I. : "Stability tests for Two dimensional recursive filters". IEEE Trans. on Audio and Electro. Vol
AU-21, n°4, August 1973.

15.   Akaike H.:"Autoregressive model fitting for control", Annals of the Institute of Statist Math. , vol 23, 1971.

16.   Justice J.H : "A Levinson-type algorithm for two dimensional Wiener filtering using bivariate Szegö polynomials". Proc. of IEEE, vol. 65, n°6, June 1977.

17. Gambotto J.P., Gueguen C. : "A multidimensional filtering approach to pattern recognition with application to texture classification and segmentation". Int. Signal Proc. Conf. Firenze 1978.

18. Gambotto J.P. : "Méthodes d'estimation linéaire multidimensionnelle : application à la reconnaisssance et à la segmentation des textures", Thesis, Dec. 1979, ENST, Paris.

19. Gambotto J.P. : "Processus bidimensionnels et modèle autorégressif vectoriel : application à la modélisation des textures", 7e colloque GRETSI, Nice 1979.

20. Deguchi K. Morishita I. : "Texture Characterization and Texture -based image partitioning using two-dimensional linear estimation techniques". IEEE Trans. on Computer vol C-27, n°8, August 1978.

21. Gambotto J.P. Gueguen C. : "A multidimensional modeling approach to texture classification and segmentation", IEEE Inter. Conf. on ASSP, April 1979, Washington.

SEGMENTATION BY SHAPE DISCRIMINATION USING SPATIAL FILTERING
TECHNIQUES

Stephen D. Pass

Department of Physics and Astronomy
University College London

ABSTRACT

Zadeh (4) introduced the notion of fuzzy set theory and
Goetcherian (6) showed how fuzzy logic could extend a variety of
binary image processes into a multi-valued (grey image) domain.
This paper introduces the basics of fuzzy set theory and discusses
the suitability of the physically parallel CLIP machine for
implementing the multi-valued operators. A form of spatial
filtering is explained and its application to shape discrimination
and image segmentation is shown. Finally, an application to a
real image processing problem is illustrated.

INTRODUCTION

In an attempt to speed the processing of two-dimensional image
data, various researchers, inspired by Unger (1), have investigated
physically parallel computer architectures. At University College
London a series of CLIP (Cellular Logic Image Processor) machines
has been built. The latest, CLIP4 (2), is a 96 x 96 array of
processing elements (P.E.'s) which can be configured with square
or hexagonal tessellation. The basic architecture of CLIP4 is
shown in Fig. 1. It will be seen that operations can be performed
on two binary images, at most, at a time. The neighbour-connectivity
allows operations based on neighbourhood relationships to be
performed in one machine instruction. Such operations include
edge-finding, shrink/expand and labelling (where free propagation
between P.E.'s during the machine instruction allows a point in
the B operand to pick out a connected set in the A operand). More
complicated algorithms, such as skeletonisation, are executed very

231

J. C. Simon and R. M. Haralick (eds.), Digital Image Processing, 231–243
Copyright © 1981 by D. Reidel Publishing Company.

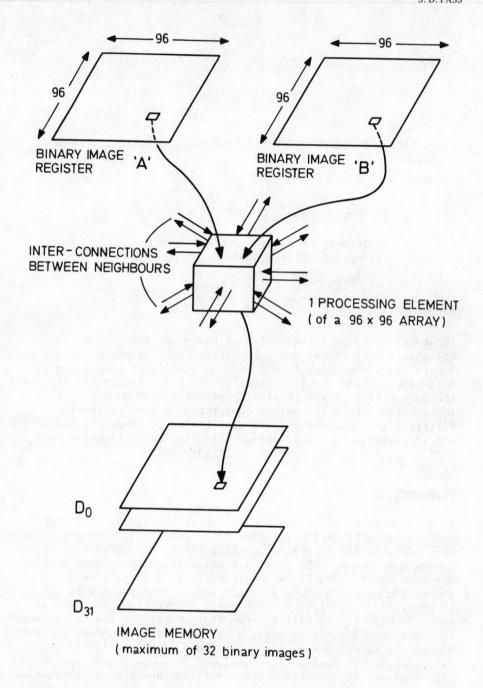

Fig. 1   Basic architecture of CLIP4.

quickly because of their dependence on the connectivity properties
of the binary image.

For processing grey value images an arithmetic approach is
usually adopted, but this is less than ideal for CLIP, where a
grey-value image is stored as a series of bit planes and the Boolean
P.E.'s perform arithmetic in a bit-serial manner. Low precision
integer addition, subtraction and multiplication are possible but
division is unfeasible since the total storage available is only
32 bit planes. Gradient and Laplacian operators can be easily
implemented on CLIP but more sophisticated segmentation algorithms
based on regional statistics (e.g. (3))present much greater
difficulties.

To utilise the power of the CLIP architecture for processing
grey images a different approach is required. If the grey image
could be treated in a similar manner to a binary image then
algorithms designed for binary images could be extended to grey
without recourse to extensive arithmetic calculation. A solution
is the adoption of some of the ideas of fuzzy set theory. Zadeh
(4) introduced fuzzy set theory and Nakagawa and Rosenfeld (5)
showed an application to image processing. Goetcherian (6) used
ideas from this short paper and showed that a series of algorithms
applied to binary images could be equally well applied to grey
images. In this paper it will be shown how the notion of spatial
filtering can be used for region segmentation of real-world
images.

BASICS OF FUZZY SET THEORY

Given a set $X = \{ x_1, ..., x_n \}$ then associated with a fuzzy
set A in X there
set A in X there is a membership function $f_A(x_i) \in [0,1]$ $1 \leqslant i \leqslant n$,

ie. each element x_i of A has a value in the interval $[0,1]$

associated with it. $f_A(x_i)$ is some estimate of the degree to
which x_i belongs to A.

The Boolean set is a restricted case of the fuzzy set where the
membership function takes the values 0 or 1,

ie. $f_A(x_i) = 0 \Leftrightarrow x_i \notin A$

        $f_A(x_i) = 1 \Leftrightarrow x_i \in A$

The algebra of fuzzy sets relates to that of Boolean sets as
shown below:

Intersection $C = A \cap B = \{x_i, f_C(x_i) \mid 1 \leqslant i \leqslant n, f_C(x_i) = MIN(f_A(x_i), f_B(x_i))\}$

Union $C = A \cup B = \{x_i, f_C(x_i) | 1 \leq i \leq n, f_C(x_i) = MAX(f_A(x_i), f_B(x_i))\}$

Complement $C = A^C = \{x_i, f_C(x_i) | 1 \leq i \leq n, f_C(x_i) = 1 - f_A(x_i)\}$

Subset $A \subseteq B = f_A(x_i) \leq f_B(x_i)$ for $1 \leq i \leq n$

                 $A \subset B = f_A(x_i) < f_B(x_i)$ for $1 \leq i \leq n$

where A, B and C are fuzzy sets in X.

AN IMAGE AS A SET OF FUZZY SINGLETONS

If each pixel of a tessellated image is considered as a fuzzy
singleton then the grey value (normalised to $[0,1]$) at that pixel
can be thought of as the degree to which the pixel belongs to the
image. The image is then a set of fuzzy singletons.

Given an image I of m x n pixels, then a pixel as a fuzzy
singleton in I is:

$$P_{ij}^I = \{p_{ij}, f_I(p_{ij}) | p_{ij} = \text{pixel at } (i,j), f_I(p_{ij}) = x_{ij}\}$$

                              where x_{ij} is the normalised grey
                              value of the pixel p_{ij} in I,
                              $1 \leq i \leq m$, $1 \leq j \leq n$

and the image is the set of fuzzy singletons covering I:

$$I = \{P_{ij}^I | 1 \leq i \leq m, \ 1 \leq j \leq n\}$$

COMBINING TWO IMAGES

Two images A and B can be combined in a point-wise manner using the
rules outlined before

     eg. $C = A \cap B = \{P_{ij}^C | 1 \leq i \leq m, \ 1 \leq j \leq n\}$

     and $P_{ij}^C = \{P_{ij}, f_C(p_{ij}) | f_C(p_{ij}) = MIN(f_A(p_{ij}), f_B(p_{ij}))\}$

            $1 \leq i \leq m$, $1 \leq j \leq n$.

In words, the new image takes values which are the minima of
corresponding points in the images A and B. The union of two
images can be similarly defined.

OPERATING ON A SINGLE IMAGE

The image I can be divided into a series of overlapping subsets
and a fuzzy operator applied to each one. For example, if I is a
square tessellated image then 3 x 3 pixels subsets can be
considered:

$$N_{ij}^{I} = \begin{array}{|c|c|c|} \hline P_{i-1,j-1}^{I} & P_{i-1,j}^{I} & P_{i-1,j+1}^{I} \\ \hline P_{i,j-1}^{I} & P_{ij}^{I} & P_{i,j+1}^{I} \\ \hline P_{i+1,j-1}^{I} & P_{i+1,j}^{I} & P_{i+1,j+1}^{I} \\ \hline \end{array}$$

where N_{ij}^{I} is the subset of I centred at (i,j).

The LOCAL MINIMUM operator applied to N_{ij}^{I} produces a new fuzzy singleton $P_{ij}^{I'}$ in the new image I' such that

$$P_{ij}^{I'} = \bigcap N_{ij}^{I}$$

$$= \{p_{ij}, f_{I'}(p_{ij}) \mid f_{I'}(p_{ij}) = \underset{kl}{MIN}(f_{I}(p_{kl})), \begin{array}{l}(i-1) \leqslant k \leqslant (i+1)\\(j-1) \leqslant l \leqslant (j+1)\end{array}\}$$

The new image is then $I' = \{P_{ij}^{I'} \mid 1 \leqslant i \leqslant m,\ 1 \leqslant j \leqslant n\}$.

The LOCAL MAXIMUM can be similarly defined. The choice of subset for applying the local operators is, of course, not limited to a 3 x 3 pixels neighbourhood. Numerical examples of the local minimum and maximum operators using 3 x 3 pixels subsets are shown in Fig. 2. The image values have not been normalised but instead $x_{ij} \epsilon [0,7]$. Subsequent examples will also use non-normalised image values. To avoid edge-effects the image edge is implicitly assumed to take the minimum value for a local max. operation and conversely the maximum value for a local min. operation. Figs. 2(d) to (f) illustrate that local min. and max. operations are duals when operating on complemented images.

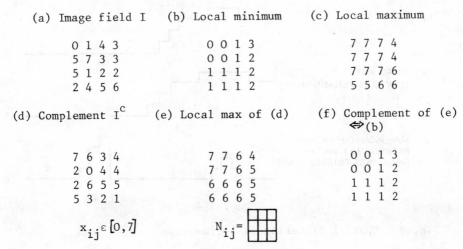

(a) Image field I   (b) Local minimum   (c) Local maximum

```
0 1 4 3            0 0 1 3            7 7 7 4
5 7 3 3            0 0 1 2            7 7 7 4
5 1 2 2            1 1 1 2            7 7 7 6
2 4 5 6            1 1 1 2            5 5 6 6
```

(d) Complement I^c (e) Local max of (d)   (f) Complement of (e) $\Leftrightarrow$ (b)

```
7 6 3 4            7 7 6 4            0 0 1 3
2 0 4 4            7 7 6 5            0 0 1 2
2 6 5 5            6 6 6 5            1 1 1 2
5 3 2 1            6 6 6 5            1 1 1 2
```

$$x_{ij} \epsilon [0,7] \qquad N_{ij} =$$

Fig. 2   Local minimum and maximum operators

SPATIAL FILTERING USING LOCAL MIN. AND MAX. OPERATORS

The manner in which the local minimum and maximum operators
can be used for spatial filtering is shown with a 1-dimensional
example in Fig. 3. The image subset used is 3 x 1 pixels. Two
local min. operations remove structures up to 4 pixels wide while
the two subsequent local max. operations return larger structures
to their previous form. The effect is thus one of low pass spatial
filtering. The numerical difference between the source and low
pass filtered images yields a high pass filtered image. It should
be noted that iterated operations on a small subset achieve the same
result as one operation on a larger subset which covers the union
of the smaller subsets, eg. in the 1-dimensional case shown, the
same result could have been achieved by operating once on 5 x 1
pixels subsets instead of twice on 3 x 1 pixels subsets.

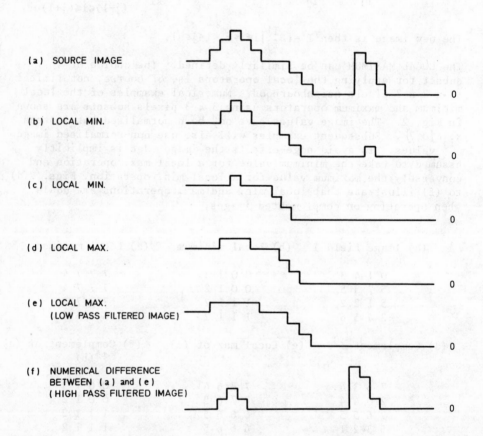

Fig. 3   Spatial filtering in 1-dimension

APPLYING FILTERING OPERATIONS TO IMAGES

For simplicity an image will be considered to consist of objects
with grey value higher than the surrounding background. The local
min. and max. operators can then be compared with the binary
operators of shrink and expand respectively since the objects
reduce in size for a local min. and increase in size for a local
max.

  ie. LOCAL MINIMUM $\rightarrow$ GREY SHRINK (S)

    LOCAL MAXIMUM $\rightarrow$ GREY EXPAND (E)

The low pass spatial filtering of an image I can then be written
as:

$$E^n S^n (I)$$

and the high pass filtering written as:

$$I - E^n S^n (I)$$

where n is the number of times the operation is applied. The size
of feature removed on low pass filtering is determined by n and
the size of the subset over which the local operators are applied.
Taking the analogy with binary images, a feature has a "depth" of
n if it takes n shrinks to remove it. Thus, $E^n S^n (I)$ removes all
features with depth up to and including n. Referring to Fig. 3,
a low pass filter of $E^2 S^2 (I)$ has been applied and structures of
depth 2 and less (up to 4 pixels wide, since the subset of local
operation is 3 x 1 pixels) have been eliminated.

SHAPE DISCRIMINATION USING SPATIAL FILTERING

The low and high pass spatial filters can be used in an
isotropic manner with, say, a 3 x 3 pixels subset (implying 8-
connectivity) but they only yield information about the minimum
dimensions of the objects in the image. Some indication of shape
can only be obtained by applying anisotropic filters based on a
linear subset. If filtering is performed separately in horizontal
and vertical directions then information about the horizontal and
vertical dimensions of the objects is obtained. This information
can be combined to distinguish between the shapes of objects. For
an 8-connected image this notion can be extended from the two
principal orthogonal directions to include the two principal
diagonal directions also.

More formally, applying the operator:

$$I - (E^n S^n (I))_\delta = F_\delta^n (I)$$

(a) Image I

```
2  0  1  2  3  3  0  3  2  4  4  2
0  6  8  4  0  1  7  9 11 15 12 13
0  4  6  6  4  1  7  8  6  7 10  0
1  4  5  7  3  3 10  7 15 14  9  5
0  2  0  2  3  4  2  2  1  3  3  2
```

(b) $F_{\updownarrow}^{2}(I)$

```
2  0  0  0  3  2  0  0  0  0  0  2
0  4  7  2  0  0  5  6  9 11  8  3
0  2  5  4  1  0  5  5  4  3  6  0
1  2  4  5  0  2  8  4 13 10  5  5
0  0  0  0  0  3  0  0  0  0  0  2
```

(c) $F_{\leftrightarrow}^{2}(I)$

```
2  0  1  2  3  3  0  1  0  2  2  0
0  6  8  4  0  0  0  2  4  8  5  0
0  3  5  5  3  0  1  2  0  1  4  0
0  1  2  4  0  0  3  0  8  7  2  0
0  2  0  0  1  2  0  0  0  1  1  0
```

(d) $\pi(I)=F_{\updownarrow}^{2}(I)\cap F_{\leftrightarrow}^{2}(I)$

```
2  0  0  0  3  2  0  0  0  0  0  0
0  4  7  2  0  0  0  2  4  8  5  0
0  2  5  4  1  0  1  2  0  1  4  0
0  1  2  4  0  0  3  0  8  7  2  0
0  0  0  0  0  2  0  0  0  0  0  0
```

(e) $E^{1}S^{1}(\pi(I))$

```
0  0  0  0  0  0  0  0  0  0  0  0
0  1  1  1  0  0  0  0  0  0  0  0
0  1  1  1  0  0  0  0  0  0  0  0
0  1  1  1  0  0  0  0  0  0  0  0
0  0  0  0  0  0  0  0  0  0  0  0
```

Fig. 4   Segmentation by shape discrimination

The square and rectangle are differentiated from the
background by using italic numerals.

where $\delta \in \Delta = \{$ principal directions of tessellation$\}$,

yields features with maximum depth n in direction δ. For each δ a different n can be specified, depending on the dimensions of the object.

The resulting set of images

$$\Pi(I) = \{F_\delta(I) | \delta \in \Delta\}$$

can be combined by point-wise minimum operations to yield a single image $\pi(I)$ which contains features with a specified maximum size in each direction $\delta \in \Delta$,

ie. $\pi(I) = \bigcap \Pi(I)$ where $\bigcap$ is a fuzzy intersection.

Low pass filtering operations can then be applied to $\pi(I)$, leaving features with some minimum dimensions specified by the low pass filter(s). The net result is an image containing features of the image I which have specified dimensions along the $\delta \in \Delta$.

A contrived example showing the separation of a noisy square from a noisy rectangle, both on a noisy background, is shown in Fig. 4. The subset for the local operators is 3 x 1 pixels and $\Delta = \{$ horizontal dir.($\leftrightarrow$), vertical dir.($\updownarrow$)$\}$. The square is 3 x 3 pixels and the rectangle is 5 x 3 pixels, giving horizontal depths of 2 and 3 respectively, and the same vertical depth of 2. Fig. 4(b) shows the result of applying $F_\updownarrow^2(I)$. As would be expected, both the square and the rectangle produce a large response. Fig. 4(c) shows the result of $F_\leftrightarrow^2(I)$. The square produces a good response but the rectangle is somewhat broken because there is structure within the rectangle of depth 2 or less but the rectangle itself has a depth of 3. Fig. 4(d) shows $\pi(I) = F_\updownarrow^2(I) \cap F_\leftrightarrow^2(I)$. The square still shows a good response since its maximum depth is 2 for each $\delta \in \{\leftrightarrow, \updownarrow\}$ but the rectangle is broken up. Finally an isotropic low pass filter based on a 3 x 3 pixels subset is applied to $\pi(I)$ to remove any features with a depth of 1 in any direction since the square has a depth of 2 in all directions on an 8-connected tessellation. The result is the square segmented from the noisy image. Of course, filtering in the vertical direction is superfluous in this example since the square and rectangle have the same depth but it illustrates the technique.

SEGMENTATION OF REAL IMAGES

Fig. 5 shows a series of filtering operations applied to a lunar image. This forms part of an algorithm being developed to perform measurements on the moon craters. Since the craters are generally circular, a hexagonal tessellation was chosen to provide a better

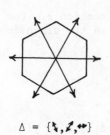 

$$\Delta = \{ \searleft, \swarrow, \leftrightarrow \}$$

(a) Principal directions of the   (b) Source image (I)
    hexagonal tessellation.

(c) $F_{\searleft}^6(I)$ (d) $F_{\swarrow}^6(I)$

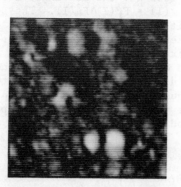

$$\Pi(I) = \{ F_{\searleft}^6, F_{\swarrow}^6, F_{\leftrightarrow}^6 \}$$

(e) $F_{\leftrightarrow}^6(I)$

Fig. 5   Spatial filtering applied to a real image.

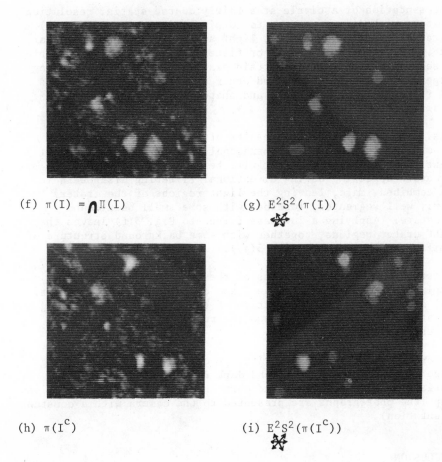

(f) $\pi(I) = \bigcap \amalg(I)$

(g) $E^2S^2(\pi(I))$

(h) $\pi(I^c)$

(i) $E^2S^2(\pi(I^c))$

Fig. 5   Spatial filtering applied to a real image. (contd.)

representation of a circle at a fairly coarse spatial resolution
(craters with diameters down to 10 pixels are to be segmented).
A crater consists of adjacent light and dark regions of similar
size. By a suitable choice of filter parameters it is possible
to segment craters within certain size bounds while eliminating
larger and smaller craters and other artifacts (such as ridges
and rifts) since their sizes and shapes do not fit the
requirements.

Figs. 5(c) to (e) show the results of filtering in the three
principal directions of the hexagonal tesselation, and Fig. 5(f)
shows their combination. It will be seen that much of the
structure appearing during directional filtering disappears in
the combined image, leaving the light regions of the craters
fairly well segmented, together with some small background
structure. Applying a low pass filter to Fig. 5(f) leaves the
light crater regions, together with some background structure of
similar shape and size (Fig. 5(g)).

Filtering for the dark regions of the craters is performed by
complementing the source image I and operating on this. Figs.
5(h) and (i) show the combination of the directional filtering
operations for dark regions and the result after low pass
filtering respectively.

The crater segmentation algorithm proceeds after the filtering
stage by relating the light and dark regions since a crater
consists of adjacent light and dark areas in a specific relationship (the photographs are presented to the camera with a constant
orientation).

CONCLUSION

A class of operators which exploits the physically parallel
architecture of CLIP has been described. The grey shrink and
expand operations execute very quickly on CLIP since they can be
decomposed into a series of binary shrink/expands, which are two
of the basic operations available on CLIP. The success of the
local operators depends only on local contrast (where "local"
is related to the subset size and number of times the operator is
applied), thus overcoming problems of shading and threshold
selection which occur in many real-world images.

REFERENCES

1. Unger, S. H.: *A computer oriented towards spatial problems*,
    1958, Proc. IRE 46(10), pp. 1744-1750.

2. Duff, M. J. B.: *CLIP 4: A large scale integrated circuit array parallel processor*, 1976, Proc. 3rd IJCPR, pp. 728-733.

3. Yakimovsky, Y.: *Boundary and object detection in real world images*, 1975, Advance papers 4th IJC on AI, pp. 695-704.

4. Zadeh, L. A.: *Fuzzy sets*, 1965, Inform. Contr. 8, pp. 338-353.

5. Nakagawa, Y., and Rosenfeld, A.: *A note on the use of local min and max operations in digital picture processing*, 1978, IEEE Trans. SMC-8, pp. 632-635.

6. Goetcherian, V.: *From binary to grey tone image processing using fuzzy logic concepts*, 1980, Patt. Recog. 12, pp. 7-15.

RELAXATION AND OPTIMIZATION FOR STOCHASTIC LABELINGS

Marc BERTHOD

INRIA, Domaine de Voluceau, 78150 LE CHESNAY (France)

This paper contains a survey of a new family of parallel iterative algorithms that allow for using context to improve the results
obtained in separately classifying the units of a global set. The
essential characteristic of these methods is that they transform
the general exponential context problem in a tractable polynomial
one, by looking for a continuous (stochastic) labeling rather than
a discrete one.

1. LABELING ENHANCEMENT : A SOLUTION TO THE PROBLEM OF CONTEXT

It is now commonly admitted by people working in the field of pattern recognition that further significant advances, in most applications, will rely on a generalized use of contextual information.
The reason for that is that man, who is by now the most efficient
known "classifier", relies highly on this type of information,
when performing perception tasks.

Among all possible definitions of context, we shall deal in that
paper with the following one : a contextual pattern recognition
problem consists of the global recognition of a set of patterns
(we shall call them "units") such that the class of any of these
patterns depends not only of the pattern itself, but also of other
patterns in the whole set. This influence of the other patterns is
the "context".
Such a situation arises in most pattern recognition applications :

- character recognition (the set of units is a word or a sentence)
- image segmentation (the set of units is an image, the units are
  the pixels)

245

- each time a complex pattern is decomposed into more elementary
  units, which are to be separately recognized and then combined
  (which leads to the graph matching problem).

Corresponding problems, and settings, may be very different, but a
large variety of them can be covered by a simple framework summarized below.

Let U be the set of units u_i i = 1, ..., N to be recognized.

Let L be a set of labels ℓ_1, ..., ℓ_M. A label is the name of a
class :u_i belongs to the class k is equivalent to say tha u_i has
label k.

A discrete labeling (D.L.) associates to every u_i a (possibly empty) subset of L, i.e., a D.L. incompletely classifies a unit among
a few classes.

An unambiguous discrete labeling (U.D.L.) is a D.L. which associates
one label, and one only, to every u_i.

This notion can be readily extended from discrete to continuous
classification, in order to deal with Bayesian classification
for exemple. So we define a continuous (or stochastic, or probabilistic) labeling (C.L.) as an application from $U \times L$ in $[0,1]$,
which associates to any $(u_i, \ell_i), P_i(\ell_i) \in [0,1]$, such that :

$$\forall i \sum_{\ell \in L} P_i(\ell) = 1$$

So a continuous labeling associates to every unit u_i a probability
vector $\overline{P}_i$: $P_i(\ell)$ may be interpreted as the probability (determined by the classifier) that unit u_i belongs to class ℓ ; we shall
call it the valuation of label ℓ, at unit u_i.

Given a C.L., one may easily achieve recognition by selecting, for
every u_i, the class (or label) with the highest valuation. The U.
D.L. so defined will be said the U.D.L. associated to $\overline{P}$.

An important characteristic of a C.L. is its ambiguity, which also
measures the quantity of information brought by the corresponding
probability vectors, or, in some sense, the degree of certainty
of the final classification. Coming from information theory, several measures of entropy may be proposed, among which we select
(for computational reasons) the quadratic entropy :

$$E = \sum_{i=1}^{N} \sum_{\ell \in L} P_i(\ell)(1-P_i(\ell)) = \sum_{i=1}^{N} (1 - ||\overline{P}_i||_2^2)$$

Clearly, E is minimal when $\overline{P}$ is unambiguous, i.e. :

$$\forall i, \; \exists \; \ell_i \quad P_i(\ell_i) = 1.$$

The fundamental problem of pattern recognition is to find an U.D. L. When no contextual information is used, this can be done by separately and independantly classifying the different units. On the contrary, when this U.D.L. is considered as a whole entity to be determined, then contextual information may be brought in the recognition of every unit, by taking in consideration whether a given label, for a given unit, is consistent with hypothesized labels, for the other units.

So contextual information is available if one knows, for any U.D. L., how much consistent it is ; and we shall say that the definition of consistency reflects the model of the world.

For example, discrete consistency may be defined by a subset L of L_i^N an U.D.L. element of L will be said to be consistent, while an U.D.L. which is not in L will be said to be inconsistent (which means that it would not be a realistic solution for the pattern recognition problem under consideration).

A continuous measure of consistency is a mapping from L^n into R, which measure, for any U.D.L. how much it is consistent, i.e. how much good a solution it is for the given P.R. problem.

There can be different contextual problems, depending on whether one starts from an initial discrete or continuous labeling, and whether the world model is given by discrete or continuous consistency.

For example, Harclick [14-17] proposes discrete relaxation as a realistic way to find an U.D.L. starting form another U.D.L., and using discrete consistency.

On the opposite, the Bayesian contextual problem as described by Toussaint [24], is to find the U.D.L. $(\ell_1, \ldots, \ell_N)$ which maximizes :

$$P_a(\ell_1, \ldots, \ell_N) \prod_{i=1}^{N} \frac{P(\overline{X}_i \mid \ell_i)}{P_a(\ell_i)}$$

where ;- $\overline{X}_i$ is the parameter vector for unit u_i

- $P(\overline{X}_i | \ell_i)$ is the probability of occurence of $\overline{X}_i$, if u_i be-
  longs to class ℓ_i
- $P_a(\ell_i)$ is the non-contextual a priori probability for u_i
  to belong to class ℓ_i
- $P_a(\ell_1, \ldots, \ell_N)$ is the a priori probability that simulta-
  nemously $u_1, \ldots, u_N$ belong to classes $\ell_1, \ldots, \ell_N$.

The general scheme to solve that problem is that of Figure 1.

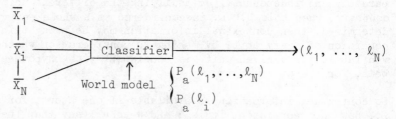

Figure 1.

Unfortunately, the general solution to the context problem requires
computation time of the order of M^N (the total number of
possible U.D.L.).The fundamental idea of labeling enhancement is to
cut-off this exponential number of operations, by replacing the discrete problem by a continuous one, following the scheme of Figure 2.

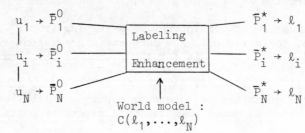

Figure 2.

In a first step, a continuous labeling (C.L.) $\overline{P}^0$ is determined by
separately classifying the units. At this stage, no contextual in-
formation is used. Starting from $\overline{P}^0$, no matter how $\overline{P}^0$ was obtained,
the labeling enhancement takes place : it incorporates contextual
information to determine an improved C.L. $\overline{P}^*$ hopefully more consistent and less ambiguous. This contextual information consists
of the knowledge of the consistence of any C.L.. Then, for every u_i
the best label in $\overline{P}^*$ is selected. The following underlying assumption is made : the original C.L. $\overline{P}^0$ is already fairly good, and

the purpose of the contextual post-processing is to correct a few errors, or to raise a few ambiguities.

The fundamental advantage of this approach is that there exist parallel iterative methods, which necessitate of the order of $N \times M^{\alpha}$ (generally $\alpha = 2$, sometimes 3), operations, to find $\bar{P}^*$, instead of M^N to find $(\ell_1, \ldots, \ell_i)$ in the discrete problem.

The first method devised to solve the labeling enhancement problem is known under the name of "probabilistic relaxation" (Rosenfeld [23]). More recently we proposed in Faugeras [9,11] and Berthod [2] another family of algorithms based on optimization.

These two general methods have in common that, starting from $\bar{P}^0$, they iteratively update, by a totally parallel process, the current labeling $\bar{P}^n$, replacing at each iteration $\bar{P}^n$ by $\bar{P}^{n+1}$ in such a way that $\bar{P}^{n+1}$ is - more consistent
                                              - less ambiguous
than P^n. They also have in common that the consistency of a C.L. relies on the notion of compatibility vector.

2. DETERMINATION OF A COMPATIBILITY VECTOR

It is always possible to assume that the consistency of an U.D.L. is known, as it directly corresponds to a solution to a pattern recognition problem : this solution may be evaluated by an expert, or its probability can be modelized. Unfortunately, this does not directly yield a measure of the consistency of a continuous labeling, which may potentially generate any of the M^N different D. L.

Rather than generalizing consistency of D.L.'s to C.L's , another possibility is to consider that, in the contextual recognition of a set of units, the labeling of a unit u_i is a function of :

- the observations made on that unit $(\bar{P}_i^0)$

- the labelings of a small subset of units : V_i, called the "neighboors" of u_i

This simply means that only a few units in U (those of V_i) have a contextual influence on the class of u_i, and this is true in most applications.

The information brought in u_i, by the labelings of the units of V_i is then gathered under the form of a compatibility vector $\bar{Q}_i$:

$Q_i(k)$ may be thought of as a prediction for the valuation of label k et unit u_i, by considering the context of u_i.

This compatibility vector will be essential either for the measure of consistency (optimization) or directly for the determination of an updated labeling (relaxation).

There can be many ways to determine $\overline{Q}_i$, and we give here only the most frequent and usefull determinations. For example, a compatibility vector is computed for every neighbour $u_j \epsilon V_i$ by :

$$Q_{ij}(k) = \sum_\ell c_{ij}(k,\ell) \, P_j(\ell) \tag{1}$$

The coefficients $c_{ij}(k,\ell)$ may be, for example, conditional probabilities that $u_i \epsilon$ class k, given that $u_j \epsilon$ class ℓ.

Couples of units may be simultaneously considered :

$$Q_{i_{j_1 j_2}}(k)= \sum_{\ell_1} . \sum_{\ell_2} c_{i_{j_1 j_2}}(k, \ell_1, \ell_2) P_{j_1}(\ell_1) P_{j_2}(\ell_2)$$

All these different compatibility vectors are then combined into a single one, attached to u_i. Again, different combinations may be considered, for example :

the weighted sum : $Q_i(k)= \sum_{j \epsilon V_i} \alpha_j Q_{ij}(k)$ (2)

or the product : $Q_i(k) = \prod_{j \epsilon V_i} Q_{ij}(k)$ (3)

Whatever combination rule is chosen, it generally results that $\forall i,k \quad Q_i(k) \geq 0$, but generally :

$$\sum_{k \epsilon L} Q_i(k) \neq 1.$$

As it may be convenient, especially for comparison with $\overline{P}_i$, that $\overline{Q}_i$ be also a continuous labeling, a new normalized compatibility vector may be determined by replacing $Q_i(k)$ by :

$$\frac{Q_i(k)}{\sum_{\ell \epsilon L} Q_i(\ell)}$$

3. RELAXATION METHODS

Relaxation methods get their name from the well-known mathematical technique. The idea is to combine, for every u_i, the current labeling, $\overline{P}_i$, with the prediction yield by the compatibility vector, into a new labeling.

Starting from $\overline{P}^0$, a sequence $\overline{P}^n$ is iteratively defined by :

$$\forall i,k \quad P_i^{n+1}(k) = \frac{P_i^n(k)Q_i^n(k)}{\sum_{\ell \in L} P_i^n(\ell)Q_i^n(\ell)} \tag{4}$$

It is trivial that $P_i^{n+1}(k) \geq 0$

$$\text{and} \quad \sum_\ell P_i^{n+1}(k) = 1$$

So $\overline{P}^n$ is a sequence of C.L. The process is terminated when the maximal difference between $P_i^{n+1}(k)$ and $P_i^n(k)$ is below some threshold.

It may be proven (Zucker [30]) that if, for some iteration, for every u_i, the label with the hightest valuation does not change, then the sequence will converge towards the corresponding unambiguous labeling. This can be used as a stopping criterion.

On the other hand, the fixed points of the process are such that :

$$\forall i \; ; \text{either} : \forall k, \; Q_i(k) = \text{Constant (i)}$$
$$\text{or} : \exists k : P_i(k) = 1$$

Experiments show that the second case is, fortunately the most frequent. Another experimental observation is that the best results are obtained before convergence, so this poses the problem of finding an experimentaly satisfying criterion.

Another problem is that many iterations (possibly more than 100) are needed to reach a fixed point. Attempts to define faster updating schemes were proposed by Zucker [28,29], for example by raising the two terms of the ratio in (4) to a power ;

$$P_i^{n+1}(k) = \frac{[P_i^n(k) \; Q_i^n(k)]^\alpha}{\sum_{\ell \in L} [P_i^n(k) \; Q_i^n(k)]^\alpha} \quad (\alpha > 1)$$

Of course, there is no way to prove that the process will reach the same fixed point depending on α.

It is worth mentionning a work of Peleg [22], who considers that, for every unit, the C.L. $\overline{P}_i$ is a random event, the probability distribution of which depends on the actual class of u_i. Necessary (but arbitrary) independency hypothesis allow then to estimate the probabilities for any u_i, to belong to any class, starting from a

labeling, and so to reestimate a new labeling. In the simplest
case, the updating formula is obtained by (1) (3) and (4), and
by substituting $P_{ij}(k,\ell)$ to $c_{ij}(k,\ell)$ where $P_{ij}(k,\ell)$ is the joint
probability that u_i has label k and u_j has label ℓ.

There has been, in the past few years a lot of succesfull applications of relaxation processes, which show their efficiency and
more generally the potential power of labeling enhancement . Most
of theses applications were in image processing/scene analysis.
but they concerned problems as varied as, for example :

- noise cleaning [5, 6, 18, 19]
- cursive script analysis [17]
- deciphering [21]
- template matching [3]
- line, curve, edge enhancement [4, 20, 24, 27, 31].

4. OPTIMIZATION METHODS

Relaxation methods were primarily designed starting from heuristic
or pragmatic considerations. Theoretical aspects were studied later on. On the contrary, the optimization approach was set up in
an attempt to give labeling enhancement stronger theoretical ba-
ses. The basic idea is to define a criterion C , which maps
$(\mathbb{R}^{NxM})^{+}$ in $\mathbb{R}$, and measures a trade-off between :

- consistency
- entropy

Entropy is defined by the quadratic entropy of $\bar{P}$. Consistency is
defined as a similarity between $\bar{P}$ and the compatibility vector $\bar{Q}$.

Starting from $\bar{P}^{0}$, the "closest" local optimum of $C(\bar{P})$ is then
iteratively searched.

The only optimization technique widely applied so far is the projected gradient method : it is not hard to see that corresponding
algorithms run in parallel on the whole set of units.

Depending on whether the compatibility vector is normalized or not,
and on what type of similarity measure is used, several different
criteria may be defined, from the same world model. For example,
the Euclidean distance between $\bar{P}$ and a normalized version of $\bar{Q}$
was proposed in [2, 9, 11]. More recently, it has been shown in
[1], that the inner product could work well even with a non-nor-
malized version of $\bar{Q}$, and moreover proved to be less sensitive
to the relative influence of entropy. These different criteria may
be summarized in the following one :

$$C = \sum_{i=1}^{N} \sum_{k \in L} (\alpha P_i(k) . Q_i(k) + \beta P_i^2(k) + \gamma Q_i^2(k))$$

where $Q_i(k)$ may or may not be normalized. For example :

$$Q_i(k) = \frac{\sum\limits_{j \in V_i(k)} \alpha_j(k) \sum\limits_{m \in L} c_{ij}(k,m) P_j(m)}{\sum\limits_{\ell} \sum\limits_{j \in V_i(\ell)} \alpha_j(\ell) \sum\limits_{m \in L} c_{ij}(\ell,m) P_j(m)}$$

When compared to relaxation, each iteration of the optimization process is of the same order of complexity, but is approximately three times longer. Experiments show that convergence (which is theoretically proven as long as C is differentiable) is generally obtained after very few iterations (10 to 40). Besides, results have proven be very satisfactory on several applications :

- semantic matching of serial images [12]
- edge enhancement [2, 11]
- multispectral image classification [9, 11].

We shall briefly describe, for example, an application to multi-spectral image classification, where the inner product alone was used as a criterion.

We conducted experiments on the Landsat picture 214-26 (July 11, 1976) which represents the Paris aera. We selected a rectangle 30 x 40 kilometers centered on the cathedral of the Notre-Dame, which represents a 512 x 512 image. A study conducted in 1975 to determine a precise description of land use in this aera provided us with ground-truth data. Each pixel of the Landsat image was attributed one among 19 possible classes. Because the description available was sometimes too functional and not physical enough, we had to merge some classes and ended with a final total of five to eight classes as follows :

λ_1 : gardens

λ_2 : agricultural aeras

λ_3 : water

λ_4 : housing

λ_5 : industry

λ_6 : cemeteries

λ_7 : public aeras

λ_8 : streets, freeways, parking lots.

In the stochastic labeling formalism, objects are pixels and conditional probabilities $P_{ij}(\lambda_k|\lambda_\ell)$ are estimated from the ground-truth data. The neighbourhoods V_j contained the eight nearest neighbors of pixel u_i.

The feature vectors $\overline{X}_i$ were the four multispectral values. We computed for every object the Mahalanobis distances d_k to each class for k = 1, ..., M

$$d_k = (\overline{X} - \overline{\mu}_k)^T \; \Sigma_k^{-1} \; (\overline{X} - \overline{\mu}_k)$$

where μ_k and μ_k are the mean vector and covariance matrix of class k, respectively. Initial probabilities were then computed using formula :

$$P_i(\lambda_k) = \frac{\dfrac{1}{d_k}}{\displaystyle\sum_{\ell=1}^{L} \dfrac{1}{d_\ell}}$$

thus yielding an initial stochastic labeling input to our optimization program.

Results on two 32 x 32 images are summarized in table 1. An increase of 18 % in the percentage of correct classification may be observed.

Table 1.

	Initial Recognition rate	Final Recognition rate
Image 1 (5 classes)	56,9 %	64 %
Image 2 (8 classes)	26,8 %	46,3 %

5. CONCLUSION

Results obtained so far in a wide variety of applications by relaxation as well as optimization techniques prove, beyond   dispute,

that labeling enhancement is worth being considered each time contextual information is to be meaningfull. New advances should be
expected anyway, particularly in the optimization domain : more
efficient optimization techniques could probably be devised [10],
and theoretical issues would be of interest for hierarchical labeling enhancement processes, for which relaxation has succesful-
ly been tried [31]. No doubt that this domain will remain open
for a while.

REFERENCES

[1]   Berthod, M., and Faugeras, O.D. : Using context in the glo-
       bal recognition of a set of objects : an optimization ap-
       proach. Proc. of IFIP Congress, Tokyo, PP. 695-698, 1980.

[2]   Berthod, M., and Faugeras, O.D. : Etiquetage par optimi-
       sation et application. 2ème congrès A.F.C.E.T.-IRIA de Re-
       connaissance des Formes et Intelligence Artificielle. Tou-
       louse, pp. 267-275, Septembre 1979.

[3]   Davis, L., and Rosenfeld, A. : Applications of relaxation,
       2 : spring loaded template matching. Proceedings of the
       3rd I.J.C.P.R., San Diego, 1976.

[4]   Davis, L., and Rosenfeld, A. : Curve segmentation by rela-
       xation labeling. I.E.E.E. Trans-on Comp. 26, pp. 1053-1057,
       1977.

[5]   Davis, L., and Rosenfeld, A. : Noise cleaning by iterated
       local averaging. I.E.E.E. Trans-on S.M.C., 8, pp. 705-710,
       1978.

[6]   Davis, L., Rosenfeld, A. : Iterative histogram modifica-
       tion. I.E.E.E. Trans-on S.M.C., 8, pp. 300-302, 1978.

[7]   Davis, L. : Shape matching using relaxation techniques.
       I.E.E.E. Trans-on P.A.M.I., Vol. 1, n° 1, pp. 60-72, Jan-
       uary 1979.

[8]   Eklundh, J.L., Yamamoto, M., Rosenfeld, A. : Relaxation me-
       thods in multispectral pixel classification. To be published
       in I.E.E.E. Trans-on P.A.M.I., 1980

[9]   Faugeras, O. and Berthod, M. : Scene labeling : an optimi-
       zation approach. Proceedings of the I.E.E.E. Conf. on Pat-
       tern Recognition and Image. Processing, Chicago, pp. 318-
       326, 1979.

[10]   Faugeras, O. : Decomposition and decentralization techni-
       ques in relaxation labeling. To be published in Computer
       Graphics and Image Processing, 1980.

[11]   Faugeras, O., and Berthod, M. : Improving consistency and
       producing ambiguity in stochastic labeling : an optimiza-
       tion approach. To be published in I.E.E.E. Trans-on P.A.M.
       I., 1980.

[12]   Faugeras, O., and Price, K. : Semantic description of aerial
       images using stochastic labeling. 5th I.J.C.P.R., Miami,
       1980.

[13]   Haralick, R.M. : Structural pattern recognition, homomor-
       phonisms and arrangements. Pattern recognition, Vol. 10,
       pp. 223-236, 1978.

[14]   Haralick, R., Kartus, J. : Arrangements homomorphisms and
       discrete relaxation. I.E.E.E. Trans-on S.M.C., 8, pp. 600-
       612, August 1978.

[15]   Haralick, R., and Shapiro, L. : The consistent labeling pro-
       blem : Part I. I.E.E.E. Trans-on P.A.M.I., 1, n° 2, pp. 173-
       183, April 1979.

[16]   Haralick, R., and Shapiro, L. : The consistent labeling pro-
       blem : Part II. I.E.E.E. Trans-on P.A.M.I., 2, n° 3, pp.
       193-203, 1979.

[17]   Heyes, K. : Reading handwritten words using hierarchical
       relaxation. T.R. 783, Computer Science Center, Univ. of
       Maryland, College Park, 1979.

[18]   Lev, A., Zucker, S.W., and Rosenfeld, A. : Iterative enhan-
       cement of noisy images. I.E.E.E. Trnas-on S.M.C., 7, pp.
       435-442, 1977.

[19]   Peleg, S. : Iterative histogram modification : 2. I.E.E.E.
       Trans-on S.M.C., 8, pp. 555-556, 1978.

[20]   Peleg, S., and Rosenfeld, A. : Determining compatibility
       coefficients for curve enhancement relaxation processes.
       I.E.E.E. Trans-on S.M.C., 8, pp. 548-554, July 1978.

[21]   Peleg, S. and Rosenfeld, A. : Breaking substitution ciphers
       using a relaxation algorithm. TR-271. Comp. Science Center,
       Univer. of Maryland, College Park, MD, January 1979.

[22]   Peleg, S. : A new probabilistic relaxation scheme. Procee-
        dings of the I.E.E. E. Conf. on Pattern Recognition and Image
        Processing, Chicago, PP. 337-343, 1979.

[23]   Rosenfeld, A., Hummel, R., and Zucker, S. : Scene labeling
        by relaxation operations. I.E.E.E. Trans-on S.M.C., 6 , pp.
        420-433, 1976.

[24]   Schachter, B.J., Len, A., Zucker, S.W., and Rosenfeld, A. :
        An application of relaxation methods to edge reinforcement.
        I.E.E.E. Trans-on S.M.C., 7, pp. 813-816, 1977.

[25]   Toussaint, G. : The use of context in pattern recognition.
        Invited address, Proceedings of the I.E.E.E. Conf. on Pat-
        tern Recognition and Image Processing, Chicago, 1977.

[26]   Ullmann, S. : Relaxation and constrained optimization by
        local processors. Computer Graphics and Image Processing,
        10, pp. 115-125, 1979.

[27]   Zucker, S., Hummel, R., and Rosenfeld, A. : An application
        of relaxation to line and curve enhancement. I.E.E.E. Trans-
        on Computer, c-26, pp. 393-403, 922,929, 1977.

[28]   Zucker, S.W.,Krishnamurthy , E.V., and Haar, R.: Relaxation
        Processes for scene labeling : convergence, speed and sta-
        bility. I.E.E.E. Trans-on S.M.C., 8, pp; 41-48. January
        1978.

[29]   Zucker, S.W., Mohammed, J.L. : Analysis of probabilistic
        relaxation labeling processes. I.E.E.E. Conf. on Image
        Processing and Pattern Recognition, Chicago, pp. 307-312,
        1978.

[30]   Zucker, S.W., Leclerc, Y.G., and Mohammed, J.L. : Conti-
        nuous relaxation and local maximum selection : conditions
        for equivalence. TR. 78-15 T, Computer Vision and Graphics
        Laboratory, Mac Gill Univ., Montréal, Canada, 1978.

[31]   Zucker, S.W., and Mohammed, J.L. : A hierarchical relaxa-
        tion system for line labeling and grouping. Proceedings
        of the I.E.E.E. Conf. on Pattern Recognition and Image
        Processing, Chicago, pp. 410-415, 1979.

COOPERATION AND VISUAL INFORMATION PROCESSING

S.W. Zucker
Computer Vision and Graphics Laboratory
Department of Electrical Engineering
McGill University
Montreal, Quebec, Canada

ABSTRACT

This paper is an attempt to develop a class of parallel,
synchronous mechanisms into the notion of cooperative algorithms,
and to study a number of their applications. The applications
range from practical, computer vision tasks to the modeling of
human capabilities in certain visual information processing tasks.
Our most fundamental goal is one of cross fertilization: to show
how the design and structure of computer vision systems can benefit
from the principles underlying human perception; and to show how
the study of human perception can benefit from the practical and
theoretical experience acquired in building computer vision systems.
Cooperation provides the common mechanism for accomplishing this.

PREFACE

This paper consists of three parts, all related by the common
theme of cooperation and its use in modeling human visual information processing. The first part serves the introductory roles
of (i) proposing a problem in human psychophysics; (ii) explicat-
ing the kind of contribution that computer vision can hope to make
to the study of human visual information processing; and (iii)
proposing an explanation of the observed perceptual phenomena.
This explanation emphasizes the roles of representations and of
algorithms that manipulate those representations.

The second part of the paper suggests one approach to structuring the algorithms. The approach is cooperative, and amounts to
building them as networks of 'simple' processors. After a brief

259

J. C. Simon and R. M. Haralick (eds.), Digital Image Processing, 259–288.
Copyright © 1981 by D. Reidel Publishing Company.

introduction to the concept of cooperation as we shall be using it,
we report some new results, obtained in collaboration with R.
Hummel, on the development of cooperative algorithms for certain
optimization problems (cf. also the paper by Faugeras in these
proceedings).

The final part of the paper ties the first two together by
presenting several examples of cooperative modeling of visual
information processes. Some of these are motivated primarily by
computer vision applications, such as finding and labeling lines
(such as veins or roads) in (biomedical or satellite) imagery.
This part of the paper will also lead to a discussion of multi-
level (and sometimes hierarchical) relaxation systems. Furthermore,
we shall present several applications of cooperative modeling
techniques to the study of human visual information processing
problems. The first of these, developed by S. Ullman, sucessfully
models human competence in a (dot) correspondence task underlying
certain of our abilities to interpret visual motion. This
correspondence process models an ability to group (visual) tokens
in time. The next application models our ability to group tokens
in space. We show several examples of 'random-dot Moire Patterns',
a grouping phenomenon discovered by L. Glass, which demonstrate
this ability quite strikingly. Structural similarities such as
these, in what appear behaviourly to be drastically different
tasks, begins to suggest common mechanisms useful to the study of
human as well as computer vision. Understanding why these mechanims
function, and how they arose evolutionarily, raise questions that
may turn out to be fundamental for scientists interested in either
computer or human vision, or both.

PART I

MOTION AND THE MUELLER-LYER ILLUSION

1. INTRODUCTION

The Mueller-Lyer illusion is a classical demonstration of the effects of context on the perceived lengths of lines. The illusion has been shown to appear stronger in blurred, as opposed to focused, presentations (see Fig. 1). The psychophysics of such effects have been documented by Ward and Coren [1976], and by Coren, Ward, Porac, and Fraser [1978] who have shown that the subjective strength of the illusion is effectively doubled with extensive blurring. Now, suppose that a subject views a sequence of images in which the illusion gradually becomes more out of focus. Will he see motion? One might be tempted to answer yes, because the difference in strength between the statically focused and the statically unfocused (i.e., blurred) cases suggests a change in the lengths of the lines comprising the illusion. Do these changes appear as a movement of the internal vertex defining the illusion, that is, the tip of the internal arrow, or do they take different form? Our experiments show that several different motion effects are perceived, none of which seem to involve the internal vertex explicitly. The reason for this, we believe, is a consequence of early visual information processing. Particular support for this conjecture is obtained by showing that our data are consistent with certain current notions about the content and character of low-level visual computations.

The effects of blur on geometric illusions, such as the Mueller-Lyer illusion, were previously studied in order to locate the causal basis for the illusion. Helmholtz [1962], for example, discussed the distortions introduced by optical imperfections of the eye. Such distortions can fill in a portion of the acute angle formed by converging lines, thereby changing the intensity array in a manner consistent with the illusion. More recently, Glass [1970] has modeled such distortions in the spatial domain (for the related poggendorf illusion), and Ginzburg [1971] has modeled them in the spatial-frequency domain, with similar conclusions about the direction of distortion. It should be noted that Ginzburg's model was motivated by a different source of blurring: the bandpass-limited channels early in the visual system, whose existence was first hypothesized to explain the threshold visibility of spatial frequency gratings [Campbell and Robson, 1968; Graham and Nachmias, 1970]; and also that it is his position that these channels are the primary cause of the illusion. We shall use a similar model to structure the blur in our experiments, and shall see that it does   cause a.static distortion in the direction of the illusion. However, additional constraints, which can be stated in terms of the representation of low-level visual information, seem to be necessary to account for the observed motion effects.

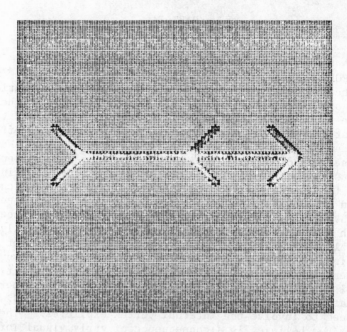

Figure 1a: Display of the Mueller-Lyer illusion just out of focus.

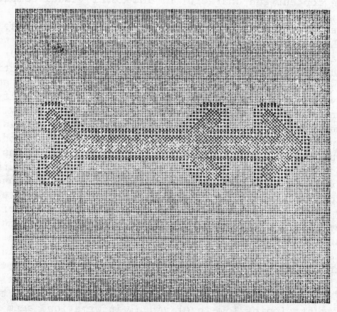

Figure 1b: Display of the Mueller-Lyer illusion substantially
out of focus.

2.   EXPERIMENT 1:   real-time blurring

To experimentally evaluate the effects of motion and blurring
on the Mueller-Lyer illusion, a number of stimuli were computed
and then presented on a real-time raster graphics display system.
The display has a resolution of 256 x 256 image points.   All
displays were computed from the Brentano form of the Mueller-Lyer
illusion, and spanned a length of about 7 inches across the monitor.
Subjects viewed the displays from a distance sufficient to remove
discrete image sampling and quantization effects (i.e., 5 to 20
feet from the monitor).   No variations in results were observed
as a function of viewing distance within this range.   All observer
reports were obtained informally.

A Gaussian function was chosen to create the blur in our
experiments because both the optical blurring function for foveal
vision [Campbell and Gubisch, 1966; Gubisch, 1967] and the spatial
smoothing underlying the bandpass-limited channels [Wilson and
Bergen, 1979] can be modeled in this way.   Mathematically, a
blurred image should be obtained by convolving such a Gaussian with
an image of the perfectly focused illusion.   However, to enable
our implementation to run in real time, this process was approximated by modeling the intensity profile across the illusion as a
Gaussian function.   To specify this function, consider the perpen-
diculars at each point along the lines comprising the figure.
Letting x denote distance along a particular perpendicular, we
have

$$Int(x) = k1 \quad exp \ \{-x/[k2 \ sigma \ ]\}^2$$

where Int(x) denotes the intensity at x, k1 and k2 are scaling
constants (for the display), and sigma, a function of time, is the
parameter that controls the extent of the blurring. We shall refer
to the position at which the profile returns to the background
value as the physical profile edge. Each subject viewed a cyclic
sequence of images presented rapidly enough for apparent motion,
beginning with the illusion in focus, then gradually going out of
focus, and finally returning back into focus. This cycle was
repeated for several minutes. Furthermore, the speed at which
sigma changed was controlled manually, so that the rate of blurring
could also be varied.

More than 20 subjects (graduate students or visitors to the
laboratory) were asked to view the cycling display, and to describe
any perceived motion effects. If, in fact, they were re-estimating
the relative lengths of the lines comprising the illusion as it
was being blurred, then one might expect that the perceived position
of the vertex formed by the internal 'arrow' (i.e., the position

at which the two lines meet) would move in the direction of
statically-described change. However, no one saw such a distinct
left to right movement of the internal vertex at any speed. The
only kind of movement observed within the illusion was a weak
'filling-in' of the convex (or, as shown in Fig. 1, the right-
hand-side) of the illusion. In other words, as the blur increased,
the inside edges of the central and rightmost arrowheads appeared
to be moving toward one another, while filling in the area between
them. No equivalent effect was observed on the other side of the
illusion.

A second, and often more powerful motion percept was that of
the entire illusion moving toward each subject in depth as it
became more blurry, and away as it became more focused. This latter
kind of motion is consistent with what would occur if the focal
plane of the eye were fixed and an object were moving in space:
it would appear to change in size and in focus. Since no perspective transformations were incorporated in the algorithms that
computed our displays, and furthermore, since the blurring algorithm
kept the focused-image template constant, the result was a perceptual increase in size and, hence, perceived motion toward the
observer. Further discussion of this issue, together with a
possible mechanism for obtaining the size increase, are presented
in Sec. 4.1.

3.   EXPERIMENT 2: Vertex movement

One explanation for the lack of perceived vertex movement
is that the vision system is incapable of capturing such movement.
For example, the position of the vertex may be redetermined for
each frame, and it may be changing, but the vision system may not
be capable of using this change to trigger a motion percept. The
second experiment was designed to test this hypothesis by simulat-
ing a different kind of motion. The illusion now remained in
focus throughout the presentation, but the angle of the arrows
defining it were varied in time (see Fig. 2). Since the subjective strength of the illusion is proportional to the cosine of
this angle [Woodworth and Schlosberg, 1962] one would expect a
strong perceived motion of the central vertex. This is precisely
what is observed. Moreover, such a result is not at all surprising,
since it is known that there are mechanisms early in vision systems
capable of responding to the motion of lines [Hubel and Wiesel,
1968].

Benussi [1921], in a classic experiment described by Woodworth
and Schlosberg [1962, p. 421], performed a similar, but coarser
vision of this experiment. His subjects observed a stroboscopic
presentation of three images: one in which the angle of the
oblique lines forming the arrows was 45 degrees with respect to
the horizontal, one in which the angle was 90 degrees, and one

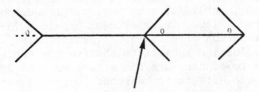

Figure 2: The angle, 0, varied in time to produce motion of the central vertex. The moving vertex is indicated by arrow.

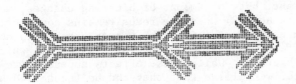

Figure 3a: Thresholded image showing the zeroes obtained by convolving a small ∇^2 G operator with the image in Fig. la.

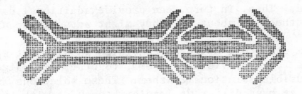

Figure 3b: Thresholded image obtained by convolving the large ∇^2 G operator with Fig. la.

in which it was 135 degrees. His subjects reported the same
effect that we observed; that is, a strong, continuous motion of
the vertex of the arrow. Thus his experiment provides independent
support for our presentation methodology. (Benussi also showed
that it was necessary for the subject to observe the entire
illusion, rather than to concentrate his attention on a particular
portion of it, for these effects to be seen. Our observations
concur on this point: if one's attention is concentrated entirely
on the internal vertex, then it can be seen to remain stationary.)

4.   THE COMPUTATIONAL APPROACH TO VISION

The experiments described thus far have demonstrated two
different motion effects: (1) when the illusion configuration
remains constant, but the amount of blurring changes, motion is
seen in depth; and (2) when the focus remains constant, but the
configuration changes, the internal vertex appears to move. A
filling-in effect was also apparent for the case in which the
illusion was blurred. While these effects appear quite different
on the surface, we believe that they can be reconciled by, and can
lend support to, the kinds of representations that are being
developed within the computational approach to early visual information processing. And, in so doing, they provide further evidence
for the competence (or sufficiency) of these representations.

We should like to stress that although the computational
approach requires making concrete models in order to carry out
experiments, it does not imply that these are the precise mechanisms functioning in humans. Nor do we wish to imply that these
are the only ways in which the computations can be posed. Our
intention, rather, is to demonstrate one way in which they can be
accomplished. This, in turn, can provide additional data for
formulating theories about how they might actually occur.

The computational approach to vision is aimed at specifying
the problems faced by vision systems, together with algorith-
mically plausible solutions to them. These solutions must reflect
constraints at many different levels, beginning with the manner
in which the abstract problem is posed all the way to a characterization of the hardware on which solutions should run. Marr
[1976] has characterized one of the earliest problems in vision
as building a symbolic representation of the intensity changes
in images, which he refers to as the (raw) primal sketch. Since
there are many possible ways of doing this in a complete, and
hence reversible fashion, another necessary property of this
representation is that the information described explicitly should
facilitate subsequent computations.

The completeness requirement for the primal sketch has resulted
in edge descriptions that parametrically specify their position,

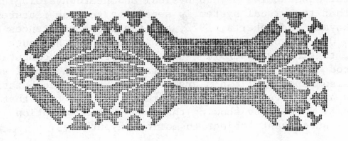

Figure 4a: Thresholded image obtained by convolving the small $\nabla^2 G$ operator with Fig. 1b.

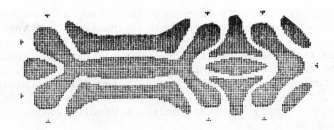

Figure 4b: Thresholded image obtained by convolving $\nabla^2 G$ operator with Fig. 1b.

oreintation, contrast, and width (or profile extent). There is,
furthermore, substantial psychophysical and neurophysiological
evidence that the visual system is sensitive to such features
[e.g., Nakayama and Roberts, 1972, and the other references cited
therein]. Our purpose in this section is to show that such descriptions, together with basic mechanisms for motion computations,
readily produce the psychophysical data reported in Sec. 2 and 3.
We shall also support Ullman's argument that motion computations
presuppose a low-level description more structured than that of
raw intensity values, by showing that one such description greatly
facilitates subsequent 'object in motion' hypotheses.

The process for computing edge descriptions that we shall use
is the one put forth by Marr and Hildreth [1979] because of its
basis in current neurophysiology. Briefly, an operator, similar
to a Laplacian but derived from the second derivative of a Gaussian,
is convolved with filtered versions of the original image. The
zero-crossings of the convolved images are then interpreted into
assertions describing the edges, with parameters as specified
above. In particular, these zero-crossings define the positions
with which the edges are associated. The filtered versions of the
image simulate the effects of the bandpass-limited channels mentioned in connection with Ginzburg's explanation of the illusion
(Sec. 1). For more detail, see the paper by Marr and Hildreth.

Once the edge features have been described, a low-level motion
computation can be performed. Ullman [1979] has shown that this
computation is decomposable into two separate tasks: (1) establishing a correspondence relation between tokens (i.e., symbolic
descriptions such as edge labels) representing components of the
objects in temporally successive images; and (2) determining
the motion of objects in space whose projections would give rise
to these tokens. Both of these tasks are readily accomplished
for our examples.

4.1 Edge-profile and object movement in Experiment 1
Figs. 3 and 4 show the result of convolving two operators,
one small (Fig. 5a) and the other larger (Fig. 5b), with the
illusion images in Fig. 1. The figures actually show thresholded
images, so that the convolution zeroes are more clearly visible--
values significantly different from zero have been set to maximal
darkness. Note, in particular, how the zeroes outline the illusion,
thereby specifying the positions of the edges in its internal
description.

The first effect caused by blurring becomes clear when the
convolutions with the larger operators are compared (Figs. 3b and
4b): the sizes of the enclosed regions grow with blurring. Such
changes in zero crossing locations in time could be very useful to
an organism, in that they could indicate objects moving in its

Figure 5a: Absolute value display of the small $\nabla^2 G$ operator.
           The central portion of the "Mexican hat" is negative.

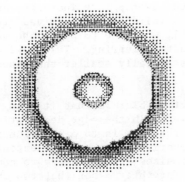

Figure 5b: Absolute value display of the large $\nabla^2 G$ operator.

field of vision. Also, they are straightforward to obtain computationally, and they may actually be computed very early in the
human visual system [Marr and Ullman, 1979]. A process that
simply looked for relative maxima among these changes would
immediately capture the motion of the profile edges within the
convex half of the illusion, because the smoothing of the acute
angles would cause these to move the fastest. Such a process
could account for the filling-in effect observed during Experiment
1.

An explanation for the lack of perceived vertex movement in
Experiment 1 can also be put forth on the basis of the zero-crossing positions. Since the arrows are represented as thick regions
rather than thin lines, there is ambiguity about where the arrows
end and the lines begin. To estimate the lengths of the horizontal
line segments, one could either discount the arrows entirely,
perhaps by subtracting them out and considering only the remaining line segments, or one could hypothesize a vertex in the center
of the arrow region. In the limit as the arrows become thin, of
course, these two schemes yield identical results. However, to
obtain illusion strengths comparable to those reported psychophysically, the former scheme must be used. The latter, when
measured carefully, yields an illusion increase of only 7% under
extreme blurring, as compared with the doubling reported psychophysically. If the arrowheads are ignored and only the remaining
line segments compared, appropriate increases in magnitude are
obtained. Moreover, a third scheme, derived from a suggestion
of Coren et al. [1968], would be to place the vertex at the point
of maximal darkness after blurring. This,however, also produces
changes that are substantially smaller than those reported psychophysically.

The second effect of blur is that it changes the width, or
profile extent, parameter in the edge descriptions. That is, it
quantifies how close the edge is to a step function, or in the
case of ramp-shaped profiles, the spatial extent of the ramp.
At least two operator sizes are necessary to compute the width
parameter, and it is for this reason that two operators were
shown in Figs. 3, 4, and 5. For our purposes, we can define the
width parameter in terms of the size of the smallest operator
whose zeroes are consistent with the larger operators evaluated
over equivalently placed sub-images. Examining Figs. 3 and 4 in
more detail demonstrates why this is the case. The smaller
operators, which are sensitive to higher spatial frequencies,
show some of the detailed irregularities that arise from the
blurred display. The response of the larger operators, which are
more restricted in their spatial-frequency response, result in
much smoother zero-crossing curves. For a given amount of blurring,
the curves are consistent when they outline equivalent illusion
figures. Now, if we define the illusion in terms of the thin,

elongated regions defined by the (zero-crossing) edges, it becomes clear that blurring changes the width parameters in the descriptions. For the less blurred image, both size operators produce consistent zeroes while, for the strongly blurred image, the width has to be defined in terms of the larger operator.

Ullman's [1979] basic approach to the interpretation of motion can now be used to obtain the second kind of motion seen in Experiment 1 - - that of an object moving in depth. The first task, establishing a correspondence between the tokens comprising the descriptions of successively blurred images, is straightforward -- they are identical in content except for the width and position parameters. It is easily within the capabilities of Ullman's [1979] linear programming scheme, which establishes the correspondence relations that minimize a function of the total distance separating tokens in successive images. To accomplish the second task, i.e., to infer a solid object moving toward the viewer, we shall appeal to the physics of the situation. If an object were indeed approaching the observer, its description would remain essentially constant, except for changes in the edge locations and profiles. Recall we are assuming that the viewer's plane of focus either remains constant or changes slowly with respect to the object's movement. In the case of an orthographic (i.e., parallel) projection, it is important to note, the blurring caused by a single Gaussian lens would be identical for object motions both toward and away from the observer. In Experiment 1, however, everyone reported motion toward them as the object was blurred, and away as it became more focused. These data have two perceptual implications.

(1) Since no perspective effects were supplied directly by our algorithms, one must inquire what they were inferred from. We believe that two separate effects conspired to make the perceived depth change the only possible one. The first effect is a negative one: if the object were moving away, then perspective changes would have to have caused it's image to shrink. Perspective must be operative because, for the focus of an object to change as drastically as it did in the experiment, the object must be relatively close to the lens. For distances far from the lens, the depth of field is large. The second effect is a positive one. When an object moves toward an observer, perspective should cause its image to grow. As we have already shown, such growth is implied by the changes in the zero-crossing positions. Thus, it would appear that these changes are quanlitatively sufficient to subsume the perspective size increase.

(2) It is possible that the observer's plane of focus was changing, contrary to our assumption that it remained constant. This would produce identical blurring effects, provided the object remained stationary. However, no one realized the perceptual

consequences of this interpretation. That is, no one saw himself
moving with respect to a stationary object. Moreover, the observer's plane of focus was probably fixed by factors external to the
experiment, such as the frame imposed by the display monitor.

In summary, Experiment 1 provides evidence that Ullman's
approach to the interpretation of motion is appropriate for components of low-level descriptions different from those that he
studied. It further suggests how simple computations on these
descriptions can imply certain perspective effects. In general,
the explanation of Experiment 1 is an example of inferring structure-from-motion; i.e., of inferring a solid (and, in this case,
flat) object moving through space on the basis of more primitive
motion computations carried out on low-level image descriptions.
It requires the assumption that the object be rigid.

4.2 Deformable Object Motion in Experiment 2:

Establishing a correspondence between the tokens comprising
successive descriptions of Experiment 2 is as straightforward as
for Experiment 1 - - the descriptions are identical in all
parameters except orientation and, to a lesser extent, length.
However, the object's motion is now very different, because no
rigid body hypothesis is tenable. Rather, the object appears to
be both fluid and articulated, and recovering it is more an
example of motion-from-structure than of structure-from-motion.

The motion-from-structure computation begins with the requirement that the 3-D structure of the perceptually moving objects
be clear from the individual images, perhaps through monocular
shape cues. Then their motions through space can be inferred in
such a way that they are consistent with the static descriptions.
In particular, Kolers [1972] has shown that when a square and a
circle are viewed in alternation, an apparent motion between them
is perceived in which the square seems to transform smoothly into
the circle, and vice versa. In both Kolers' example and in
Experiment 2, the objects seem to be flat, like wire-frames lying
in the plane. The motion percepts seem similar as well, like
flexible objects smoothly deforming one into the other.

5. DOES THE MUELLER-LYER ILLUSION ARISE CENTRALLY?

One of the most intriguing questions about visual illusions
is how centrally they arise. Theories for the Mueller-Lyer
illusion have ranged from Helmholz's ocular imperfections to
Gregory's [1963] theory of inappropriate constancy scaling. The
computational results in Sec. 4 suggest that very early processing
is sufficient for describing the illusion as well as several
derived motion effects. More particularly, they suggest that the
position and width variables carry the essential information.

Since these parameters are expressed in retinal-or image-based coordinates, which also provide the basis for stereo-fusion, one would also expect the illusion to be present in stereoscopic presentation. Indeed, Julesz [1971] has shown that it is. It does not follow, however, that it must be computed entirely after fusion.

An alternative to the deformable object motions observed in Experiment 2 might have been predicted from Gregory's [1963] theory that the Mueller-Lyer illusion derives from inappropriate constancy scaling in depth. If the lines comprising the illusion are interpreted as the contours of corners in 3-space, then the apparent depth of these corners from the viewer can be estimated as a function of the angles in the illusion [Gregory, 1968]. This implies a full 3-D surface representation. However, no one saw motions of such objects, even when they were instructed to attempt to do so. One reason for this could be that such motions were in fact attempted, but, since the perspective transformations were now incorrect, the motions were discarded. A simpler, and to our view more appealing, explanation is that the motion computation takes place separately from the computation of such 3-D representations.. This latter view is further supported by the fact that a qualitatively invariant 3-D corner can be visible for a range of angles in Experiment 2, especially when the illusion is presented in two parts: one with convex arrows and the other with concave ones. Also, if subjects concentrate their attention only on the internal arrow, it can be seen as a solid, L-shaped figure rotating back and forth. However, both of these 3-D figures should be recoverable from the structure-from-motion computation already described.

6. CONCLUSIONS

Visual illusions often provide insight into the constraints involved in visual information processing, and this paper was an attempt to explore one of these in a new way. The Mueller-Lyer illusion was presented undergoing two different kinds of real-time distortion: one in which it was blurred in and out of focus, and the other in which the angles of the arrows defining it were changed from acute to obtuse and back again. It was shown that the motion percepts resulting from these displays were different from what might be expected solely on the basis of static data, and, furthermore, that these percepts are consistent with certain current ideas about low-level visual computations. In particular, we have shown that the width and position parameters in edge descriptions can support an 'object in motion' hypothesis, and that such objects moving in depth are often what people see. Our basic approach was to show that particular computations that could give rise to these effects were also in agreement with the physics of the perceived configuration and viewing scenario. Another computation showed how the edge profiles defining a perceived filling-in effect could be isolated. While it may be possible for both of these tasks to be implemented by a third, and different computation, those chosen were already hypothesized for other reasons. Furthermore, the decomposition is supported by the psychophysical fact that both motion effects are perceivable simultaneously and independently.

An examination of the low-level description of the blurry illusion showed that illusion strengths comparable to those reported psychophysically could be obtained by subtracting the full arrowheads from the figure, and then considering only the relative lengths of the remaining line segments. The arrowheads could be defined, e.g., by the positions of the zero-crossings resulting from a particular operator convolution. These experiments thus suggest that discounting the arrowheads is a possible strategy underlying subjective reports about the illusion.

The second experiment, which was designed to demonstrate that the vertex of the internal arrow defining the illusion is capable of motion, also implied that the position of this vertex is derived from the descriptions of the lines forming it. Perhaps it is derived from their common endpoints. Computational experiments with the first blurred configuration suggested that no such vertex description is explicitly maintained, because of ambiguities in placing it. Furthermore, if it were explicit, then certain other motion effects would be expected, analogous to those in Experiment 2, and these never occured.

The internal description of the blurry Mueller-Lyer illusion was fixed by the edge-position and the edge-width parameters, which are extremely low-level constructs. It is not clear that other geometric illusions are determined this early, since at least some, such as the Poggendorf, can be induced by subjective figures. The motion blurring methodology presented in this paper is currently being applied to the other geometric illusions, to test the generality of the effects reported here.

REFERENCES:

1. Benussi, V., Stroboskopische Scheinbewegungen und geometrisch-hoptische Gestalttauschungen, Arch ges Ps, 1912, 24, 31-62.

2. Campbell, F. W. and Robson, J.G., Applications of Fourier analysis to the visibility of gratings, J. Physiol. (London), 1968, 197, 551-556.

3. Campbell, F.W. and Gubisch, R.W., J. Physiol. (London), 1966, 186, 558.

4. Coren, S., Ward, L., Porac, C., and Fraser, R., The effect of optical blur of visual-geometric illusions, Bulletin of the psychonomic Soc., 1978, 11(6), 390-392.

5. Ginzburg, A., Psychological correlates of a model of the human visual system, Proc. National Electronics Conf, 1971, 283-290. Available from IEEE.

6. Glass, L., Effect of Blurring on a simple geometric pattern, Nature, 1970, 228(5278), 1341-1342.

7. Graham, N., and Nachmias, J., Detection of grating patterns containing two spatial frequencies: a comparison of single-channel and multiple-channel models, vision Research, 1971, 11, 251-259.

8. Gregory, R.L., Distortion of visual space as inappropriate constancy scaling, Nature (London), 1963, 199, 628.

9. Gregory, R.L., Perceptual illusions and brain models, Proc. Roy. Soc., B, 1968, 171, 279.

10. Gubisch, R.W., Optical performance of the human eye, J. Opt. Soc. Amer., 1967, 57, 407.

11. Helmholtz, H. Von, Treatise on Physiological Optics, Dover, New York, 1963 reprint, Southall, ed.

12. Hubel, D.H. and Wiesel, T.N., Receptive fields and functional architecture of monkey striate cortex, J. Physiol. (London), 1968, 195, 215-243.

13. Julesz, B., Foundations of Cyclopean Perception, University of Chicago Press, Chicago, 1970.

14. Kolers, P., Asepcts of Motion Perception, Pergamon Press, New York, 1972.

15. Marr, D., Early Processing of visual information, Proc. Roy. Soc. London, B, 275, 483-534.

16. Marr, D. and Hildreth, E., Theory of edge detection, A I Memo 518, M.I.T., April, 1979.

17. Marr, D. and Ullman, S., Directional selectivity and its use in early visual processing, A I Memo 524, M.I.T., June, 1979.

18. Nakayama, K. and Roberts, D., Line length detectors in the human visual system: Evidence from selective adaptation, Vis. Res., 1972, 12, 1709-1713.

19. Ullman, S., The Interpretation of Visual Motion, MIT Press, Cambridge, 1979.

20. Ward, L. and Coren, S., The effect of optically induced blur on the Mueller-Lyer illusion, Bull. Psychon. Soc., 1976, 7, 483-484.

21. Wilson, H. and Bergen, J., A four mechanism model for threshold spatial vision, Vis. Res., 1979, 19, 19-32.

22. Woodworth, R. and Schlosberg, H., Experimental Psychology, rev. ed., Holt, Rinehart, and Winston, New York, 1962.

PART II

COOPERATIVE RELAXATION PROCESSES

1. WHAT ARE COOPERATIVE ALGORITHMS?

The Oxford English Dictionary defines cooperation as "Working" together to the same end". This definition has a number of pre-suppositions necessary for understanding its application to algorithms. First and foremost, it implies a community of distinct entities capable of communicating with one another. Such communication is necessary for achieving the common end, or, in our case, the solution to a prespecified problem. Problems are amenable to cooperative solutions when they can be posed in terms of a network of subproblems, and a cooperative algorithm embodies this structure in a particular architecture -- that of a network of processors. It should be noted that the subproblem may or may not imply meaningful problems on their own; the requirement is that their solution contributes to the solution of the original problem. Thus, without further restrictions, the first pre-supposition essentially amounts to one of distribution.

The second definitional presupposition, that each entry is incapable of solving the entire problem, begins to focus our understanding more on the notion of cooperation as we intend to use it. It has two separate implications: (i) that the processor be simple, perhaps not even Turing universal, and (ii) that each processor be restricted in the information available to it. Both of these restrictions will be applicable to the cooperative algorithms that we shall be considering, which contain identical processors at each node of the network. Each of these processors is very simple, and implements a prespecified sequence of numerical operations. Furthermore, each one is restricted to communicate directly only with a sparse set of "nearest" neighbors. Such cooperative algorithms solve problems in an iterative fashion, propagating information, or partial results, only along network interconnections. We shall call these specific algorithms relax-ation labelling processes, and shall study them both in the abstract and in relation to the problem of finding lines in digital imagery. When the processors implement a numerical computation, as above, we shall refer to them as continuous relaxation processes. When they implement a symbolic computation, such as comparing patterns against a list or data base of patterns, we shall refer to them as discrete relaxation processes.

Two additional, but less well-known properties typically attributed to cooperation are hysteresis and abrupt order/disorder transitions. Both of these properties have their origins in the first coherent model for a cooperative phenomenon -- the Ising [1925] model of ferromagnetism. Briefly, this model viewed

ferromagnetic materials as consisting of dipole arrays in which
neighboring dipoles were coupled into having either parallel or
antiparallel orientations. The strength of the coupling varied
monotically with temperature, and gave rise to an abrupt disorder
(randomly oriented dipoles) to order (aligned dipoles) transition
at a specific temperature (the Curie point). Such behaviour had
been observed in ferromagnetic materials. Moreover, hysteresis is
present in such models because the ordered state is more difficult
to leave than to enter; loosely, the dipoles, once aligned,
attempt to pull the system back. For reasons described below,
however, these latter two properties may provide too much constraint on our algorithms, and shall not play an integral role in
our notion of cooperation. Thus, in summary, when we refer to
cooperative algorithms, we shall have sparse networks of simple
processors in mind (see also Ullman, 1979b).

2. WHY STUDY COOPERATIVE ALGORITHMS?

There are both practical and theoretical reasons for studying
cooperative algorithms in general, and for studying cooperative
approaches to line finding in particular. As we shall now attempt
to show, the general reasons suggest a methodology based, in part,
on particulars, and vice versa. The reason is similar to the need
for both experimental and theoretical physics: the experiments
suggest constraints on what the theories must account for, and
provide the data with which they are ultimately tied to reality.
Theories, on the other hand, unify large amounts of data, and,
while they are being refined, suggest key experiments.

In addition to interacting, however, theoretical and experimental physics must be conducted partially in isolation of one
another. For example, it was necessary to discover differential
equations, or the mathematical theory of groups, so that they
could be developed as the language for expressing theories of
electromagnetism, or of quantum mechanics, respectively, [Dyson;
1969; see also Wigner, 1960]. One such a language has been
established, the interaction can take place most fruitfully. The
mathematical analysis of the Ising model of ferromagnetism took
many years [Newell and Mantrell, 1953] and such cooperative notions
are now used in modeling a wide range of physical problems [Haken,
1975].

The increasing prominence of very large scale integration
(VLSI) makes the practical attraction of cooperative algorithms
clear, in that they are easily implementable in hardware [Kung,
1979]. Given this practical attraction, we must ask when can we
make use of it. Or, in other words, what can be computed cooperatively? This is a deep theoretical issue, related to the
issues of decomposability of large systems in general, and, more

particularly, to the meaning of local and global for theories of computation. These are very difficult questions, which have, so far, resisted serious attempts to answer them.

Progress has been made toward understanding questions of decomposability and locality in restricted areas, however, which helps to place them in our theoretical vocabularies as well as possibly sharpening our intuitions about them. We have already seen one example from Physics. Cooperation arises in economics when a community of individuals acts to set commodity prices to make the "best" use of limited resources, or to maximize the total of their profits [Dorfman, Samuelson, and Solow, 1958]. The study of such particular economic situations has resulted in models for cooperation that are precise enough for analysis and has led, among other results, to a distributed theory of linear programming [Arrow, Hurwicz, and Uzawa, 1958], to game theory [Von Neumann and Morgenstern, 1947], and to an economic theory of teams [Marschak and Radner, 1972]. Each of these results now stands on its own, amenable to further analysis and to application in areas different from its economic origin. Two cases in point are the application of distributed linear programming to motion perception [Ullman, 1979], and the application of economic team theory to program structuring [Fox, 1979]; see also [Simon, 1969].

Cross-disciplinary applications, while often useful, are, in many other cases, dangerous, especially when details from one problem carried over to another. A particular point is the current controversy over the correct form for models of human stereopsis. In particular, Julesz [1971], on the basis of an apparent hysteresis effect [Fender and Julesz, 1967], has postulated a cooperative model very similar to that proposed by Ising for ferromagnetism. Marr and Poggio [1979], however, have reinterpreted this effect into one of a simple memory that has drastically different implications for stereopsis models. While the debate between them will probably be decided in the neurophysiological laboratory, the essential point for us is that we do not loose sight of computation as our home base, and of low-level visual information processing as our application area. The two are tied together through the computational approach to vision, which is where we shall now turn.

2.1 Cooperative modeling of low-level visual computations

The computational approach to vision attempts to formulate algorithmic solutions to problems in the processing of visual information. Once a problem has been formulated, the algorithm designed to implement its solution must reflect constraints all the way from the abstract problem down to the hardware on which it is destined to run [Marr, 1977]. In the first part of this section we attempted to argue that (i) for an approach to algorithm design

to be available to researchers, either it must already be in their
vocabulary or they must invent it, and we might add that we believe
invention is usually much harder than identification; and (ii)
that there are technical problems associated with a class of
algorithms whose solutions are required by the users of the
algorithms. Moreover, like in Physics, the study of algorithms
should be both experimental and theoretical.

All problems in low-level visual information processing seem
to have one point in common: the vastness of the data. For
engineering problems, when efficiency or time enter into the choice
of an algorithm, distribution seems to be a necessity. And, given
the practical advantages of networks of simple processors, cooperation seems desirable. For the study of human perception,
such networks appear to be a reasonable computational form within
which the neural machinery of low level vision can be modeled.
Therefore understanding them may facilitate inferences about how
human visual information processing might be carried out. In
particular, a more precise understanding of cooperation could
indicate whether it is indeed an appropriate form of modeling
for human vision.

A second, pervasive aspect of low-level visual computations
is the conceptual need to organize them into levels. This need
arises from the complexity of the computations, as well as from
the fact that certain computations have explicit prerequisites.
Consider, for example, the problem of designing a template for
the recognition of a pattern. If the pattern is large, i.e., if
it consists of many inter-related pieces, then the computational
effort required to perform the match is large. Furthermore, this
entire effort must be repeated for each position and orientation
(or other relevant variable) in the match space -- there is no way
to take advantage of partial results. In short, the advantage
of such large templates is that they permit detailed and precise
pattern definitions; their disadvantage is computational expense.

The time honored solution to such problems is decomposition
and organization [Simon, 1969] -- consider, e.g., what the
Dewey decimal system does (and does not do) for libraries. In the
template example, the full template should be partioned into a
network of smaller ones, and attempts made to match these. When
successful, they can be dealt with as named entities, and reformed
into the global pattern. Often these sub-templates can make
implicit properties explicit, such as the orientation associated
with an edge, which are then useful in matching the full pattern.
Dewey uses subclassifications as the named entities at the
different levels. The trade-off, of course, is that putting the
sub-patterns together is again a combinatorial problem, unless
additional assumptions can be brought to bear. Relaxation labeling processes essentially transform this combinatorial problem

into one of optimization, with the proviso that it be solved cooperatively - (cf. Davis and Rosenfeld, 1976; Kovalevsky, 1979, Faugeras and Berthod, 1979).

In specifying the computational solution to a vision problem, two separate issues must be discussed: representations and mechanisms for manipulating those representations. These issues can be treated both separately and interactively, and, again both theoretical and experimental results are desirable. There-fore, in order to address these issues more concretly, in the lecture we shall choose the problem of finding lines in images in some detail. This problem is of practical interest, because it arises in many application areas (e.g., handwriting analysis [Hayes, 1979]; remote sensing; biomedical image analysis; indus-trial image processing and quality control). It would also seem to be an intuitively clear domain in which to explore the exper-imental physics" of cooperative solutions to vision problems, because, as we shall show, it leads to a conceptually homogeneous framework for modeling them. This framework is, furthermore, sufficiently well defined to permit mathematical analysis for some of its properties.

3. THE LABELING PROBLEM:

Suppose that we are given a set of entities, numbered (arbitrarily) $i = 1,2, \ldots, N$, and a set of labels $\lambda \varepsilon \Lambda_i$, attached

to each of these entities. The problem is to find a best label for each entity, where best will be specified as an optimization problem defined below. As an example, the entities may correspond to picture points and the labels to assertions about the presence (or absence) of oriented unit line segments at those points. Thus, for example, there are nine labels attached to each pixel.

To motivate the optimization problem, suppose that there is a unique label λ at each position i. Then we can speak of the compatibility (or consistency) between a pair of labels (λ,λ') located at two respective positions (i,j) as a function whose value is $r_{ij}(\lambda,\lambda')$. We shall require that $\infty \leq r_{ij}(\lambda,\lambda') < \infty$,

where negative values indicate incompatible relationships and positive values indicate compatible ones. In our line example, these compatibilities can represent a model of orientation good continuation: line segments whose orientations align into a smooth curve are preferable to those whose orientations align into a jagged curve. If we refer to the assignment of each unique label to every entity as a labelling $\mathcal{L}$, then the quality (or cost) of the labeling can be expressed concisely by the global functional

$$C(\mathcal{L}) = \Sigma \; r_{ij} \; (\lambda, \lambda') \tag{1}$$

where the sum is taken over all pairs (λ, λ') on all entities (i,j). Clearly, to find the best quality labeling, we wish to maximize a form such as $C(\mathcal{L})$. That is, we wish to find a label assignment $\lambda \rightarrow i$, $i=1,2, \ldots, N$, that maximizes $C(\mathcal{L})$.

The relationship between the label finding problem and a formal notion of consistency can now be obtained by introducing a function $\Lambda(i)$ that assigns a unique label ℓ to node i, and by specifying the functional $C(\cdot)$ individually for each node. That is,

$$C_i(\ell) = \Sigma_{\lambda'} \; r_{ij} \; (\ell, \lambda')$$

indicates the support that the unique labels at all nodes neighboring i contribute to ℓ. Clearly, we desire the labeling $\mathcal{L} = \Lambda(i)$ such that

$$C_i(\Lambda(i)) \geq C_i(\lambda) \qquad \forall \lambda \; , \; \forall \; i. \tag{2}$$

The difficulty with the above formulation, however, is that it is expressed under the assumption that there are unique label assignments to each entity. In general, this is not the case -- a set of labels is more often appropriate to express the uncertainty associated with the labeling of each entity. In the line example, uncertainty arises because the response of label line operators is ambiguous. That is, it is not clear, on the basis of operators evaluated over a single position in the image, whether they are indicating a line segment or not. This ambiguity can be expressed by initially considering each possible interpretation as a label potentially assigned to the pixel's position.

The potential labels attached to each entity can be ordered further. For example, it is the case that line operators are loosely matched with their task, so that a strong response can be taken as a reasonable first estimate that a line segment is present. And, on the other hand, a weak response can be taken as an estimate that no line is present. Thus, these estimates can be used to order the labels according to their likelihood, and we shall represent the position of each label on this likelihood scale by a number $p_i(\lambda)$, indicating how likely label λ is

for entity i. Historically, these numbers were thought of as indicating the probability that label λ is correct for node i, although this interpretation is no longer required. We shall require, however, that they be comparable; i.e., that

$$0 \leq p_i(\lambda) \leq 1 \qquad \forall \; i, \; \lambda \tag{3}$$

$$\sum_{\lambda \epsilon \Lambda_i} p_i(\lambda) = 1 \quad \forall \ i$$

The ordering of the labels at each node affects our optimization problem. If we view this ordering as reflecting the information initially available to the process computing the optimal assignment, we certainly want it to appear in the problem statement. This can be accomplished by generalizing our previous argument involving unique assignments to considering all possible assignments weighted by their likelihoods. Thus,

$$C(P) = \sum r_{ij}(\lambda,\lambda') \ p_i(\lambda) \ p_j(\lambda'). \tag{4}$$

The P now indicates an uncertain labeling. Note that, in the original argument, each label was unique, i.e., $p_i(\lambda) = 1$ for all λ and i, so that (4) reduces to (1).

If we view the likelihoods $p_i(\lambda)$ as indicating whether λ is assigned to i in an optimal labeling, then, beginning with their initial values $p_i^o(\lambda)$, we should like to adjust them in a direction

that increases $C(P)$. Thus, we can think of an iteration of the form:

$$p_i^{k+1}(\lambda) = f(p_i^k(\lambda), \ q_i^k(\lambda)) \tag{5}$$

where $q_i^k(\lambda)$ is an updating component that specifies how to change $p_i^k(\lambda)$ at the kth iteration. In the end, a value of $p_i^\infty(\lambda) = 1$

indicates that λ should be associated with i, while $p_i^\infty(\lambda) = 0$

indicates that it should not.

A standard method of increasing the functional (4) is gradient ascent, and it is this method that we shall now apply.

$$\text{grad } C(P) = q_k(\ell)$$

$$= \frac{\partial}{\partial p_k(\ell)} \left(\sum_{i,\lambda, \ j,\lambda'} r_{ij}(\lambda,\lambda') p_i(\lambda) p_j(\lambda') \right)$$

$$= \sum_{j,\lambda'} [\ r_{kj}(\ell,\lambda') + r_{jk}(\lambda',\ell)] \ p_j(\lambda')$$

which becomes, for symmetric compatibilities,

$$q_k(\ell) = \sum_{j,\lambda'} 2 \ r_{kj}(\ell,\lambda') \ p_j(\lambda'). \tag{6}$$

What remains, to complete the updating formula (5), is to combine
the updating vector (6) with the current estimate p_i^k (λ) to yield
the gradient ascent updating formula:

$$p_i^{K+1} (\lambda) = p_i^K (\ell) + \varepsilon^K q_i^K (\lambda) \qquad (7)$$

where ε^K is a step size for the kth iteration.

4. NEW RESULTS

4.1 A Projection Algorithm:

The updating terms can be viewed geometrically as defining
a direction of movement. In order for the update p_i^{K+1} ($\cdot$) to

still be in the labeling space defined by (3), q_i^k ($\cdot$) must

correspond to the elements of vectors in the tangent space $T_p k$,
where

$$T_{\overline{p}k} = \{ \overline{q} = (\overline{q}_1, \ldots, \overline{q}_N) : \sum_{\lambda =1}^{M} q_i(\lambda) = 0 \; \forall \; i ,$$

$$\text{and } q_i(\lambda) \geq 0 \text{ if } p_i (\lambda) = 0 \}.$$

As long as q lies above the interior of this space, the update
can be found by a simple projection (see, e.g., Rosen's gradient
algorithm); however, when it lies along an edge, the projection
is much more complex. The following algorithm, developed with
J.L. Mohammed, is sufficient for the task:

Step 1: Construct $\overline{q}_i'$ according to:

$$q_i' = q_i - \frac{1}{M} \sum_j q_j$$

Step 2: Construct the following sets:

$$S = \{ \} , D = \{i : p_i = \phi \}$$

and set FLAG = FALSE
        THRESHOLD = ϕ

Step 3: $\forall$ i ϵ D, do:

If $q_i' <$ THRESHOLD, then set

$$D = D \setminus \{i\}$$

$$S = S \cup \{i\}$$

FLAG = TRUE

Step 4: If FLAG is TRUE, then
(set FLAG=FALSE,

$$\text{THRESHOLD} = -\sum_{i \in S} q_i \frac{1}{M - |S|}$$

and repeat 3 and 4)
Else

$$\forall i \text{, set } v_i = \begin{cases} \phi & i \in S \\ q_i' - \text{THRESHOLD, } i \notin S \end{cases}$$

and return $\overline{V}$ as the — projection.

4.2 The Fundamental Variational Inequality:

The development of the updating formula (7) was based on a gradient ascent technique for maximizing C(P). We shall now derive an alternate formulation of it, in terms of a variational inequality. The reason for this is that the alternate formulation does not require the assumption of symmetric compatibilities.

Suppose that $\overline{p}$ is at a local maximum of C(P). Let $\overline{v}$ be any other probabilistic labeling assignment. We consider the function of one real variable

$$g(t) = C((1-t)\overline{p} + t \overline{v}), 0 \leq t \leq 1.$$

Note that the convex combination $(1-t)\overline{p} + t \overline{v}$ is a probabilistic labeling assignment for $0 \leq t \leq 1$, since the assignment is convex. Because $\overline{p}$ is at a local maximum, g(t) is non-increasing in t at t = 0. That is $g'(0) \leq 0$. But

$$g'(0) = \text{grad } A(\overline{p}) \cdot (\overline{v} - \overline{p})$$

$$\sum_{i,\ell} q_i (\ell) \cdot (v_i(\ell) - p_i(\ell))$$

$$\sum_{i,\ell} \sum_{j,\ell} 2 r_{ij}(\ell,\ell') p_j(\ell') (v_i(\ell) - p_i(\ell)),$$

Now, if we let P denote the convex space of certainty factors defined by (3), $P = \{\overline{p} = (\overline{p}_1, \ldots, \overline{p}_N), p_i(\lambda_1), \ldots, p_i(\lambda_m))$ the problem of finding local maxima of (4) becomes equivalent to finding $\overline{p} \in P$ such that

$$\sum_{i,\lambda} \sum_{j,\lambda} r_{ij}(\lambda,\lambda')p_j(\lambda') \, [v_i(\lambda) - p_i(\lambda)] \le 0 \quad \forall \, \bar{v} \, \varepsilon \, P$$

This statement can be reformulated as the variational inequality for consistency. Let

$$[\bar{q}\,(\bar{p})]_{i\lambda} = \sum_{j,\lambda'} r_{ij}(\lambda,\lambda') \, p_j(\lambda').$$

Then the problem is to find $\bar{p} \, \varepsilon \, P$ such that

$$\bar{q}\,(\bar{p}) \cdot \bar{t} \le 0 \quad \forall \, \bar{t} \, \varepsilon \, T_{\bar{p}}(P),$$

where $T_{\bar{p}}(P) = \{\, \bar{t} = r(\bar{r} - \bar{p}): \bar{v} \, \varepsilon P, \, r > 0\}$.

The importance of this variational inequality is that it defines the stopping criteria for our gradient ascent, whether or not the compatibilities are symmetric. In particular, the solution $\bar{p} \, \varepsilon \, P$ satisfies

    (i)   If $0 < p_i(\lambda) < 1$

         and $0 < p_i(\lambda') < 1$

         then $q_i(\lambda) = q_i(\lambda')$

   (ii) if $p_i(\lambda) = 0$

         and $p_i(\lambda') > 0$

         then $q_i(\lambda) \le q_i(\lambda')$.

4.3 Relationship to consistency inequality:

It can be proved that the solutions to the fundamental variational inequality correspond to those of the consistency inequality (1).

4.4 Relationships to previous work:

The updating formula (7) is similar to that used by Faugeras and Berthod, although its development from the labeling problem is quite different. Furthermore, the classical relaxation formula of Rosenfeld, Hummel, and Zucker can be shown to be an approximation to (7), which leads us to conjecture that its empirical successes occurred when finding local maxima of (4) was appropriate for the problem under consideration. This conjecture further implies that the probabilistic approach is not necessary for explaining relaxation (cf. Peleg). Finally it should be noted that the results in this paper can be extended to arbitrary orders of compatibilities, one special case of which is Ullman's algorithm for computing correspondences. In particular, he is finding maxima of the functional (in our terms)

$$C(P) = \Sigma \; r_i \; (\lambda) \; p_i \; (\lambda).$$

PART III

RANDOM DOT MOIRE PATTERNS

Suppose that we are given a random collection of dots distributed over a planar surface. Now, suppose that we superimpose
a second distribution of dots onto the first. If the two distributions are uncorrelated, then the result will appear as a
single, flat distribution with roughly twice the density as the
original. If the second distribution is a related to the first,
however, then a number of interesting patterns emerge. Several
of these patterns are shown on this page, each of which is obtained
by making the second distribution a modification of the first.
In particular, in Fig. (a), the second distribution is simply a
copy of the first that has been translated along the vertical axis;
in (b) the second distribution is an expansion of the first; in (c)
an expansion and rotation; and in (d) an expansion along the
X-axis and a contraction along the Y-axis. These examples are
adaptations of those in Glass and Perez, Nature (246), p. 360,
1973. During the lecture we will discuss algorithms for extracting the perceived structure in these patterns. The algorithms
are both cooperative and local, thereby providing a common basis
for comparison between these two classes.

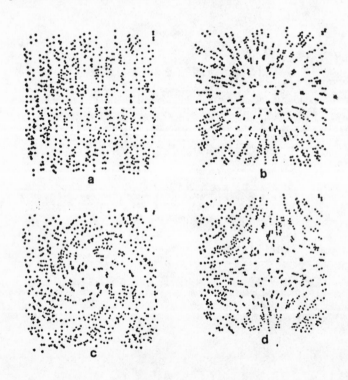

SHAPE DESCRIPTION*

Theodosios Pavlidis

Dept. of Electrical Engineering
& Computer Science
Princeton University
Princeton, NJ 08544, U.S.A.

ABSTRACT:
A review of methodologies for Shape Description is
presented. The emphasis is on simple methods such as
contour analysis, thinning, and projections.

1. INTRODUCTION

Shape analysis is one of the fundamental problems
in pattern recognition and is of interest also in
interactive graphics. It is relevant where one must
make a decision on the basis of the form of objects
that he sees. The precise psychophysical definition
of what shape is outside the scope of this paper.
Certainly, terms such as "elongated", "sharp corner",
etc. refer to shape. One could define shape by default by stating that it is the information contained
in a binary image without considering the color of
the regions. This definition limits us to the shape
of silhouettes but it is adequate for many applications
and in particular the recognition of alphanumeric
characters.

One can distinguish two modes of shape recognition.
In one, a person looks at the total object and makes
a decision on the basis of the overall structure.
This is commonly the case in the recognition of hand-

---

*Portions of this paper will appear as part of a book
on Graphics and Image Processing by T. Pavlidis.
All rights reserved. The research leading to this
paper was supported by NSF Grant ENG 76-16808.

J. C. Simon and R. M. Haralick (eds.), Digital Image Processing, 289–310.

printed letters, especially Chinese, where one identi-
fies strokes or other basic building blocks. In another mode one examines the contour of the silhouette,
usually looking for corners, protrusions, intrusions
and other points of <u>high curvature</u>. An example is the
recognition of silhouettes of human profiles, or the
check for defects of the outlines of the circuit elements on printed wiring boards. Of course, there are
many situations where both modes must be used: An engineering drawing contains lines and letters recogniz-
able by their structure, as well as circles or hexagons
which must be distinguished from each other on the
basis of their contours. Most methodologies in the
past have been oriented towards one of the two modes,
and since it is theoretically possible to apply them
on any objects they have been used in cases where they
were not well suited. Ideally, one would like to have
a mixed method which could adapt to the mode most appropriate for the object examined.

<u>Example</u>: A thinning algorithm can be used to trans-
form an alphenumeric character into a graph where most
branches would correspond to strokes. The same algorithm would also reduce a rectangle to a graph and it is
possible to discriminate between two kinds of rectangles on the basis of the relative length of the branches
of the graph. However, it is easier to check directly
the relative sizes of the sides, and the cost of the
thinning process cannot be justified.

        Curvature is an important feature in shape analysis,
not only when it is used directly, but also indirectly.
Unfortunately, its direct measurement is not always
feasible because of noise. The formula given in most
texts of calculus requires taking a second derivative
and thus it cannot be used in any practical situation.
On the other hand it is possible to estimate the radius
of curvature by the geometrical construction shown in
Figure 1. Indeed, if A, B, and C are points on a curve,
M and N are the midpoints of the intervals AB and BC
respectively, and K the point where the normals at M
and N intersect, then R, the length of BK equals the
radius of a circle passing through A, B and C.

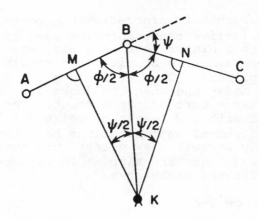

Figure 1.   Construction used for the estimation of the
             radius of curvature.

    If the angle ABC equals ϕ, and Ψ denote its com-
plement, and the lengths of AB and BC are each equal
to 2e, then a straightforward calculation shows that

$$R = \frac{e}{\cos\left(\frac{\phi}{2}\right)} = \frac{e}{\sin\left(\frac{\Psi}{2}\right)} \qquad (1)$$

It is clear from Figure 1 and equation (1) that R is an
increasing function of ϕ. If Ψ is small, as it is
often the case, then the <u>curvature</u> <u>c</u> which is defined
as the inverse of the radius of curvature will be given
approximately by

$$c = \frac{\Psi}{2e} = \frac{\pi - \phi}{2e} \qquad (2)$$

It is possible to eliminate some of the effects of
noise by calculating the curvature not on successive
pixels bit on triplets where the distance <u>e</u> is large
compared with the spatial noise frequency. There exist
many variations of this technique but we shall not
deal with them here. An indirect determination of

curvature maxima can be achieved by polygonal approximations because corners will tend to be placed near
such maxima [11].

2. CONTOUR ANALYSIS

It is possible to proceed with a contour analysis
based on a primitive representation such as a chain
code. Or with a higher level representation consisting of an approximation by pieces of smooth curves
(e.g. by B-splines). The latter is preferable when
the data are noisy and when one looks for features
involving a large part of the contour. The former is
best for data with low levels of noise and localized
features. Polygonal approximations have been used
often not only because they detect curvature maxima,
but also because they are simpler to implement than
other curve fitting techniques.

2.1. Chain Code Techniques

The chain code has been used widely in image
processing and a very common form uses eight directions,
as shown in Figure 2. If the points were chosen from
among adjacent elements of an image matrix (along rows,
columns, or diagonals), then representing them by their
matrix indices will require 4 bits per point (in order
to represent the values -1, 0, and +1 for each coordinate) while the chain code requires only 3 bits
per point.

An even more efficient form is offered by the
differential chain code where each point is represent
by the difference between two successive absolute codes.
In this case we still have eight values (0, +1, +2,
+ 3, and 4), but their occurence is not equally likely.
For relatively smooth curves we expect the values 0,
and +1 to occur more frequently than all the others,
while 4 will be very rare. Then we can use a variable
length code to represent each direction. Table 1
shows one possible assignment:

If we assume that the direction 0 occurs 40% of
the time, the directions +1 20% each, the directions
+2 8%, the directions +3 I.5%, then the average number of

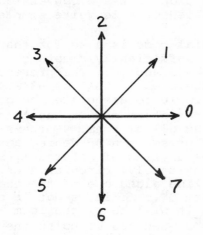

Figure 2. Definition of the basic chain code. The
  broken lines delineate the set of directions which
  are assigned to a particular element of the code.

Table 1

Direction	Code
0	0
+1	01
−1	011
+2	0111
−2	01111
+3	011111
−3	0111111
4	01111111

bits per point will be 2.395, considerably less than
that of the plain chain code. Whether a chain code
representation is desirable for curve manipulation
depends on the application, but the conversion to
x-y coordiantes and vice versa is quite simple.

　　　　If the differential code is used for shape analysis
and if the contour is sufficiently smooth the only
symbols present will be 0 and ± 1. The presense of ± 2
implies a 90 degree angle while that of ± 3 a 45 degree
angle. It would be simple to associate the occurence
of such symbols with curvature maxima but this is not
sufficient for locating all of them. For example, the
sequence 01110 also represents a 45 degree angle.
There are four classes of arcs of interest: Straight
lines, corners, (approximately) circular arcs, and
notches. A straight line along one of the chain code
directions has the form 0^n. If it is not in one of
these directions then it will be of the form
$(0^m+1-1)^k$ or $(0^m-1+1)^k$. Because of noise the actual
code may not be as regular but will always be characterized by the occurence of +1-1 or -1+1 pairs. A circular
arc on the other hand will be of the form $(0^m+1)^k$ or
$(0^m-1)^k$, the main feature being that 1's will occur
singly and they will be of the same sign. A corner
will either contain one of the high values of the code
or a sequence of 1's with the same sign. A notch is a
sharp intrusion or protrusion and it is of interest
in certain applications where one wants to have very
smooth contours so that the presence of a notch will
indicate a defect in terms of the chain code a notch
is a sequence of high values of the code adding up to
0 (or at most ± 1). Figure 3 shows some examples.

　　　　One can come up with simple algorithms for detecting such features and their applicability depends on
how noise free the data are. One useful technique is
based on the definition of regular expressions or,

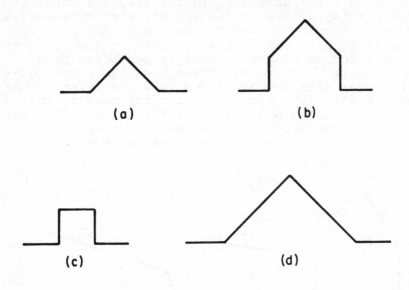

Figure 3.   Examples of notches. The respective
    differential chain codes are as follows:
(a)...0+1-2+10...,
(b)...0+2-1-2-1+20...,
(c)...0+2-2-2+20...,
(d)...0+10-20+10... .

(equivalently) <u>finite</u> <u>automata</u> which match such features
For example, the following expression will recognize
some of the notches shown in Figure 3:

0(+1 or +2)(a string of at most two symbols)(-1 or -2)0
                                              ⌃   ⌃
                                             +1   +2

2.2. Polygonal Approximation Techniques

If the contour is first processed by an algorithm
which finds a polygonal fit to it, then the corners
found in this way are candidates for curvature maxima.
Because curve fitting algorithms check only for error
bounds it is not true that any corner of the polygon
is also a corner of the contour as perceived by a
human observer. There are three possible exceptions:
(a) A linear side may be broken by a corner as shown
in Figure 4a. (b) A circular arc has been approximated by a regular polygon (Fig. 4b). (c) A contour
corner has generated two or three polygonal corners
with short sides between them (Figs. 4c and 4d)

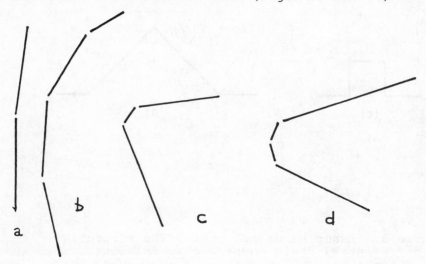

Figure 4. Examples of polygonal corners which do not
    correspond to contour corners.

These cases can be treated by examining the following:
(a) Absolute size of the polygonal corner. If it is
near 180 degrees then the corner is considered spurious and removed from any further consideration.
(b) Variance in length between successive sides and
corners. A sequence "Long-Short-Long" or "Long-Short-
Short-Long" can be interpreted as denoting a contour
corner near the location of the short sides. On the
other hand, a sequence of sides of approximately equal
length suggest the possiblity of a fit to a circular
arc. In that case the angles must also be checked and
if they are also approximately equal, then one can conclude that an arc is present, otherwise, the corners
are considered to be "true" corners. This kind of
analysis can be expressed in a syntactic formalism and

one can derive regular expressions applicable to polygonal sequences. (See Bibliographical Notes).

2.3. Shape Features

The identification of corners, lines, arcs, and,
possibly, notches on a contour can form the basis for
further analysis. One may look for more complex
features or study the <u>attributes</u> of the features found.
The following are some possible attributes:
(1)   Length of arc or line, or take length of the
      sides forming a corner.
(2)   Depth of concave arc or concave corner (see Fig. 5).
(3)   Orientation of the bisecrtix of the arc on the
      corner.
(4)   Position of the corner or the center of the arc.
      This can be useful in the case of shapes where
      orientation is important (e.g. alphenumerics).

Figure 5. (See Figure   at end of paper).

3. STRUCTURAL ANALYSIS

Structural analysis must identify the elementary
building blocks of an object whose silhouette is
available as part of a binary image. This can be done
directly by decomposition, or by finding a thinned
version of the object. The latter method is theoretically applicable to any objects but in practice it
yields reasonable results only for objects which are
"thin" to start with. Decomposition techniques tend
to be rather expensive computationally and we shall
not discuss them here. Instead we shall present some
simple thinning methods which could be used for the
recognition of line drawings or alphanumeric characters,
objects which are likely to appear in an interactive
graphics environment.

3.1. General Thinning

There exist numerous thinning algorithms in the
literature. One recently proposed method proceeds by
successive contour peelings while maintaining certain
"multiple" pixels. We proceed to define these formally.

<u>Definition 1.</u> The <u>c-neighbors</u> of a pixel belonging to
a contour C are the defined as the previous and next
elements to it along the traversal defined by a contour
tracing algorithm.

Note that the two c-neighbors need not be distinct.
The have this property only when the contour is a simple
path.

<u>Definition 2</u>. A pixel is said to be <u>multiple</u> if one
or more of the following conditions hold:
- (a) It is traversed more than once during contour
tracing.
- (b) It has no neighbors in the interior of the
region.
- (c) It has at least one direct neighbor which
belongs to the contour, but which is not one
of its c-neighbors.

See [11] for the definition of a direct neighbor.

In the case of binary images, the first two
conditions can be checked easily by a using markings
during the contour tracing. Indeed, at the end of the
tracing simple pixels of the contour will have value
exactly 2, while multiple pixels satisfying condition
(a) will have values greater than 2. A second traversal
can be used to check condition (b) (no neighbor with
value 1) and condition (c) (a d-neighbor with value 2
or greater which is not a c-neighbor). Figure 6 shows
an interpretation of multiple pixels in terms of sampling of the analog region and its boundary.

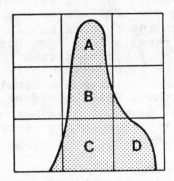

Figure 6. Intuitive illustration of the multiple pixels.
A corresponds to a boundary folding (condition b), B
to two disjoint arcs mapped on the same pixel (condition
a), and C and D are adjacent pixels where disjoint arcs
of the boundary are mapped (condition c).

If one insists on connected skeletons, then it is
necessary to keep not only multiple pixels, but a few
others as well:

Definition 3.   A <u>skeletal pixel</u> is one for which one
of the following conditions is true:

    (a)   It is a multiple pixel.

    (b)   It has a d- or i-neighbor which has been
         identified as a skeletal pixel during an
         earlier tracing.

   This leads to Algorithm 1.

Algorithm 1.   Thinning Algorithm for the set Q.

1.   Set S = $\emptyset$. [Empty set.]

2.   <u>While</u> Q is not empty do steps 3-6.

       <u>Begin</u>.

3.   Find B(Q) by a contour tracing algorithm.

4.   Find L(Q) and M(Q) by retracing B(Q) while
       checking the conditions of Definition 2. (The
       retracing can be done with the help of a queue
       containing the chain codes).

5.   Find K(Q) by examining all pixels in B(Q) for
       neighbors in S

                              [Steps 4 and 5
       may be combined for more efficiency.]

6.   Set S = S U M(Q) U K(Q) and Q = Q - B(Q).

       <u>End</u>.

7.   End of Algorithm.

    A complete analysis of this algorithm is presented
elsewhere [14].   Figure 7 shows an example of its
application

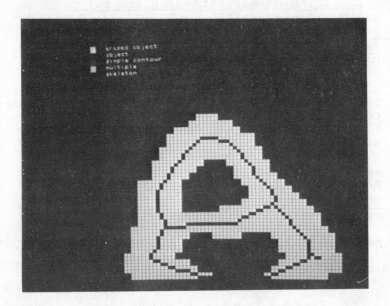

Figure 7.   An example of the application of Algorithm 1.
The resulting skeleton was further edited by an algorithm described in [14].

3.2.   Simple Thinning and Applications to Shape
       Analysis
     The general algorithms, such as the one given
above tend to be rather expensive computationally.
A "quick and dirty" method is listed as Algorithm 2.

Algorithm 2.   Fast Thinning Algorithm.

1.   Form the Line Adjacency Graph (LAG) and label each
     node with the location of the center and the width
     of the respective interval. {Let d denote the
     spacing between scan lines.}

2.   Utilize a graph traversal algorithm to find all
     paths consisting of nodes with degree (1,1), and,
     possibly, (1,0) or (0,1).

3.   While any of the conditions of steps 5, 6, and 8
     have been found to be true do:
          Begin.

4.   For each path found in step 2 do:
               Begin.

5.   If a path contains a subpath where the
               width of all nodes is below a given
               threshold w_0 and the number of nodes
               exceeds a given value N_0, then a vertical
               branch has been found and all nodes of
               the subpath should be labeled as belong-
               ing to such a branch.

6.   If a path contains a subpath where the
               width of all nodes exceeds a threshold
               $N_0 \cdot d$ and the number of nodes is below
               w_0/d, then a horizontal branch has been
               found and all nodes of the subpath should
               be labeled as belonging to such a branch.

7.   If neither of the conditions stated in
               steps 5 and 6 are true, then do:
                    Begin.

8.   Find a subpath where all centers are
                    approximately colinear and estimate
                    the slope of the line ϕ.
                    If the width of all the nodes is
                    below $w_0 \sin\phi$ and their number ex-
                    ceeds $N_0 \cos\phi$, then a branch with
                    slope ϕ has been found and all nodes
                    of the subpath should be labeled
                    accordingly.
                    End.

               End.

          End.

9.   End of Algorithm.

     Algorithm 2 does a very simple form of thinning
and it is expected that for many objects it will leave
significant parts unlabeled. It does not attempt to
partition paths into subpaths in any optimal way and
this may contribute to an increase in the number of
"leftovers". However, it is much faster though than
any of the general algorithms and it may not be a bad

idea to use it as a preprocessor, even when the objects
do not consist primarily of thin lines. In that case
one could apply one of the general thinning algorithms
on the parts of the silhouette that have been left unlabeled. Since our goal now is structural analysis it
is reasonable to relax the connectivity requirement
for the skeleton. In particular, we may define as
skeletal pixels only those which are multiple pixels
and omit step 5 of Algorithm 1. The following examples
illustrate these points.

Example: Figure 8a shows a silhouette and the respective LAG. It is seen that three branches are clearly
identified and found by Algorithm 2. In Figure 8b the
same algorithm fails to find the curved stroke.

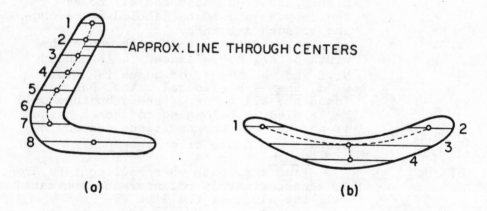

(a)   (b)

Figure 8. Examples of silhouettes where Algorithm 2.
   operates successfully (a) and unsuccessfully (b).

Example: Removing the connectivity requirement of
Algorithm 1 results in a skeleton where isolate branches
correspond to connected components.

         Additional methods for decomposition can be found
in the literature cited in the Bibliographical Notes.
A major obstacle in the application of structural
techniques has been the lack of convenient classification techniques. The complete treatment of the subject
is beyond the scope of this paper but the following is
a summary of what are the major problems.
         (a) The results of structural analysis are ex-
pressed best as a graph. It is possible to describe
these graphs in terms of graph grammars but this does

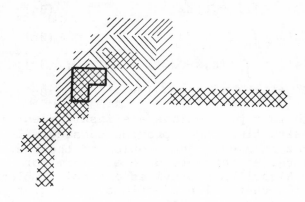

Figure **9.** Decomposition of a silhouette effected by
   a modified thinning algorithm.

not seem to be a fruitful method in spite of its theoretical attraction.   There are at least two reasons for
this: (a) It is very difficult to infer grammars from
samples of data.   (b) It is very difficult to parse
graph grammars. See [11] for a discussion of these
questions.
       (b) An alternative to graph grammars is to define
certain logical functions about the (labeled) graph
properties. For example, "there exist two vertical
strokes". Then one can establish a decision tree or
define binary vectors whose components correspond to
these logical functions. Such vectors can then be
classified according to appropriate statistical rules.

4.   PROJECTIONS
       In certain applications, it is convenient to use
projections of the kind that has become well known in
crossaxial tomography [3]. This is particularly true
for alphanumeric characters which are written with
strokes along a few directions: vertical, horizontal,
and two diagonals. Such strokes will appear as peaks
in the projection along their direction and could be
detected easily in a number of ways. We proceed with
the analysis by writing the projection equations for
ϕ = 0 degree, 90 degree, 45 degree, and 135 degree.

$$p(0°, t) = \int_R f(x, t)\, dx \tag{3a}$$

$$p(90°, t) = \int_R f(t, y)\, dy \tag{3b}$$

$$p(45°, t) = \int_R f(x, x - \sqrt{2}t)\, dx \tag{3c}$$

$$p(135°, t) = \int_R (x, \sqrt{2}t - x)\, dx \tag{3d}$$

Even though four projections are insufficient for a good reconstruction, they provide considerable information about strokes. One problem in their application is finding the appropriate discrete form for eqs. (3c) and (3d). Algorithm 3 provides a simple implementation which assures that all projections are vectors of the same size. It assumes that N is even.

The operations performed in steps 6-13 have the effect of grouping the 2N-1 diagonals of the NxN grid into N pairs, each consisting of two diagonals, except the two middle ones which share the main diagonal. Figure 10 shows some examples.

In order to visualize the amount of information present in these projections, we attempt to reconstruct the characters by simple backprojections. The results are shown in Figures 11 and 12. The examples demonstrate that indeed, considerable information about the gross appearance of the character is contained in the backprojections.

The detection of peaks in the projections can be achieved by any technique which is appropriate in waveform analysis. Usually, the level of the noise is low and simple techniques such as thresholding or even differentiation are applicable. The method was used in the past for the recognition of alphebetic characters with reasonable success, even though only two projections were calculated. (See Bibliographical Notes.)

The examples of reconstruction in Figures 11 and 12 suggest some of the power and limitations of the method. Details on the contour is circular rather than rectilinear, as in the case of the letter R. However, because of this lack of sensitivity, it may be appropriate for multifont character recognition.

Algorithm 3. Calculating Projections for Shape Analysis.

1. For ℓ = 0 to N-1 set $p(0,\ell) = \sum_{i=0}^{n-1} f(i,\ell)$

2. For ℓ = 0 to N-1 set $p(1,\ell) = \sum_{i=0}^{n-1} f(\ell,i)$

3. Initialize p(2,*) and p(3,*) to zero.
4. For j = 0 to N-1 do:
   Begin.
5.   For i = 0 to N-1 do:
       Begin.
6.   Compare i+j to N-1
7.   If greater, then add f(j,i) to $p(2,\frac{i+j+1}{2})$

8.   If less, then add f(j,i) to $p(2,\frac{i+j}{2})$

9.   If equal, then add f(j,i)/2 to $p(2,\frac{N}{2}-1)$

       and f(j,i)/2 to $p(2,\frac{N}{2})$

10.   Compare j-i to 0.
11.   If greater, then add f(j,i) to $p3,\frac{j-i+N}{2})$

12.   If less, then add f(j,i) to $p(3,\frac{j-i+N-1}{2})$

13.   If equal, then add f(j,i)/2 to $p(3,\frac{N}{2}-1)$

       and f(j,i)/2 to $p(3,\frac{N}{2})$

       End.
    End.
14. End of algorithm

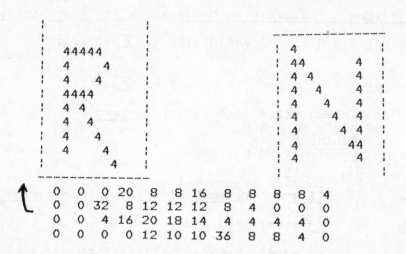

```
44444              4
4      4           44          4
4      4           4 4         4
4444               4  4        4
4 4                4    4      4
4 4                4      4    4
4      4           4       44  4
4      4           4        44 4
   4               4          4
```

```
0   0   0  20   8   8  16   8   8   8   8   4
0   0  32   8  12  12  12   8   4   0   0   0
0   0   4  16  20  18  14   4   4   4   4   0
0   0   0   0  12  10  10  36   8   8   4   0
```

Figure 10.   Binary matrices representing printed characters and the respective projections listed in the
order they are found by Algorithm 3.

```
3                    4        4
3                    5        4
3                    5        4
533333               5        4
4                    5        4
4                    5        4
5 333                5        4
4                    5        4
4                    4        4
4                    4        3
4                    4        3
4                    4        3
```

Figure 11.   Reconstruction of the examples of Fig. 10
using only vertical horizontal backprojections.

```
:                   :      : 4                    :
:                   :      : 43          3        :
:                   :      : 3 3         3        :
: 3333333           :      : 3  3        3        :
:  5                :      : 3   3       3        :
:  54               :      : 3    3      3        :
:  4443             :      : 3     3 3   3        :
:  3 43             :      : 3       3 3 :
:  3  33            :      : 3        33 :
:  3    33          :      : 3         4 :
:  3     3          :      :            3 :
:                   :      :              :
```

Figure 12.   Reconstruction of the examples of Fig. 10
using all four backprojections.

5.   BIBLIOGRAPHICAL NOTES

The literature on shape analysis is enormous.
See [12] and [13] for a general review. We concentrate
here on that part which is relevant to man made objects
and, in particular, character recognition. A review
of the early work on that subject can be found in [19].
Examples of contour analysis can be found in many
papers. The use of concavities for the description of
written symbols is emphasized in [1], [2], and [20].
The first paper also discusses the syntactic analysis
of polygonal approximations of contours and the use
of regular expressions in that environment.

Examples of simple structural descriptions can be
found in [2], [5], [6], [7], and [9]. [6] is probably
the earliest paper where the Line Adjacency Graph (LAG)
is used (although implicitly) and it forms the basis
for a decomposition procedure. The method described
in [9] performs, in effect, an analysis of the LAG in
order to find the skeleton. [2] describes a direct
thinning algorithm (of the same general type as
Algorithm 1) and uses it for numeral recognition. It
is interesting to note that although structural descriptions were proposed quite early (as far back as
1959), they have found limited use in practice. The
most likely reason is that, even the simplest ones,
tend to have more extensive computational requirements
than purely heurestic techniques. As the price of

computing decreases though one expects their implementation to become "inexpensive" and therefore should be
kept in mind as potential candidates in any system
design. See [4] and [17] for some recent decomposition
algorithms. See [14] for a discussion of the thinning
problem including references to earlier work. [16] and
[18] are good places to look for parallel thinning
algorithms.

        Projections offer a cheap way of performing some
structural analysis. Their use for character recognition was first described in [10] where a recognition
of 98% was achieved on a testing set consisting of
1880 multifont characters (samples of a,c,d,e,f,h,i,
k,l, and t). Only vertical and horizontal projections
were used but the integration was carried out separately
after each gap. Strokes were determined by a combination of thresholding and slope detection and their
presense in certain locations was used to generate
binary features. Then standard statistical techniques
were used for the final classification. Another application of projection is described in [8] where the
Fourier transforms of four projections were used for
the recognition of Chinese characters.

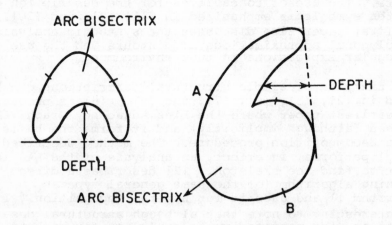

Figure 5. Definition of Attributes for Shape Analysis.

6. LITERATURE

[1]   Ali, F. and T. Pavlidis, "Syntactic Recognition
      of Handwritten Numerals", IEEE Trans. Systems,
      Man, Cybernetics, SMC-7 (1977), pp. 537-541.

[2]   Beun, M. "A flexible method for automatic read-
      ing of handwritten numerals", Philips Technical
      Review, 33, (1973), Part I: pp. 89-101. Part
      II: pp. 130-137.

[3]   Budinger, T. F. "Computed Tomography: Three-
      Dimensional Imaging with Photons and Nuclear
      magnetic Resonance", in Biomedical Pattern
      Recognition and Image Processing, (K.S. Fu and
      T. Pavlidis, editors), Chemie Verlag, 1979, pp.
      179-212.

[4]   Bjorklund, C.M. and T. Pavlidis "Global Shape
      Decomposition Using the k-Syntactic Similarity
      Approach", Proc. IEEE Conf. on Patter Recognition
      and Image Processing, Chicago, August 1979, pp.
      599-603.

[5]   Fu, K.S. (editor), Syntactic Patter Recognition,
      Applications, Springer, 1977.

[6]   Grimsdale, R. L., F.H. Summer, C.J. Tunis, and
      T. Kilburn, "A System for the automatic recognit-
      ion of patterns", Proc. IEE, 106B, 1959, pp. 210-
      221.

[7]   Hattich, W., M. Tropf, and G. Winkler, "Combinat-
      ion of statistical and syntactical pattern recog-
      nition - applied to classification of unconstrain-
      ed handwritten numerals", Proc. Fourth Intern.
      Joint Conference on Pattern Recognition, Kyoto,
      November 1978, pp. 786-788.

[8]   Nakimano, Y., K. Nakato, Y. Uchikura, A. Nakajima
      "Improvement of Chinese Character Recognition
      Using Projection Profiles" Proceedings First
      International Joint Conference on Pattern Recog-
      ition, Washington, D.C. (1973) pp. 172-178.

[9]   Naito, S., H. Arakawa, and I. Masuda, "Recognit-
      ion of handprinted alphanumerics and symbols on
      centroid lines". Proc. Fourth Intern. Joint
      Conference on Pattern Recognition, Koyoto,
      November 1978, pp. 797-801.

[10] Pavlidis, T. "Computer Recognition of Figures
      Through Decomposition" Information and Control
      14 (1968) pp. 526-537.

[11] Pavlidis, T. Structural Pattern Recognition,
      Springer, 1977.

[12] Pavlidis, T. "A Review of Algorithms for Shape
      Alanysis", Computer Graphics and Image Processing,
      7, (1978), pp. 243-258.

[13]   Pavlidis, T. "Algorithms for Shape Analysis of
        Contours and Waveforms", IEEE Trans. Pattern
        Analysis and Machine Intelligence, (accepted for
        publication).
[14]   Pavlidis, T. "A Thinning Algorithm for Discrete
        Binary Images", CGIP (in press).
[15]   Pavlidis, T. and F. Ali, "Computer Recognition
        of Handwritten Numerals by Polygonal Approximat-
        ions" IEEE Trans. Systems, Man, Cybernetics,
        SMC-5 (1975), pp. 610-614.
[16]   Rosenfeld, A. "A Characterization of Parallel
        Thinning Algorithms" Information and Control vol.
        29 (1975), pp. 286-291.
[17]   Shapiro, L.G. and R.M. Haralick, "Decompositon
        of Two-Dimensional Shapes by Graph-Theoretic
        Clustering", IEEE Trans. Pattern Analysis and
        Machine Intelligence, 1, (1979) pp. 10-20.
[18]   Tamura, H., "A Comparison of Line Thinning
        Algorithms from Digital Geometry Viewpoint",
        Proc. Fourth Int. Joint Conf. Pattern Recogn.
        Kyoto, Japan, Nov. 7-10, 1978, pp. 715-719.
[19]   Ullmann, J.R. "Picture Analysis in Character
        Recognition", Digital Picture Analysis, (A.
        Rosenfeld, editor), Springer 1976, pp. 295-343.
[20]   Yamamoto, K. and S. Mori, "Recognition of hand-
        printed characters by outermost point method",
        Proc. Fourth Intern. Joint Conference on Pattern,
        Kyoto, November 1978, pp. 794-796.

STRUCTURAL SHAPE DESCRIPTION FOR TWO-DIMENSIONAL AND
THREE-DIMENSIONAL SHAPES

Linda G. Shapiro

Department of Computer Science
Virginia Polytechnic Institute and State University

ABSTRACT

Shape analysis consists of producing a description of a shape
that can be used for identification, grouping, or further
processing of the shape. In this paper we discuss structural
shape analysis techniques including syntactic techniques and
relational techniques. Examples are given of a shape grammar, a
two-dimensional relational shape description, and a three-
dimensional relational shape description.

In shape analysis problems, we are given a two-dimensional or
three-dimensional shape and are required to produce a description
of the shape that can be used to identify the shape or to group
it with similar shapes. The shape may be given as a gray-tone
image, a binary image, a sequence of boundary points, a polygonal
approximation, or in the three-dimensional case as a 3D array.
The description may consist of a single label, a vector of
features, or a more complex hierarchic or relational structural.

Shape analysis techniques can be divided into structural and
nonstructural techniques. Nonstructural techniques describe a
shape as a vector of scalar features. For example, for a
discrete binary picture function $f(x,y)$, the p,q'th moment is
given by

$$M_{p,q} = \sum_{f(x,y) = 1} x^p y^q.$$

311

J. C. Simon and R. M. Haralick (eds.), Digital Image Processing, 311–326.
Copyright © 1981 by D. Reidel Publishing Company.

A vector of moment invariants of the picture function $f(x,y)$ can be used to describe the shape [2,6]. As a second example of a nonstructural technique, the boundary of the shape can be expressed as a function, the function approximated by a Fourier series, and the coefficients of the Fourier series used for shape descriptors [5,11,12,18].

A structural description of an object consists of 1) primitive pieces of the object, 2) properties of the pieces and of the whole, and 3) relationships among the pieces. Thus structural descriptions tend to be more robust than nonstructural descriptions and can describe more complex objects. We can divide structural descriptions further into two classes: syntactic or grammatical descriptions and relational descriptions.

SYNTACTIC DESCRIPTIONS

A phrase structure grammar is a four-tuple $G = (V_T, V_N, P, S)$ where V_T is a set of terminal symbols or primitives, V_N is a set of nonterminal symbols, $S \subseteq V_N$ is the start symbol, and P is a set of productions or rules. Starting from the start symbol, the production rules can be used to generate strings of terminal and nonterminal symbols. The set of all strings of only terminal symbols that can be generated by grammar G is called the language of G (denoted $L(G)$). For example, for the grammar $G = (\{a,b\}, \{S\}, P, S)$ where P is given by

$$\{S \longrightarrow ab$$
$$S \longrightarrow aSb\},$$

the language $L(G)$ is the set of strings $\{a^n b^n \mid n > 0\}$.

Grammars generate the strings of a language. For each phrase structure grammar G, there is a machine called a "parser" that recognizes and analyzes the strings of the language $L(G)$. Thus if we can describe a language by a grammar, then we can write a program that can recognize legal strings of that language. In particular for the simple right and left linear grammars, efficient programs may be devised to recognize strings of the language.

Picture grammars are an extension of phrase structure grammars to two-dimensional data. As for strings, if classes of pictures can be described by grammars, then it should be possible to construct picture parsers to recognize and analyze such pictures. Kirsch [7] suggested a two-dimensional picture grammar in 1964 with productions of the form

<array configuration> --> <array configuration>

whose language was a set of digitized right triangles with V at one vertex, W at the opposite vertex, and R at the right angle, H's along the hypotenuse, B's along the base, L's along the leg, and I's in interior pixels.   Such "array grammars" have been further explored by Dacey [4] and by Milgram and Rosenfeld [8]. They seem to be adequate for generating polygonal shapes, but not terribly useful in general recognition problems.

Most picture grammars have had one-dimensional productions describing the connection of two-dimensional symbols.   For example, in the more recent work of Pavlidis and Ali [10], a general shape grammar was given by production rules similar to:

$$
\begin{aligned}
&\text{BDY} \longrightarrow \text{BDY + TRUS} \mid \text{BDY + QUAD} \mid \text{TRUS} \mid \text{QUAD} \\
&\text{TRUS} \longrightarrow \text{STROKE} \mid \text{CORNER} \\
&\text{STROKE} \longrightarrow \text{LINE1 + BREAK + LINE2} \mid \\
&\qquad\quad \text{LINE1 + LINE2 where} \\
&\qquad\quad \mid 180 - \text{angle(LINE1,LINE2)} \mid < \epsilon \\
&\text{CORNER} \longrightarrow \text{LINE1 + BREAK + LINE2} \mid \\
&\qquad\quad \text{LINE1 + LINE2 where} \\
&\qquad\quad \text{angle(LINE1,LINE2)} > \epsilon \\
&\text{LINE} \longrightarrow \text{LINE + V} \mid \text{V where} \\
&\qquad\quad \text{angle(LINE,V)} < \epsilon \\
&\text{QUAD} \longrightarrow \text{QUAD + V} \mid \text{V where} \\
&\qquad\quad \text{variance(V-angle)} < d \text{ and} \\
&\qquad\quad \text{variance(V-length)} < d,
\end{aligned}
$$

where BDY is the entire contour of the shape, TRUS is a sharp protrusion or intrusion, CORNER is a corner, STROKE includes two antiparallel segments, LINE is a long, linear segment, QUAD is an arc that can be approximated by a quadratic curve, BREAK is a short segment, and V is one of the vectors into which the shape has been segmented.   This grammar and associated stochastic parser has been successfully used in character recognition experiments.

There are two general methods for parsing a picture: 1) create a string describing the picture and parse the string, and 2) use the grammar to guide the search for certain relationships in the picture.   Method 1) is the commoner of the two.   It suffers from problems with poor segmentations of the shape causing erroneous parses.   Its more serious shortcoming is the belief that a two-dimensional picture can be captured by a one-dimensional string description.   Method 2 seems more reliable, but harder to carry out. A syntactic pattern recognition system using this principle was described in Shapiro and Baron [15].

Relational Descriptions

A relational description of an object or shape is a set D =
{R1,R2,...,RK} of N-ary relations. For each relation Rk, there
is a positive integer Nk and a sequence of domain sets
$S(1,k),...,S(Nk,k)$ such that Rk $\subseteq$ $S(1,k)X...XS(Nk,k)$. The
elements of the domain sets may be atoms (nondecomposable) or
complex relational descriptions. There is usually one relation,
the attribute-value table that stores global properties or
features of the object. Other relations describe the
interrelationships among parts of the object or interactions of
the object with other objects. A phrase structure grammar is a
simple relational description where the only relationship is
symbol adjacency.

Relational descriptions for scene analysis have been around
for some time. (See Barrow, Ambler, and Burstall [3].) Only
recently with the increased use of relaxation operators in
relational matching have they been used for shape analysis. We
will describe two relational shape models -- one for two-
dimensional shape description and one for three-dimensional
object descriptions.

A Two-Dimensional Relational Shape Model

Given a two-dimensional shape represented by an ordered
sequence of points around its boundary, we have shown that the
shape can be decomposed into a set of near-convex simple parts by
applying a graph-theoretic clustering algorithm to the binary
relation LI = { (p1,p2) | the line segment joining p1 and p2 lies
interior to the boundary of the shape} [16]. The set of
nontrivial intrusions into the boundary of the shape can be
produced by applying the same clustering algorithm to the second
relation LE = { (p1,P2) | the line segment joining p1 and p2 lies
exterior to the boundary of the shape}.

Let S = {s1,...,sn) be the set of simple parts of a shape,
and let I = {i1,...,im} be the set of intrusions. Then P = S U I
is the set of shape primitives. Let d: P x P ---> $[0,\infty)$ be a
distance function which gives a relative measure of the distance
between two simple parts, two intrusions, or a simple part and an
intrusion with the property that if p1 and p2 touch or overlap,
then d(p1,p2) = 0. Let D_3 be a nonnegative real number. Define
the Boolean function p: P^3 ---> {true,false} by

$$p(p1,p2,p3) = \begin{cases} \text{true if } d(p1,p2) \leq D, \\ \quad d(p2,p3) \leq D, \text{ and} \\ \quad \text{it is possible to draw a straight} \\ \quad \text{line from a boundary point of p1} \\ \quad \text{through p2 to a boundary point of} \\ \quad \text{p3 such that there is a significant} \\ \quad \text{area of p2 on both sides of the line} \\ \\ \text{false otherwise.} \end{cases}$$

The predicate p(p1,p2,p3) is true if both p1 and p3 touch or almost touch p2 and a part of p2 lies between p1 and p3.

Using the predicate p, we can define two relations Rp and Ri that characterize the structure of the shape. The ternary relation Rp is defined by Rp = {(i1,s,i2) ∈ I x S x I | p(i1,s,i2) = true}. Thus Rp consists of triples of the form (intrusion 1, simple part, intrusion 2) where the simple part protrudes between the two intrusions. The second ternary relation Ri is defined by Ri = {(s1,i,s2) ∈ S x I x S | p(s1,i,s2) is true and d(s1,s2) ≤ D}. Ri consists of tuples of the form (simple part 1, intrusion, simple part 2) where the two simple parts touch or nearly touch and form part of the boundary of the intrusion. Figure 1 shows a block letter E which has been decomposed into simple parts 1, 2, 3, and 4 and intrusion 11 and 12 and gives the Rp and Ri relations for this shape. The two-dimensional shape descriptions have been utilized by a program that matches noisy character shapes to stored models. For more details of the matching techniques, see Shapiro [14].

A Three-Dimensional Relational Shape Model

A major task in scene analysis is the recognition of three-dimensional objects from two-dimensional perspective projections. The two-dimensional data is often noisy and distorted and, since edge finding and region growing techniques are not perfect, objects can have missing parts or extra parts. For example, an image of a table may appear to have only three legs, because the fourth leg is shadowed and is missed by the region grower. It is not until the object is recognized as a table that the existence of the fourth leg can be hypothesized. One way to handle this problem is to use a database of rough descriptions of the three-dimensional objects. The rough descriptions together with inexact-matching techniques (see Shapiro and Haralick, [17]) can allow tentative identification of the two-dimensional view as one of a small number of three-dimensional objects. Low-level routines may then use more specific high-level knowledge about an object to achieve a better segmentation of the image.

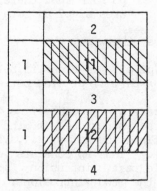

$$R_p = \{(11,3,12)\}$$

$$R_i = \{(1,11,2),$$
$$(1,11,3),$$
$$(1,12,3),$$
$$(1,12,4)\}$$

Figure 1 illustrates the protrusion relation R_p and the intrusion relation R_i of a block letter 'E'.

Relational descriptions can be used as follows to represent three-dimensional objects. Each object has a collection of near-convex parts. The relations describe the constraints on how the parts fit together. The attribute-value table contains global properties of the object to be used in grouping three-dimensional objects for quicker access. The parts of the object themselves are represented by relational descriptions. Since the parts are intended to be simple and since we are constructing only rough descriptions, the part descriptions only contain attribute-value tables describing their properties. However, the recursive relational structure gives us the option of adding refinements in the future to more precisely describe the objects.

Probably the most important three-dimensional object representation used in computer vision is the generalized cylinder. A generalized cylinder consists of an axis and

arbitrary shaped normal cross-sections along the axis.   A three-
dimensional description of an object consists of descriptions of
the generalized cylinders and of the joints connecting them.
(See Agin and Binford [1] and Nevatia and Binford [9].)   The
descriptions we suggest in this paper are related to the
generalized cylinder approach, but are at a grosser level.

Sticks, Plates, and Blobs

   Most man-made and many natural three-dimensional objects can
be broken into parts in an obvious and often unambiguous way.   In
our gross descriptions,   there are three kinds of three-
dimensional parts: long thin parts, flatish wide parts, and
parts that are neither thin nor flat.   The long thin parts will
be called sticks their identifying characteristic is that one
dimension is significant and two dimensions are not. Thus we can
picture a stick as having significant length, but very small
width and depth.   The flatish wide parts will be called plates
plates have two nearly flat surfaces connected by a thin edge
between them.   They also have two significant dimensions, which
may be thought of as length and width, and an insignificant third
dimension.   Finally, the parts that are neither thin nor flat
will be called blobs.   Blobs have three significant dimensions.
All three kinds of parts are 'near-convex'; that is a stick
cannot bend very much, the surfaces of a plate cannot fold too
much, and a blob can be bumpy, but cannot have large concavities.
Figure 2 shows several examples of sticks, plates, and blobs. We
will assume that all three-dimensional objects of interest to us
can be approximated by these three primitives.   We grant that
there are some objects that will have awkward decompositions with
respect to this model, for instance, a cup might require four
plates to approximate it. For such objects, the model would have
to be extended to include both positive and negative primitives;
such an extension is beyond the scope of this preliminary paper.

   Formally, we define the parts of a three-dimensional object
as follows. A stick is a 4-tuple ST = (En,I,Cm,L) where En is
the set of two ends of the stick; I is the interior of the stick
(the surface points of the stick except for its ends); Cm is the
center of mass of the stick; and L is its length. A plate is a
4-tuple PL = (Eg,S,Cm,A) where Eg is the edge separating the two
surfaces; S = {S1,S2} is the set of surface points, partitioned
into the points S1 on surface one and the points S2 on surface
two; Cm is the center of mass of the plate; and A is its area.
Finally, a blob is a triple BL = (S,Cm,V) where S is the set of
surface points of the blob, Cm is its center of mass, and V is
its volume.

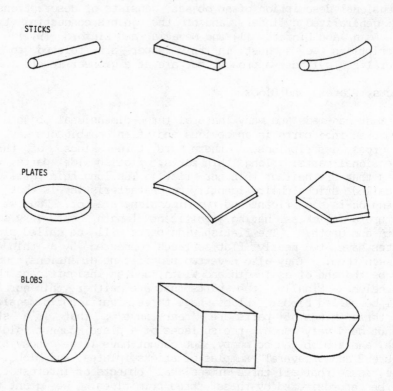

Figure 2 illustrates several examples each of sticks, plates, and blobs.

Relations

In describing an object, we must list the parts, their types (stick, plate, or blob), and their relative sizes; and we must specify how the parts fit together. For any two primitive parts that connect, we specify the type of connection and up to three angle constraints. The type of connection can be end-end, end-interior, end-center, end-edge, interior-center, or center-center where "end" refers to an end of a stick, "interior" refers to the interior of a stick or surface of a plate or blob, "edge" refers to the edge of a plate, and "center" refers to the center of mass of any part.

For each type of pairwise connection, there are one, two, or three angles that, when specified as single values, completely describe the connection. For example, for two sticks in the end-

end type connection, the single angle between them is sufficient to describe the connection. For a stick and a plate in the end-edge type connection, two angles are required: the angle between the stick and its projection on the plane of the plate and the angle between that projection and the line from the connection point to the center of mass of the plate. For two plates in the edge-interior type connection, three angles are required.

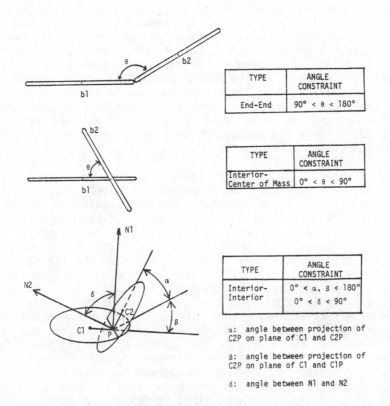

TYPE	ANGLE CONSTRAINT
End-End	90° < θ < 180°

TYPE	ANGLE CONSTRAINT
Interior-Center of Mass	0° < θ < 90°

TYPE	ANGLE CONSTRAINT	
Interior-Interior	0° < α, β < 180°	
	0° < δ < 90°	

α: angle between projection of C2P on plane of C1 and C2P

β: angle between projection of C2P on plane of C1 and C1P

δ: angle between N1 and N2

Figure 3 illustrates three examples of the constrained connections of two simple parts.

Requiring exact angles is not in the spirit of our rough models. Instead we will specify permissible ranges for each required angle. Figure 3 gives three examples of the constrained connections of two parts, showing the type of connection and the constrained angles. The angle constraints given in Figure 3 actually show the entire range of values permitted for these angles. In our relational model, binary connections are described in the CONNECTS/SUPPORTS relation which contains

10-tuples of the form (Part1, Part2, SUPPORTS, HOW, VL1, VH1, VL2, VH2, VL3, VH3) where Part1 connects to Part2, SUPPORTS is true if Part1 supports Part2, HOW gives the connection type, VLi gives the low-value in the permissible range of angle i and VHi gives the high value in the permissible range of angle i, i = 1, 2, 3.

The CONNECTS/SUPPORTS relation is not sufficient to describe a three-dimensional object. One shortcoming is its failure to place any global constraints on the resulting object. We can make the model more powerful merely by considering triples of parts (s1,s2,s3) where s1 and s3 both touch s2 and describing the spatial relationship between s1 and s3 with respect to s2. Such a description appears in the TRIPLE CONSTRAINT relation and has two components: 1) a boolean which is true if s1 and s3 meet s2 on the same end (or surface) and 2) a contraint on the angle subtended by the center of mass of s1 and s3 at the center of mass of s2. The angle constraint is also in the form of a range and is illustrated in Figure 4.

The relational description for an object consists of ten relations. The A/V relation or attribute-value table contains global properties of the object. Our A/V relations currently contain the following attributes: 1) number of base supports, 2) type of topmost part, 3) number of sticks, 4) number of plates, 5) number of blobs, 6) number of upright parts, 7) number of horizontal parts, 8) number of slanted parts. The A/V relation is a simple numeric vector, including none of the structural information in the other relations. It will be used as a screening relation in matching; if two objects have very different A/V relations, there is no point in comparing the structure-describing relations. We are also using the A/V relations as feature vectors to input to a clustering algorithm. The resulting clusters represent groups of objects which are similar. Matching can then be performed on cluster centroids instead of on the entire database of models. Other relations include SIMPLE PARTS, PARALLEL PAIRS, PERPENDICULAR PAIRS, LENGTH CONSTRAINT, BINARY ANGLE CONSTRAINT, AREA CONSTRAINT, VOLUME CONSTRAINT, TRIPLE CONSTRAINT and CONNECTS/SUPPORTS.

Access Mechanism

Given a two-dimensional shape, we do not want to try to match it against all three-dimensional models in the database. Therefore, we require a method for quickly narrowing the search. This problem is similar to the problem of document retrieval. In some document retrieval systems (see Salton [13]), the user supplies a query vector of weighted terms that should appear in the documents he wishes to access, where the weights designate importance. Each document is also represented by a vector of

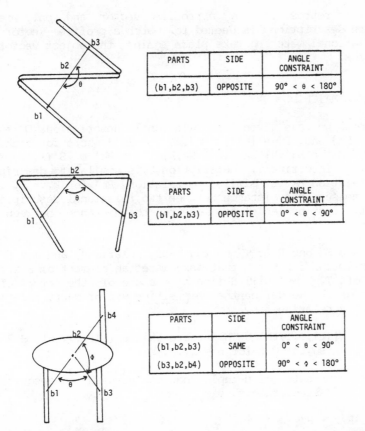

PARTS	SIDE	ANGLE CONSTRAINT
(b1,b2,b3)	OPPOSITE	90° < θ < 180°

PARTS	SIDE	ANGLE CONSTRAINT
(b1,b2,b3)	OPPOSITE	0° < θ < 90°

PARTS	SIDE	ANGLE CONSTRAINT
(b1,b2,b3)	SAME	0° < θ < 90°
(b3,b2,b4)	OPPOSITE	90° < φ < 180°

Figure 4 illustrates three examples of constrained connections among three simple parts.

weighted terms, where the weights designate frequency of occurrence. The documents are grouped into clusters of similar documents based on their vectors. Each cluster is described by a profile vector. The query vector is matched first against each profile vector, using a similarity measure such as a correlation coefficient. Only those documents in high ranking clusters have their document vectors matched against the query vector. This eliminates much unnecessary searching and matching. If we consider the attribute-value table as a weighted vector describing an object, we can develop a similar clustering scheme for the three-dimensional objects.

We have encoded attribute-value tables for a fairly large sample of office furniture objects and experimented with several different clustering algorithms to group the objects into classes. As in the document retrieval systems, each class of objects is represented by a profile vector and only when a candidate description is deemed to match a profile vector will full relational matching take place against the object vectors in that class.

RELATIONAL MATCHING

Suppose we are given two relational descriptions $D = \{R1, R2, \ldots, RK\}$ and $D' = \{R1', R2', \ldots, RK'\}$ where for each $k = 1, \ldots, K$, $Rk \subseteq S(1,k) \times \ldots \times S(Nk,k)$ and $Rk' \subseteq S'(1,k) \times \ldots \times S'(Nk,k)$. Intuitively, description D is similar to description D' if relation Rk is similar to relation Rk' for $k = 1, \ldots, K$. Thus to measure the distance between two relational descriptions, we must first be able to measure the distance between two relations.

Let $R \subseteq S^N$ and $R' \subseteq T^N$ be two N-ary relations, and let f be a binary relation $f \subseteq S \times T$ that associates an element of S with an element of T. We will define a measure of the error of the association f. We define the <u>composition</u> R∘f of N-ary relation R with binary relation f by

$$R \circ f = \{(t1, \ldots, tN) \in TN \mid \text{there exists } (s1, \ldots, sn) \in R \text{ with } (sn, tn) \in f \text{ for } n = 1, \ldots, N\}$$

There are four sets of N-tuple that can be used to describe the error of the association f.

1) $R \circ f - R'$

    This set consists of N-tuples that arise when an N-tuple of relation R is transformed by f to an N-tuple of T^N but this new N-tuple is not a part of R'.

2) $R' \circ f^{-1} - R$

    This set consists of N-tuples that arise when an N-tuple of relation R' is transformed by f^{-1} to an N-tuple of S^N but this new N-tuple is not a part of the relation R. This set is the symmetric equivalent of set 1) and is used here because we are interested in two-way matching.

3) $R - R' \circ f^{-1}$

    This set consists of N-tuples of R that are not included in the group of N-tuples obtained by applying f^{-1} to each N-tuple of R'.

4) R' - R ∘ f

This set consists of N-tuples of R' that are not included in the group of N-tuples obtained by applying f to each N-tuple of R.

Example

Consider the two chairs C and C' shown in Figure 5 and the corresponding simplified binary connection relations

$$R = \{(1,2),(2,3),(2,4),(2,5),(2,6)\} \text{ and}$$
$$R' = \{(A,B),(B,C),(B,D),(B,E),(B,F),$$
$$(C,G),(D,G),(E,H),(F,H)\}.$$

Suppose we wish to measure the error of the association f given by

$$f = \{(1,A,),(2,B),(3,C),(4,G),(6,F)\}.$$

Then the two compositions are given by

$$R \circ f = \{(A,B),(B,C),(B,G),(B,F)\} \text{ and}$$
$$R' \circ f^{-1} = \{(1,2),(2,3),(2,6),(3,4)\},$$

and the four sets of interest are

SET	NUMBER ELEMENTS
$R \circ f - R' = \{(B,G)\}$	1
$R' \circ f^{-1} - R = \{(3,4)\}$	1
$R - R' \circ f^{-1} = \{(2,4)(2,5)\}$	2
$R' - R \circ f = \{(B,D),(B,E),(C,G),$ $(D,G),(E,H),(F,H)\}.$	6

One method of defining the error of f is by a normalized convex sum of the number of elements in each of the four sets.

$$E_{R,R'}(f) = \frac{a|R \circ f - R'| + b|R' \circ f^{-1} - R| + c|R - R' \circ f^{-1}| + d|R' - R \circ f|}{a|R \circ f| + b|R' \circ f^{-1}| + c|R| + d|R'|}$$

This measure will be 0 when R' is an isomorphic image of R and f is the isomorphism; and it will be 1 in the worst possible case when $R \circ f \cap R' = R' \circ f^{-1} \cap R = \emptyset$. It has the advantage of simplicity and the disadvantage of counting all N-tuples of a relation equally when some relationships may be more important than others.

Once such a measure of relational error has been defined, we can define the structural error SE(D,D') of two descriptions D and D' by

$$SE(D,D') = \min_{f} \sum_{k=1}^{K} wk \ E_{Rk,Rk'}(f).$$

where wk is the weight assigned to relation Rk.

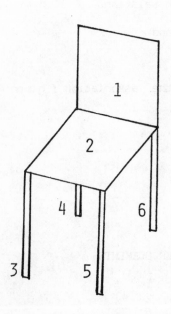

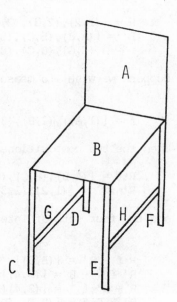

R = {(1,2) (2,3) (2,4)   R' = {(A,B), (B,C), (B,D),
      (2,5) (2,6)}   (B,E), (B,F), (C,G),
                                        (D,G), (E,H), (F,H)}

Figure 5 illustrates two similar chairs and their binary connection relations. Two shapes match when their structural error is sufficiently low.

CONCLUSIONS

We have described several structural shape description techniques. The syntactic techniques give us a mathematical formalism plus a well-known kind of recognition algorithm for each kind of grammar. They suffer from problems of erroneous segmentations and are often limited when they try to express a two-dimensional shape as a one-dimensional string. Relational descriptions are more robust, but also more complex. Relational matching algorithms tend to be NP-complete problems. The recent popularity of relaxation algorithms to reduce tree search time has helped make relational shape matching a feasible approach to shape analysis.

REFERENCES

1. Agin, G.J. and Binford, T.O., "Computer Description of Curved Objects", IEEE Transactions of Computers, Vol C-25, No. 4, 1976.

2. Alt, F.L., "Digital Pattern Recognition by Moments", JACM, April 1962, pp. 240-258.

3. Barrow, H.G., Ambler, A.P. and Burstall, R.M., "Some Techniques for Recognizing Structure in Pictures", in Frontiers of Pattern Recognition, S. Watanabe, ed. Academic Press, New York, 1972, pp. 1-29.

4. Dacey, M.F., "The Syntax of a Triangle and Some Other Figures", Pattern Recognition 2, 1970, pp. 11-31.

5. Granlund, G.H. "Fourier Preprocessing for Hand Print Character Recognition", IEEE Transactions on Computers, Vol. C-21, Feb. 1972, pp. 195-201.

6. Hu, M.K., "Visual Pattern Recognition by Moment Invariants", IRE Transactions on Information Theory, Feb. 1962, pp. 179-187.

7. Kirsch, R.A., "Computer Interpretation of English Text and Picture Patterns", IEEE Transactions on Electronic Computers, Vol. EC-13, 1964, pp. 363-376.

8. Milgram, D.M. and Rosenfeld, A., "Array Automata and Array Grammars", IFIP Congress 71, Booklet TA-2, North-Holland, Amsterdam, 1971, pp. 166-173.

9. Nevatia, R. and Benford, T.O., "Description and Recognition of Curved Objects", Artificial Intelligence 8, 77-98, North-Holland Publishing Co., 1977.

10. Pavlidis, T. and Ali, F., "A Hierarchical Syntactic Shape Analyzer", IEEE Transactions on Pattern Analysis and Machine Intelligence, Vol. PAMI-1, No. 1, Jan. 1979, pp. 2-9.

11. Persoon, E., and Fu, K.S., "Shape Discrimination Using Fourier Descriptors", IEEE Transactions on Systems, Man, and Cybernetics, Vol. SMC-7, No. 3, March 1977, pp. 170-179.

12. Richard, C.W. and Hemami, H., "Identification of Three-Dimensional Objects Using Fourier Descriptors of the Boundary Curve", IEEE Transactions on System, Man, and Cybernetics, Vol. SMC-4, No. 4, July 1974, pp. 371-378.

13. Salton, G., Dynamic Information and Library Processing, Prentice-Hall, Englewood Cliffs, NJ, 1975.

14. Shapiro, L.G., "A Structural Model of Shape", IEEE Transactions on Pattern Analyses and Machine Intelligence, Vol. 2, No.

15. Shapiro, L.G. and Baron, R.J., "ESP^3: A Language for Pattern Description and a System for Pattern REcognition", IEEE Transactions on Software Engineering, March 1977.

16. Shapiro, L.G. and Haralick, R.M., "Decomposition of Two-Dimensional Shapes by Graph-Theoretic Clustering", IEEE Transactions on Pattern Analysis and Machine Intelligence, Vol. 1, No. 1, Jan. 1979, pp.

17. Shapiro, L.G. and Haralick, R.M., "Structural Descriptions and Inexact Matching", to appear in IEEE Transactions on Pattern Analysis and Machine Intelligence, 1981.

18. Zahn, C.T. and Roskies, R.Z., "Fourier Descriptors for Plane Closed Curves", IEEE Transactions on Computers, Vol. C-21, No. 3, March 1972, pp. 269-281.

SHAPE GRAMMAR COMPILERS

Thomas C. Henderson

Institut für Nachrichtentechnik
Deutsche Forschungs- und Versuchsanstalt für Luft-
und Raumfahrt (DFVLR) Oberpfaffenhofen
8031 Weßling, BRD

A general shape grammar scheme is proposed for producing a
shape parsing mechanism by means of a shape grammar compiler from
a high-level shape grammar description. In particular, we show
how such a relation can be defined between a special class of
shape grammars and a parsing mechanism called a hierarchical constraint process. Thus, we propose: a shape grammar formalism
which accounts for most aspects of 2-D shape, a parsing mechanism
which uses constraints between pieces of the shape, and an automatic method to compute constraint relations between the vocabulary symbols of the shape grammar.

INTRODUCTION

Many syntactic shape models have been proposed, e.g., see
Shaw (12), Fu and Bhargava (4), and Rosenfeld and Milgram (10),
but the choice of parsing algorithms for these models has received scant attention; an exception to this is Stockman (13).
However, if any practical result is expected from the syntactic
approach, it is necessary to establish the relation between the
shape grammar and the parsing mechanism. In this paper we restrict our attention to bottom-up shape parsing, and we show how
such a relation can be defined between a special class of shape
grammars and a parsing mechanism called a hierarchical constraint
process.

In choosing the parsing mechanism for a given shape grammar,
the problem is much the same as that faced by a compiler writer
trying to choose a recognizer for a string grammar. In the traditional bottom-up parsing approach (see Gries (5) or Aho and

327

J. C. Simon and R. M. Haralick (eds.), Digital Image Processing, 327–336.
Copyright © 1981 by D. Reidel Publishing Company.

Ullman (1)), a recognizer is implemented in a general way using tables. These tables are derived from the given grammar and describe relations between the vocabulary symbols of the grammar. A constructor is designed which, given a grammar, checks it for suitability and builds the necessary tables for the recognizer. That is, to implement the recognizer for a given grammar, the constructor is run with the grammar as data, and the output is merged with the recognizer. We will show that the hierarchical constraint process (see Henderson (7)) is an extension of this approach from string grammars to shape grammars.

The general shape grammar scheme is to produce a shape parsing mechanism by means of a shape grammar compiler from a high-level shape grammar description. The shape parsing mechanism performs the actual analysis of unknown shapes and outputs an organization imposed on the shape primitives in terms of the underlying grammar. This is analogous to using a compiler compiler to produce a string grammar compiler from a high-level programming language description.

Most syntactic shape analysis methods proposed have dealt with the shape grammar (or model) at length, while the corresponding parsing algorithm has been chosen ad hoc and from a string grammar perspective, e.g., You and Fu (16) use Earley's algorithm. The shape parsing mechanism has usually been constructed manually. Finally, in most formalisms it is a tedious process to produce the shape grammar for any interesting class of shapes. This provides an impetus for developing more suitable user-oriented languages for describing a shape grammar. Thus, one problem faced is the design of suitable shape grammar description languages, and the subsequent problem is the construction of correct and efficient compilers for such languages.

As an example of this approach to syntactic shape analysis, we propose a class of constraint grammars and a method for automatically deriving the shape parsing mechanism. In particular, we

- define a shape grammar formalism which accounts for most aspects of 2-D shape,
- define a parsing mechanism which uses constraints between pieces of the shape, and
- define an automatic method to compute constraint relations between the vocabulary symbols of the grammar.

This process can be viewed as a generalization of traditional precedence grammar techniques in that the precedence grammars involve constraints between string grammar symbols. With string grammars, bottom-up parsing involves scanning from left to right until the tail of the handle is found, then scanning right to left from the tail until the head of the handle is found. This works well enough for string grammars, but shape grammars pose the problem of complicated relations between the symbols, and these relations must be accounted for and taken advantage of by the shape parsing mechanism.

CONSTRAINT GRAMMARS

Grammatical models for shape analysis have been developed and investigated by many people. With some simple modifications these models can be integrated in a natural way with constraint analysis techniques. An extension of the geometrical grammars of Vamos (14) will be used to model shapes.

A stratified context-free grammar, G, is a quadruple (T,N,P,S), where T is the set of terminal symbols, N is the set of non-terminal symbols, P is the set of productions, and S is the start symbol. Let $V = (N \cup T)$ be the set of all vocabulary symbols. Associated with every symbol $v \in V$ is a level number, $\ln(v):V \to \{0,1,\ldots,n\}$, where $\ln(S)=n$, and $\forall v \in T$, $\ln(v)=0$.

T consists of symbols which correspond to relatively large pieces of the shapes modeled by the grammar, e.g., straight-edge approximations to the boundary of the shape.

N consists of symbols each of which has a level number from 1 to n associated with it. In any rule $v := a$ (the rewrite part of a production), if $\ln(v)=k$, then every symbol in the string a is at level k-1. Furthermore, $\forall v \in V$

v = <name part> {attachment part} [semantic part], where

<name part> is a unique name by which the symbol v is known, {attachment part} is a set of attachment points of the symbol, [semantic part] is a set of predicates which describes certain aspects of the symbol.

P consists of productions of the form $(v:=a,A,C,Ga,Gs)$, and $v:=a$ is the rewrite part that indicates the replacement of the symbol v by the group of symbols a, where $v \in N$ and $a=v1v2\ldots vk$ such that $vi \in V$ and $\ln(vi)=(\ln(v)-1)$ for $i = 1,k$; A is a set of applicability conditions on the syntactic arrangement of the vi; C is a set of semantic consistency conditions on the vi and consists of various predicates describing geometric and other properties of the vi; Ga is a set of rules for generating the attachment part for v, the new symbol; Gs is a set of rules for generating the semantic part of v, the new symbol.

Stratified grammars have been described in detail elsewhere (Henderson (6)). These grammars give rise to a large set of

contextual constraints on the organization of a shape. It is
these constraints which must be derived so that a constraint
oriented parsing mechanism can be constructed.

SHAPE GRAMMAR COMPILATION

We now discuss the procedures for deriving the constraints
from the grammar. Two types of constraints, syntactic and semantic, are described. The semantic attributes of a vocabulary symbol are computed from the attributes of the symbols which produce
it. Knuth (8) gives a way to define semantics for context-free
string languages using both inherited and synthesized attributes;
however, the shape grammar formalism here is closer to property
grammars of Aho and Ullman (1) in that properties are associated
with the vocabulary symbols.

Consider a vocabulary symbol as representing a piece of the
boundary of a 2-D shape. If a vocabulary symbol is part of a complete shape, then it is adjacent to pieces of the shape which can
combine with it to produce a higher level vocabulary symbol. Suppose that the set of all possible neighbors of a vocabulary symbol, v, at any given attachment point is known, and that a silhouette boundary includes v. If at one of v's attachment points
the attached symbol is not in the neighbor set of v, then the
silhouette boundary cannot produce the start symbol. This type of
constraint is called a syntactic constraint. Without these constraints, several levels of vocabulary symbols might be built
before it can be determined that some vocabulary symbol lacks the
appropriate context. The use of constraints, however, makes it
possible to detect this much earlier.

The other type of constraint involves some (usually geometric) relation between the semantic features of two vocabulary
symbols. E.g., let $v = \langle v \rangle \{a,b\}[s]$, and let $w = \langle w \rangle \{c,d\}[t]$; then
it may be the case that s is parallel to t in any parse producing
the start symbol. These kinds of constraints are called semantic
constraints. This also makes it possible for high level information to be specified, e.g., the orientation of some high level
part of the shape, and this information can be used to eliminate
silhouette boundaries that are not consistent with this information.

Let $G = (.T,N,P,S)$, let v,w, and $x \in V$, let $at(v)$ denote the
attachment points of v, and let $av \in at(v)$. The syntactic constraints are defined by specifying for each symbol v a set of vocabulary symbols which may be attached to v. Note that, in general, the attachment points of the symbols must also be specified. To determine this set, we define:

(v,av) Ancestor (w,aw) iff ∃p∈P such that the rewrite rule
of p is v := ...w..., and ∃ aw∈at(w) such that aw is identified
with av in Ga of p. Then we say that v is an ancestor of w
through attachment point av of v and aw of w, where av and aw
represent the same physical location.

(w,aw) Descendent (v,av) iff (v,av) Ancestor (w,aw).

(v,av) Neighbor (w,aw) iff
    a) ∃p∈P such that the rewrite rule of p is x := ..v..w..,
and aw is specified as being joined to av in A of p, or
    b) ∃ x∈V with ax∈at(x), and ∃y∈V with ay∈at(y) such that
(x,ax) Ancestor (v,av) and (y,ay) Neighbor (x,ax) and (w,aw)
Descendent (y,ay).

Using matrix representations for these relations, the de-
scendents and neighbors of a symbol at a particular attachment
point can be computed. See Gries (5) for an introduction to binary relations, their representation using matrices and their
manipulation. Let s be the number of vocabulary symbols in V, and
let the Boolean matrix Amn be the square matrix of order s whose
(i,j)th entry is 1 iff symbol vi is in relation A to symbol vj
through endpoint m of vi and endpoint j of vj (consider the endpoints of vocabulary symbols to be ordered). A relation (which is
dependent on endpoints) is then fully specified by a total of
k·k matrices, where k is the number of endpoints per symbol. However, if the grammar is written so that endpoints are interchang-
able, then one matrix will define the relation, i.e., all k·k matrices are the same.

The Ancestor relation, Amn, is the transitive closure of the
matrix having a 1 in the (i,j)th position if the condition given
in the definition is satisfied. The Descendent relation, Dmn, is
just the transpose of Amn. Given Amn and Dmn, the Neighbor relation, Nmn, is computed as:

$$Nmn := Nmn + \sum_p \{Dmp * [\sum_q (Npq * Aqn)]\},$$

where + is Boolean "or" and * is Boolean "and", and Npq is just
the explicit neighbors given in the productions. This definition
of Nmn follows directly from the definition of the relation. The
first term on the right, Nmn, corresponds to part a) of the nei-
bor relation, i.e., this is the neighbor relation defined explicitly in the productions. Let DNAmn denote the second term, i.e.,

$$DNAmn(i,j) = \sum_p \{Dmp * \sum_q (Npq \ Aqn)\}$$

where Rmn(i,j) denotes (i,m) R (j,n) for relation R. Let NApn denote the second term of DNAmn, i.e.,

$$NApn(i,j) = \sum_q (Npq*Aqn).$$

Finally, fix q and consider the computation of the (i,j)th entry of the qth summand:

$$\sum_k \{Npq(i,k)*Aqn(k,j)\}.$$

Then each element of this sum is of the form:

(i,p) N (k,q) and (k,q) A (j,n)

which means that vocabulary symbol i at attachment point p is a neighbor of k at point q which is an ancestor of j at point n. Thus, NApn(i,j)=1 means that vocabulary symbol i neighbors an ancestor of j at attachment point p of i and n of j. Likewise, DNAmn(i,j)=1 means that vocabulary symbol i at attachment point m is a descendent of some neighbor (called x in the definition of the neighbor relation) of some ancestor (called y in the definition) of j at point n. This is precisely what the definition calls for. This relation defines a set of neighbors for every symbol, and if a vocabulary symbol in the silhouette boundary fails to have a neighbor from this set, then that silhouette boundary fails to produce the start symbol. This relation constitutes the syntactic constraints.

Semantic constraints can be generated in much the same way as syntactic constraints: by defining binary relations and computing their transitive closure. For example, the axes of two symbols are parallel if a production states this explicitly or by transitivity through some third symbol. Such relations also allow for semantic features to be fixed (set to some constant), e.g., the orientation of one axis of a vocabulary symbol could be set to 45 degrees, and this certain information can be propagated to other vocabulary symbols. This can be done for example, by having global information available describing known orientations of the vocabulary symbols. In this way, it is possible to determine that certain boundary silhouettes cannot be parsed. The parallel relation was computed between all vocabulary symbols. Note that some symbols may not be parallel to any symbol; moreover, some hypothesized symbols may require other hypotheses of the same symbol to exist. The parallel relation was computed using a binary-valued matrix, whose rows and columns correspond to the axes of the vocabulary symbols.

In general, a transitive relation is computed as:

    P := {P(0)+I̲}*P(0)!

where I̲ is the complement of the Boolean identity matrix, and
P(0)! is the transitive closure of P(0), the explicit parallel
relation. (An explicit relation is one found directly in the productions.) Computed this way, a symbol is only parallel to itself
if there must exist two distinct vocabulary symbols of the same
name part in the boundary silhouette. Relations which are not
transitive, e.g., perpendicular, require special procedures for
their computation, but any transitive relation can be computed
with the above form.

HIERARCHICAL CONSTRAINT PROCESSES

     The discussion so far applies to string grammars as well as
shape grammars. However, a major problem faced by any syntactic
shape analysis method, but not by string parsers, is the ambiguity of the underlying data. Usually no clear-cut decision can be
made in associating the terminal symbols of a grammar with the
shape primitives. Thus, the parsing mechanism must not only overcome the problem of parsing a complicated arrangement of symbols
(i.e., concatenation is no longer the only relation between symbols), but must also disambiguate the interpretation of a given
shape primitive.

     The hierarchical constraint process (or HCP) has been pro-
posed as a solution to this problem (Davis (2), Henderson and
Davis (6) and Henderson (7)). Given a set of shape primitives (a
set of linear segments approximating the boundary of some silhouette) and a stratified shape grammar for the class of shapes,
HCP performs the following actions:

          - associate with each shape primitive a set of possible
            interpretations, i.e., terminal symbols,
          - determine the initial network of hypotheses, that is,for
            each possible interpretation of each shape primitive,in-
            sert a node in the network; two nodes of the network are
            connected if their underlying shape primitives are phy-
            sically adjacent,
          - apply the procedures BUILD, CONSTRAIN and COMPACT to the
            network until the network is empty or the start symbol
            is produced.

     The shape primitives are generated by computing several
piecewise linear approximations to the boundary of the shape. A
modified split-and-merge algorithm (see Pavlidis (9)) fits
straight edges to the boundary using the cornerity measure proposed by Freeman and Davis (3) to choose break points.

The association of terminal symbols with shape primitives will (in the limit) be to hypothesize every terminal symbol for each primitive. However, methods for reducing the number of hypotheses include using a more global analysis to derive indications of appropriate scale, orientation, etc. from simple global properties, e.g., histogram selected features of the primitives themselves to infer properties of particular vocabulary symbols.

The network of hypotheses represents all possible sentential forms for the given shape primitives. Since our grammars define simple closed curves, every cycle in the network represents a distinct set of interpretations of the primitives and must be parsed. However, this is usually much too large a set to be parsed one after the other. The hierarchical constraint process computes a bottom-up parse of all the cycles in parallel. This is done by applying the constraints to the network and can be described by specifying three simple procedures and two sets which these procedures manipulate.

BUILD - Given level k of the network, BUILD uses the productions of the grammar to construct level k+1 nodes. Physically adjacent hypotheses are linked, and a record is kept of which nodes are used to construct each level k+1 node. All level k+1 nodes are put into the CONSTRAIN-SET, and all level k nodes are put in the COMPACT-SET (both of these sets are initially empty).

CONSTRAIN - While the CONSTRAIN-SET is not empty, CONSTRAIN examines each member of that set; if a node fails to satisfy the constraints, then its neighbors are put into the CONSTRAIN-SET, any nodes it helped produce and the nodes used to produce it are put into the COMPACT-SET, and it is deleted from the network.

COMPACT - While the COMPACT-SET is not empty, COMPACT examines each member of that set; given a node n from the COMPACT-SET, if one of the lower level nodes used to produce n has been deleted, or if n has not helped produce a node at the level above it, (and that level has been built), then n's neighbors are put into the CONSTRAIN-SET, any nodes it helped produce and the nodes used to produce n are put into the COMPACT-SET, and n is deleted from the network.

This then is the shape parsing mechanism. The constraint propagation is based on the discrete relaxation techniques developed by Waltz (15) and Rosenfeld et al (11).

DISCUSSION

In analyzing a class of shapes, we proceed as follows:

- define a stratified shape grammar for the class of shapes,
- derive the syntactic and semantic constraints between the vocabulary symbols of the grammar, and
- apply the hierarchical constraint procedure to a set of shape primitives using the constraints to eliminate incorrect hypotheses.

Successful experiments have been run for detecting airplane shapes (see Henderson (7) for results). However, several problems have been encountered. The shape grammar can have 30 to 40 productions, and a convenient means for defining a grammar has yet to be developed. Thus, at present, shape grammars are a major source of error and usually require much debugging. Furthermore, it may be more convenient to provide for attachment elements (instead of points) so that shape primitives might include polygons as well as line segments.

A major problem with the hierarchical constraint process is the initial size of the network of hypotheses. If there are 50 shape primitives and 10 terminal symbols in the grammar, then the initial network has 500 nodes. Each node has much information associated with it, and consequently a lack of storage may result. However, in such a case, it is possible to define a partial network of hypotheses by choosing one hypothesis at a time for a given primitive (or primitives) and running HCP on the resulting network separately.

It might be questioned whether stratification of the grammar is necessary. Without stratification, constraints cannot be applied usefully since a symbol will continually be rebuilt even though it fails to satisfy contextual constraints at some later point. Stratification ensures that a hypothesis will be made only once, and if it is incorrect, then it will be permanently deleted.

Thus, we have shown one possible approach to constructing a theory of shape parsing. While drawing heavily on the traditional theory of parsing, the ambiguity of the underlying data requires a different parsing strategy. We have shown how to implement a bottom-up, constraint-driven parsing mechanism and how it relates to traditional string parser theory.

REFERENCES

1. Aho, S. and J.D. Ullman (1973) The Theory of Parsing, Translation and Compiling, Vol. 2, Prentice Hall, New Jersy.
2. Davis, L. (1978) "Hierarchical Relaxation for Shape Analysis," Pattern Recognition and Image Processing Conf., Chicago, Il., June, pp. 275-279.

3. Freeman, H. and L. Davis (1977) "A Corner-Finding Algorithm for Chain Coded Curves," IEEE Trans. on Computers, Vol. C-26, March, pp. 297-303.

4. Fu, K.S. and B. Bhargava (1973) "Tree Systems for Syntactic Pattern Recognition," IEEE Trans. on Computers, Vol. C-22, Dec., pp. 1087-1099.

5. Gries, D. (1969) Compiler Construction for Digital Computers, John Wiley, New York.

6. Henderson, T. and L. Davis (1980) "Hierarchical Models and Analysis of Shape," BPRA 1980 Conf. on Pattern Recognition, Oxford, England, p.79.

7. Henderson, T. (1979) "Hierarchical Constraint Processes for Shape Analysis," Ph.D. Dissertation, U. of Texas, December.

8. Knuth, D. (1968) "Semantics of Context-Free Languages," Math. Systems Theory, Vol. 2, no. 2, pp. 127-145.

9. Pavlidis, T. and S. Horowitz (1973) "Piecewise Approximation of Plane Curves," Pattern Recognition, pp. 346-405.

10. Rosenfeld, A. and D. Milgram (1972) "Web Automata and Web Grammars," in Machine Intelligence (eds. B. Meltzer and D. Michie), 7, Edinburgh U. Press, pp. 307-324.

11. Rosenfeld, A., R. Hummel and S. Zucker (1976) "Scene Labeling by Relaxation Operations," IEEE Trans. on SMC, Vol. SMC-6, pp. 420-433.

12. Shaw, A.C. (1969) "A Formal Picture Description Scheme as a Basis for Picture Processing Systems," Inf. and Con., Vol. 14, pp. 9-52.

13. Stockman, G. (1977) "A Problem Reduction Approach to the Linguistic Analysis of Waveforms," U. of Maryland, TR-538, May.

14. Vamos, T. and Z. Vassy (1973) "Industrial Pattern Recognition Experiment - A Syntax Aided Approach," IJCPR, Wash., pp. 445-452.

15. Waltz, D. (1975) "Understanding Line Drawings of Scenes with Shadows," in The Psychology of Computer Vision, (ed. P. Winston), McGraw-Hill, pp. 19-91.

16. You, K. and K.S. Fu (1977) "Syntactic Shape Recognition," in Image Understanding and Information Extraction, Summary Report of Research, March, pp. 72-83.

A FACET MODEL FOR IMAGE DATA: REGIONS, EDGES, AND
TEXTURE

Robert M. Haralick

Virginia Polytechnic Institute
and State University
Blacksburg Virginia 24061

ABSTRACT

In this paper we present a facet model for image
data which motivates an image processing procedure
that simultaneously permits image restoration as well
as edge detection, region growing, and texture analysis. We give a mathematical discussion of the model,
the associated iterative processing procedure, and
illustrate it with processed image examples.

1.   INTRODUCTION

The world recorded by imaging sensors has order. This
order reflects itself in the regularity of the image
data taken by imaging sensors. A model for image data
describes how the order and regularity in the world
manifests itself in the ideal image and how the real
image differs from the ideal image. In this paper we
describe a facet model for image data and suggest some
procedures for image restoration, segmenting, and texture analysis using the facet model.

J. C. Simon and R. M. Haralick (eds.), Digital Image Processing, 337–356.

The facet model for image data assumes that the spatial domain of the image can be partitioned into regions having certain gray tone and shape properties. The gray tones in a region must all lie in the same simple surface. The shape of a region must not be too jagged or too narrow.

To assure regions which are not too jagged or narrow, the simplest facet model assumes that for each image there exists a $K > 1$ such that each region in the image can be expressed as the union of $K \times K$ blocks of pixels. The value of K associated with an image means that the narrowest part of each of its regions is at least as large as a $K \times K$ block of pixels. Hence, ideal images which have large values of K have very smoothly shaped regions.

To make these ideas precise, let Zr and Zc be the row and column index set for the spatial domain of an image. For any $(r, c) \in Zr \times Zc$, let $I(r, c)$ be the gray value of resolution cell (r, c) and let $B(r, c)$ be the $K \times K$ block of resolution cells centered around resolution cell (r, c). Let $P = \{P(1),\ldots,P(N)\}$ be a partition of $Zr \times Zc$ into its regions.

In the slope facet model, for every resolution cell $(r, c) \in P(n)$, there exists a resolution cell $(i, j) \in Zr \times Zc$ such that:

(1) shape region constraint
   $(r, c) \in B(i, j) \subset P(n)$

(2) region gray tone constraint
   $I(r, c) = a(n)r + b(n)c + g(n)$

The actual image J differs from the ideal image I by the addition of random stationary noise having zero mean and covariance matrix proportional to a specified one.

$$J(r, c) = I(r, c) + n(r, c)$$

where

$$E[n(r, c)] = 0$$

$$E[n(r, c)\, n(r', c')] = ks(r - r', c - c')$$

The flat model of Tomita and Tsuji (1977) and Nagao and Matsuyama (1978) differs from the slope

facet model only in that the coefficients a(n) and
b(n) are assumed to be zero. Nagao and Matsuyama also
use elongated neighborhoods with a variety of orientation. This variety of neighborhoods, of course, leads
to a more general and more complex facet model. A
second way of generalizing the facet model is to have
the facet surfaces be more complex than sloped planes.
For example we could consider polynomial or trignome-
tric polynomial surfaces. In the remainder of this
paper, we consider only the flat facet and the sloped
facet model.

To illustrate the validity of the facet model we
consider the image shown in Figure 1a. Using a 2x2
window and iterating with the slope facet procedure
described in the next section to a fixed point, there
results the image shown in Figure 1b after 20 iterations. The logarithm of the absolute value of the
difference between the original and the 2x2 slope
facet image is shown in Figure 1c. To see the effect
of the window size, Figure 1d shows the resulting
fixed point image after 16 iterations of the slope
facet procedure. The logarithm of the absolute value
of the difference between the original and the 3x3
slope facet image is shown in Figure 1e. It is clear
that as the window size increases, the error increases
and becomes more spatially correlated. Notice that
most of the error occurs around region edges. In Figure 1d one can begin to see region edges becoming just
a little jagged due to the fact that the 3x3 window is
too large of a window for the slope facet model for
this image.

These kinds of experiments have been repeated with
other kinds of images with similar results. More work
needs to be done to determine the best compromise between window size and complexity of the function fitting each facet. We will be reporting on such studies
in a future paper.

$\hat{J}(-1,-1)$

8	5	2
5	2	-1
2	-1	-4

$\hat{J}(-1,0)$

5	5	5
2	2	2
-1	-1	-1

$\hat{J}(-1,1)$

2	5	8
-1	2	5
-4	-1	2

$\hat{J}(0,-1)$

5	2	-1
5	2	-1
5	2	-1

$\hat{J}(0,0)$

2	2	2
2	2	2
2	2	2

$\hat{J}(0,1)$

-1	2	5
-1	2	5
-1	2	5

$\hat{J}(1,-1)$

2	-1	-4
5	2	-1
8	5	2

$\hat{J}(1,0)$

-1	-1	-1
2	2	2
5	5	5

$\hat{J}(1,1)$

-4	-1	2
-1	2	5
2	5	8

Figure 1 shows the filtering masks to be used for
least squares estimation of the gray
value for any position in a 3 x 3 block.
Each mask must be normalized by dividing
by 18.

Figure 1a shows the original.

Figure 1b shows the fixed point 2 x 2
      slope facet (20 iterations).

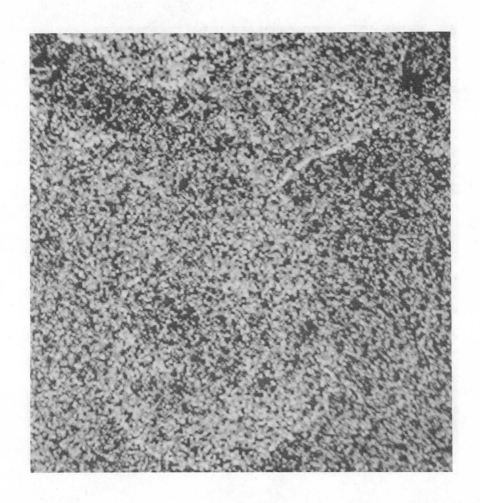

Figure 1c shows the log error at 2 x 2
        fixed point (white high).

Figure ld shows the fixed point 3 x 3
        slope facet (16 iterations).

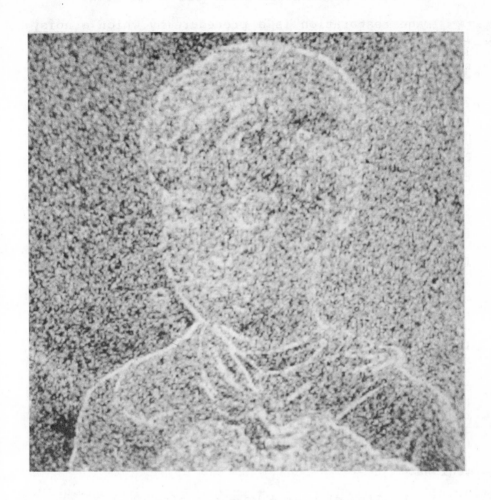

Figure 1e shows the log error at 3 x 3
        fixed point (white high).

2. IMAGE RESTORATION UNDER THE FACET MODEL

Image restoration is a procedure by which a noisy image is operated on in a manner which produces an image which has less noise and is closer to the form of the ideal image than the observed image is. Iterative and relaxation techniques have become important techniques for achieving this (Rosenfeld, 1977; Rosenfeld, 1978; Overton and Weymouth, 1979; Lev et. al. 1977). The facet model suggests the following simple non-linear filtering procedure. Each resolution cell is contained in K^2 different K x K blocks. The gray tone distribution in each of these blocks can be fitted by either a flat horizontal plane or a sloped plane. One of the K x K blocks has smallest error of fit. Set the output gray value to be that gray value fitted by the block having smallest error. For the flat facet model this amounts to computing the variance for each K x K block a pixel participates in. The output gray value is then the mean value of the block having smalllest variance.

The filtering procedure for the sloped facet model is more complicated and we give a derivation here of the required equations. We assume that the block lengths are odd so that one of the block's pixels is its center. Let the block be (2L + 1) x (2L + 1) with the upper left-hand corner pixel having relative row column coordinates (-L, L), the center pixel having relative row column coordinates (0, 0), and the lower right-hand corner pixel having relative row column coordinates (L, L). Let J(r, c) be the gray value at row r column c. According to the sloped facet model, for any block entirely contained in one of the image regions.

$$J(r, c) = ar + bc + g + n(r, c)$$

where n(r, c) is the noise.

A least squares procedure may be used to determine the estimates for a, b, and g.

Let f(a, b, g) =

$$\sum_{r = -L}^{L} \sum_{c = -L}^{L} (ar + bc + g - J(r, c))^2.$$

The least squares estimate for a, b, and g are those which minimize f. To determine these values, we take the partial derivative of f with respect to a, and b, set these to zero and solve the resulting equations for a, b, and g. Doing this we obtain:

$$a = \frac{3}{L(L+1)(2L+1)^2} \sum_{r=-L}^{L} r \sum_{c=-L}^{L} J(r, c)$$

$$b = \frac{3}{L(L+1)(2L+1)^2} \sum_{c=-L}^{L} c \sum_{r=-L}^{L} J(r, c)$$

$$g = \frac{1}{(2L+1)^2} \sum_{r=-L}^{L} \sum_{c=-L}^{L} J(r, c)$$

The meaning of this result can be readily understood for the case when the block size is 3 x 3. Here L = 1 and

$$a = \frac{1}{6} [J(-1, *) - J(1, *)]$$

$$b = \frac{1}{6} [J(*, -1) - J(*, 1)]$$

$$g = \frac{1}{9} J(*, *)$$

where an argument of J taking the value dot means that J is summed from -L to L in that argument position. Hence, a is proportional to the slope down the row dimension, b is proportional to the slope across the column dimension, and g is the simple gray value aver-

age over the block. Figure 2 illustrates the masks
that may be used to compute a, b, and g for a couple
window sizes.

The fitted gray tone for any resolution cell
(r, c) in the block is given by

$$H(r, c) = ar + bc + g$$

For the case where L=1,

$$H(r, c) = [J(1, *)-J(-1, *)]r/6$$

$$+ [J(*, -1) - J(*, 1)]c/6$$

$$+J(*, *)/9$$

Writing this expression out in full:

$$H(r, c) = \{J(-1,-1) \ (-3r - 3c + 2)$$

$$+ J(-1, 0) \ )-3r +2)$$

$$+ J(-1, 1) \ (-3r + 3c + 2)$$

$$+ J(0, -1) \ (-3c +2)$$

$$+ J(0, 1) \ (3c + 2)$$

$$+ J(1, -1) \ (3r - 3c + 2)$$

$$+ J(1, 0) \ (3r + 2)$$

$$+ J(1, 1) \ (3r + 3c + 2)\}/18$$

This leads to the set of linear filter masks shown in
Figure 2 for fitting each pixel position in the 3 x 3
block.

The sloped facet model noise filtering would examine each of the K x K blocks a pixel (r, c) belongs
to. For each block, a block error can be computed by

$$e^2 = \sum_{r = -L}^{L} \sum_{c = -L}^{L} (J(r, c) - J(r, c))^2$$

One of their K^2 blocks will have lowest error. Let (r*, c*) be the coordinates of the pixel (r, c) in terms of the coordinate system of the block having smallest error. The output gray value at pixel (r, c) is then given by H(r*, c*) where H is the affine function estimating the gray values for the block having the smallest error of fit.

Haralick and Watson (1979) prove convergence of this iteration procedure for any size or set of neighborhood slopes.

3. REGION AND EDGE ANALYSIS

Edge detection and region growing are two areas of image analysis which are opposite in emphasis but identical in heart. Edges obviously occur at bordering locations of two adjacent regions which are significantly different. Regions are maximal areas having similar attributes. If we could do region analysis, then edges can be declared at the borders of all regions. If we could do edge detection, regions would be the areas surrounded by the edges. Unfortunately, we tend to have trouble doing either: edge detectors are undoubtedly noisy and region growers often grow too far.

The facet model permits an even handed treatment of both. Edges will not occur at locations of high differences. Rather, they will occur at the boundaries having high differences between the parameters of sufficiently homogeneous facets (Haralick, 1980). Regions will not be declared at just areas of similar value of gray tone. They will be the facets: connected areas whose resolution cells yield minimal differences between of region parameters. Here, minimal means smallest among a set of resolution cell groupings. In essence, edge detection and region analysis are identical problems that can be resolved with the same procedure and in this section we describe how.

Recall that the facet model iterations produce the parameters a and b. The fact that the parameters a and b determine the value of the slope in any direction is well known. For a planar surface of the form:

$$g(r, c) = ar + bc + g$$

the value of the slope at an angle t to the row axis is given by the directional derivative of g in the direction t. Since a is the partial derivative of g with respect to r and b is the partial derivative of g with respect to c, the value of the slope at angle t is [a cos t + b sin t]. Hence, the slope at any direction is an appropriate linear combination of the values for a and b. The angle t which maximizes this value satisfies

$$\cos t = \frac{a}{a^2 + b^2} \quad \text{and} \quad \sin t = \frac{b}{a^2 + b^2}$$

and the gradient which is the value of the slope in the steepest direction is

$$a^2 + b^2 .$$

The sloped-facet model is an appropriate one for either the flat world or sloped world assumption. In the flat world each ideal region is constant in gray tone. Hence, all edges are step edges. The observed image taken in an ideal flat world is a defocussed version of the ideal piecewise constant image with the addition of some random noise. The defocussing changes all step edges to sloped edges. The edge detection problem is one of determining whether the observed noisy slope has a gradient significantly higher than one which could have been caused by the noise alone. Edge boundaries are declared in the middle of all significantly sloped regions.

In the sloped facet world, each ideal region has a gray tone surface which is a sloped plane. Edges are places of either discontinuity in gray tone or derivative of gray tone. The observed image is the ideal image with noise added and no defocussing. To determine if there is an edge between two adjacent pixels, we first determine the best slope fitting neighborhood for each of the pixels by the iteration facet procedure. Edges are declared at locations having significantly different planes on either side of them.

This model does not take into account defocussing and, therefore, does not recognize whether or not

highly sloped regions are edges. The determination of
whether a sloped region is an edge region or not may
depend on the significance and magnitude of the slope
as well as the semantics of the image. One of the
important research goals of the facet model is to work
out this kind of problem.

In either the case of the noisy defocussed flat
world, or the noisy sloped world we are faced with the
problem of estimating the parameters of a sloped surface for a given neighborhood and then calculating the
significance of the difference of the estimated slopes
of two adjacent neighborhoods. To do this we proceed
in a classical manner. We can use a least square
procedure to estimate parameters and we can measure
the strength of any difference by an appropriate
F-statistic.

In summary, the facet image restoration iteration
procedure produces more than just a restored gray
tone. For each pixel, it also produces the
a, b, and g parameters. Using these parameters we can
determine whether or not neighboring pixels lie in the
same connected facet. Because the parpmeters come not
form the pixel's central neighborhood but from the
pixel's best neighborhood, the determination of
whether two pixels lie in the same facet requires that
the parameters for each pixel be taken out of their
individual relative coordinate systems and be placed
in the same coordinate system. Linking together
neighboring pixels with the same a, b, g paramenters,
permits us to identify the facets which are characterized by the connected sets of pixels that constitute
them. These facets become the regions and edges are
the boundaries between regions.

4.   TEXTURE ANALYSIS

Following Haralick (1979) textures can be classified as being weak textures or strong textures. Weak
textures are those which have weak spatial-interaction
between the texture primitives. To distinguish between them it may be sufficient to only determine, for
each pair of primitives, the frequency with which the
primitives co-occur in a specified spatial relationship. In this section we discuss a variety of ways in
which primitives from the facet model can be defined

and the ways in which relationships between primitives
can be defined.

4.1   PRIMITIVES

A primitive is a connected set of resolution cells
characterized by a list of attributes. The simplest
primitive is the pixel with its gray tone attribute.
Sometimes it is useful to work with primitives which
are maximally connected sets of resolution cells having a particular property. An example of such a primitive is a maximally connected set of pixels all having the same gray tone or all having the same edge
direction.

Gray tones and local properties are not the only
attributes which primitives may have. Other attributes include measures of shape of a connected region
and homogeneity of its local property. For example, a
connected set of resolution cells can be associated
with its length or elongation of its shape or the variance of its local property.

Attributes generated by the facet model include
the a, b, and g parameters plus the average error of
fit for the facet. These attributes can be used by
themselves or used to generate derived attribute
images such as that created from a^2+b^2 . The relative extreme primitives can be defined in the following way:

Label all pixels in each maximally (minimally)
connected relative maxima (minima) plateau with an
unique label. Then label each pixel with the label of
the relative maxima (minima) that can reach it by a
monotonically decreasing (increasing) path. If more
than one relative maxima (minima) can reach it by a
monotonically decreasing (increasing) path, then label
the pixel with a special "c" for common. We call the
regions so formed the descending components of the
image.

4.2   SPATIAL

Once the primitives have been constructed, we have available a list of primitives, their center coordinates, and their attributes. We might also have available some topological information about the primitives, such as which are adjacent to which. From this data, we can select a simple spatial relationship such as adjacency of primitives or nearness of primitives and count how many primitives of each kind occur in the specified spatial relationship.

More complex spatial relationships include closest distance or closest within an angular window. In this case, for each kind of primitive situated in the texture, we could lay expanding circles around it and locate the shortest distance between it and every other kind of primitive. In this case our co-occurrence frequency is three dimensional, two dimensions for primitive kind and one dimension for shortest distance. This can be dimensionally reduced to two dimensions by considering only the shortest distance between each pair of like primitives.

Co-occurrence between properties of the descending components can be based on the spatial relationship of adjacency. For example, if the property is size, the co-occurrence matrix could tell us how often a descending component of one size is next fo a descending component of another size.

To define the concept of generalized co-occurrence, it is necessary to first decompose an image into its primitives. Let Q be the set of all primitives on the image. Then we need to measure primitive properties such as parameter value, region, size shape, etc. Let T be the set of primitive properties and f be a function assigning to each primitive in Q a value for each property of T. Finally, we need to specify a spatial relation between primitives such as distance or adjacency. Let SC Q x Q be the binary relation pairing all primitives which satisfy the spatial relation. The generalized co-occurrence matrix P is defined by:

P(t1, t2) =

$$\frac{\#\{(q1, q2) \in S \mid f(q1)=t1 , f(g2) = t2\}}{\#S}$$

P(tl, t2) is just the relative frequency with which
two primitives occur with specified spatial relationship in the image, one primitive having property tl
and the other primitive having property t2.

Zucker (1974) suggests that some textures may be
characterized by the frequency distribution of the
number of primitives any primitive has related to it.
This probalility p(k) is defined by:

$$p(k) = \frac{\#\{(q \in Q \mid \#S(q) = k\}}{\#Q}$$

Although this distribution is simpler than co-occurrence, no investigation appears to have used it in
texture discrimination experiments.

5. CONCLUSION

In this paper we considered the gray tones of an
image to represent the height of a surface above the
row-column coordinates of the gray tones.   The
observed image is then the surface of the underlying
ideal image plus random noise. The ideal image is
composed of a patchwork of constrained surfaces sewed
together.

We called each patch a facet and in the ideal
image, the facets must satisfy the constraints of the
facet model for image data: the facet model constrains the shape of each facet to be exactly composed
as a union (possibly over-lapping) of a given set of
neighborhood shapes and it constrains the surface to
be a sloped plane surface (or some other more general
polynomial surface).

The goal of image restoration is to recover the
ideal gray tone surface which underlies the observed
noisy gray tone surface. Although the noise prevents
recovering the precise underlying ideal surface, we
can recover that gray tone surface which is the "closest surface" to the observed noisy surface and which
also satisfies the facet model constraints.

The procedure we suggest for recovering the underlying surface consists of iterating best neighborhood least squares fits.

Associated with each given pixel is a set of all the neighborhoods of given shapes that contain it. Each one of these neighborhoods can be fitted with the best fitting surface. One of these neighborhoods will have a best fitting surface whose residual error is smallest among all the neighborhoods tried. The parallel iterative procedure consists of replacing each pixel gray tone intensity with the height of the best fitting surface in its lowest residual error neighborhood. The procedure is guaranteed to converge and actually achieves essential convergence in a few iterations. The resulting image is an enhanced image having less noise, better contrast, and sharper boundaries.

Image restoration is not the only use of the facet model. The facet model processing provides us with additional important information. By collecting together all pixels participating in the same surface facet, we transformed the pixel as our processing and analysis unit into the surface facet as our processing and analysis unit. Now edge boundaries, for example, can be defined to occur at the shared boundary of all neighboring facets whose surface paramenters are significantly different. Homogeneous regions can be defined by linking together all those neighboring surface facets whose parameters are significantly the same. Texture can be characterized by the co-occurrence statistics of neighboring primitives which are not the pixel gray tones as in the usual occurrence approach but which are the facets characterized by their boundary, shape, size, and surface parameter attributes.

Our paper has been mainly theoretical laying out a variety of uses of the facet model in image processing. Future papers will describe our experimental results.

REFERENCES

Haralick, R. M. 'Statistical and Structural Approaches
     to Texture', Proceeding of the IEEE Vol 67, No. 5,
     May 1979, p 786-804.

Haralick, R. M. and L. Watson 'A Facet Model For Image
     Data', Proceeding IEEE Conference on Pattern
     Recognition and Image Processing Chicago, Ill.,
     August 1979, p 489-497.

Haralick R. M., 'Edge and Region Analysis for Digital
     Image Data', Computer Graphics and Image Processing
     Vol. 12, No. 1, January 1980, p 60-73.

Lev, A., S. W. Zucker, and A. Rosenfeld, 'Iterative
     Enhancement of Noisy Images', IEEE Transactions on
     Systems, Man, and Cybernetics Vol. SMC-7, No. 6,
     June 1977, p 435,442.

Nagao, M. and T. Matsuyama, 'Edge Preserving
     Smoothing', Computer Graphics and Image Processing
     Vol 9, 1979, p 394-407.

Overton K. J., and T. E. Weymouth, 'A Noise Reducing
     Preprocessing Algorithm' Proceeding IEEE Conference
     on Pattern Recognition and Image Processing
     Chicago, Ill., August 1979, p 498-507.

Rosenfeld, A. 'Iterative Methods In Image Analysis',
     Proceedings IEEE Conference on Pattern Recognition
     and Image Processing Troy, N.Y., June 1977, p
     14-18.

Rosenfeld, A. 'Relaxation Methods in Image Processing
     and Analysis', Proceedings of the 4th International
     Joint Conference on Pattern Recognition Kyoto,
     Japan, November 1978, p. 181-185.

Tomita, F. and S. Tsuji, 'Extraction of Multiple
     Regions by Smoothing in Selected Neighborhoods',
     IEEE Transactions on Systems Man, and, and
     Cybernetics Vol SMC-7, No. 2, February 1977 p.
     107-109.

PATCHED IMAGE DATABASES*

Steven L. Tanimoto

Univ. of Washington, Seattle WA, USA

Large images or many images are common in LANDSAT and CAT-
scan analysis, and VLSI design. A technique is described here
for manipulating large images as collections of smaller ones.
This general method begins by the user specifying areas of interest in his image(s). The areas are then covered (automatically
or interactively) with patches of a convenient size. An index to
the collection of patches is created. This database is then used,
with the help of an image operating system, in lieu of the original data for the user's application. Data compaction and faster
access, particularly for interactive applications, are benefits
of this method.

INTRODUCTION

The management of large image databases is becoming more of
a problem than figuring out how to analyze or use the data in some
applications. For example, in VLSI design (3), the rasterized
chip designs are useful in visual inspection of a design and also
as input to design-rule checking programs. However, the size of
these images (typically 1 million by 1 million pixels) makes it
difficult to interact with them in real-time.

LANDSAT images are often used in studies of ground cover
(e.g. forest monitoring) over particular regions of irregular
shape (e.g. Mt. Baker National Forest in Washington). Consequently many of the pixels in the original images (those outside the
National Forest) are irrelevant for the application. Definition

*This work was supported under NSF Grant ENG-79-09246

J. C. Simon and R. M. Haralick (eds.), Digital Image Processing, 357–362.
Copyright © 1981 by D. Reidel Publishing Company.

of a relevant subset followed by efficiently representing the subset can reduce both the memory and time requirements for processing the relevant pixels.

Computerized axial tomographical scans threaten to overwhelm
existing image memory and processing capabilities, as not just
single slices but sets of slices (representing body geometry in
3-D), and then sequences of such sets (over time) become more
common. Methods for restricting attention to particular organs
or substructures and working efficiently with them are desirable.

These and other applications motivated the development of a
method for working with images in which data can be represented
at various resolutions, as required by the application. Yet the
method presents a logically uniform view of the data to the user
and his algorithms. The patched-image database technique is applied in two phases: (a) creation of the patched database and (b)
utilization of the database.

For the present, creation is very interesting and most of
this paper discusses the three steps involved in creation of
patched databases. Subset definition, the first of these, may be
done interactively or automatically. For example, in one of our
studies the coastal regions near Puget Sound were selected automatically (2). The second step is covering of the subset with
patches. The objective here is to cover all of the relevant pixels
using as few patches as possible (to save space). Patches are
used because they are easily derived from the original data without loss of information, and they can be easily processed by conventional image processing algorithms or sent to a display screen.
The third step is building an index for the collection of patches.
This, also, may be done automatically or interactively. The purpose of the index is not just to express the spatial relationships
among the patches, but to speed access to pixel information in the
patches. It can also provide defaults or low-resolution data for
pixels outside of the special subset.

SUBSET DEFINITION

Most users know what regions of images are relevant to their
studies. Even when "everything" is relevant, they can usually say
what areas need full detail and what areas need only coarse representation. Therefore, providing users with an interactive facility
to trace out relevant areas in their LANDSAT (or other) images, is
a suitable enough aid for building subsets.

In our experiments with the Puget Sound coastal areas (2,5),
we defined our subset through a combination of thresholding, thinning and expanding operations. Well-known classification and seg-

mentation techniques can be used successfully for defining subsets because the regions need not be exact, as long as they contain the relevant pixels. For example, one may choose a threshold
conservatively, and thus eliminate from the subset only pixels
sure to be irrelevant.

COVERING

    After the subset is defined, it is covered with patches.
Patches are typically 64 by 64 pixels. The covered portions of
the original image are then saved on the patches as a representation of the subset of interest. Even when patches are permitted
to overlap, the combined area of the patches is usually less than
the area of the original image; hence, storage space can be saved.

    Interesting design issues for a system employing covering
are (a) patch shape and size and (b) overlap vs. no overlap in
patch coverage. Square patches are usually simplest and best.
However, if an application requires display of image data on a
non-square screen, patches of a general rectangular shape might
be better there.

    In choosing patch size, one trades off the added efficiency
(in space) of working with smaller patches (for most subsets)
against the increased overhead in managing the larger collections
of patches (5). The size may be selected according to other criteria as well: (a) the amount of available main memory on the
system where the database will be used, and (b) the size of the
display screen for interactive applications, or the size of array
that can be handled by an array processor (if one is available).

    Patch placement (covering) may be done either manually, inter-
actively, or automatically. For small databases, manual covering
is best. However, automatic methods are preferable when the number of patches in the database exceeds 200 or so. Algorithms for
automatic patch placement range from linear-time algorithms that
give fair-to-good solutions, to computationally intractable approaches for the minimum number of patches. Here are discussed
the essential features of the following algorithms: tiling, banded 1-D-optimal, greedy, and optimal.

    Tiling gives a cover by the following procedure: The origi-
nal image is partitioned by a grid into patch-sized regions (except possibly along two borders). Then every region containing
part of the subset is covered by one patch. The solution is
found after only one pass over the data. In the worst case, four
times as many patches as necessary are used (2).

    The banded 1-dimensional-optimal method improves upon til-

ing (5). It permits adjustment of patch placements along one direction (e.g. horizontal). Placing patches optimally along a line
to cover points on the line can be done optimally in linear time.
Thus the banded method also requires only one pass over the data.
It produces slightly more efficient coverings (10% in our experiments) than tiling. Neither tiling nor banded covering produces
overlapping patch placements. One can improve the banded method
slightly by allowing a vertical adjustment of patch placements
after each horizontal band has been processed.

The greedy method works as follows: The first patch is
placed so as to cover the maximum number of pixels in the relevant subset. The covered pixels are then removed from the subset.
The second patch is then placed in similar fashion, pixels removed, etc., until no pixels are left in the subset. This method
occasionally produces efficient coverings. However, it may also
give arbitrarily inefficient coverings, as a result of excessive
overlap (2).

Optimal covering is an NP-complete problem (1). The demonstration of this is an interesting construction that simulates
boolean logic with configurations of points in the plane. The
result of this is that subsets containing looped chains of pixels,
that intersect one another, are very difficult to cover optimally.
Subsets of LANDSAT images that correspond to highway systems, or
subsets of VLSI designs showing certain kinds of wires, may have
such structure.

INDEXING

In order for a collection of patches to provide useful image
information, there must be an index which allows rapid access to
the right patch for a particular query (4). Clearly the positions
of the patches must be stored in order to maintain the correspondence between the patch data and the original image data in and
around the subset.

A simple index is a list of triples (patchx, patchy, patch-
address) giving the coordinates of the southwest corner of the
patch and its address in the (auxiliary -- e.g. tape or disk)
storage system. In order to retrieve information about a particular pixel, the list is scanned to find the (first) patch containing the pixel, the patch is accessed, the correct pixel located,
and its information read.

A quad-tree index (2,4) provides an average access-path
length that is considerably shorter than that provided by the list
of patches. Such an index may be built automatically or interactively. For the latter, the user views his original image

(possibly at reduced resolution) with the patch positions superimposed. He then points to regions (initially the entire picture)
that are to be "split" (partitioned into quadrants for purposes
of the index). The system builds a quad tree that corresponds to
the partition. The objective is to produce a tree whose leaves
correspond to regions of the original image that have just a few
patches impinging upon them. The quad tree can then be used (with
hierarchical search) to find a small set of patches in which the
desired pixel might be found. An automatic method for constructing quad trees performs essentially the same steps as done interactively by the user.

PAGING

    Once the database of patches plus index has been created, it
can be used with the help of an "image operating system". This system is designed to disguise the patched structure of the image
representation. Its job is to automatically page in and out the
patches of the image as they are needed by an application program.
Several concepts from computer operating systems are relevant to
image operating systems: working sets of pages (patches), marked
page lists, page replacement schemes, locality of access, and prepaging.

    Many image processing operations involve local computations on
an image. Some of these could be done by operating on 1 to 5 raster lines at a time. However, others (such as border following,
or interactive display with roaming) do not work well with pages
organized as raster lines. They work much better with pages
corresponding to patches. Furthermore there is a tendency with
these algorithms for an access to one patch to be succeeded by an
access to a neighboring patch. Such locality in access allows
efficient prepaging of data likely to be accessed, reducing de-
lays due to page faults.

REFERENCES

1. Fowler, R. J., Paterson, M. S., and Tanimoto, S. L. The Complexity of Packing and Covering in the Plane and Related Intersection Graph Problems. Tech. Rept. 80-05-02, Dept. Computer
Science, FR-35, Univ. of Wash. Seattle WA 98195, U.S.A. May 1980.

2. Tanimoto, S. L. Covering and Indexing an Image Subset. Proc.
1979 IEEE Conf. on Pattern Recog. and Image Processing, Chicago,
IL, pp. 239-245.

3. Tanimoto, S. L. Color-Mapping Techniques for Computer-Aided

Design and Verification of VLSI Systems. Computers and Graphics (1980, in press).

4. Tanimoto, S. L. Hierarchical Picture Indexing and Description. Proc. 1980 IEEE Workshop on Picture Data Description and Management, Pacific Grove, CA, pp. 103-105.

5. Tanimoto, S. L. and Fowler, R. J. Covering an Image Subset with Patches. Proc. Fifth International Conf. on Pattern Recognition, Miami Beach, FL Dec. 1980.

MAP AND LINE-DRAWING PROCESSING

Herbert Freeman
Rensselaer Polytechnic Institute, Troy, NY, USA

Abstract: Maps are line drawings used to represent data that is
intrinsically two-dimensional in nature. The most common use of
maps is for depicting geographically distributed data, though
maps also find application for representing data based on totally
different spatial coordinates. Once confined to representation
solely on paper, today's maps tend increasingly to be stored in
digital form on magnetic tape or disk, with the "paper" map being
merely a "hard copy" of the stored data. This article discusses
some of the different map forms, data structures, and algorithms
for analysis and manipulation, and describes some of the current
problems relating to the processing of all-digital map data.

1. INTRODUCTION

A map is a special kind of line drawing used to represent
data that is inherently two-dimensional in nature. Although we
tend to think of maps primarily in connection with geographically-
distributed data, maps are, in fact, much more generally used -
finding application also for such diverse forms of data as
electromagnetic antenna radiation patterns and electric-potential
distributions over parts of the human body. Maps play an ever-
increasing role in modern society - as highway maps, terrain
contour maps, maps of population density, maps showing the geographic distribution of personal income, natural resources, or
particular diseases, and maps of political or property boundaries,
to mention only some of the more important.

Until recently maps were drawn or printed on paper, and the
paper was both the storage medium and the display medium for the

363

J. C. Simon and R. M. Haralick (eds.), Digital Image Processing, 363–382.
Copyright © 1981 by D. Reidel Publishing Company.

data. Beginning about twenty years ago, interest developed in
storing map data in digital form, on magnetic tape or disk; the
paper copy was to be used only when "hard-copy" output of the
digitally stored data was desired.

The advantages of storing a map in digital form are impressive.
First of all, it is a simple matter to copy the data and transmit
it rapidly to remote locations without any loss of precision.
Secondly, alterations in the data can be easily made, whether to
correct errors or to incorporate additional data. Access to a
large map data base for information retrieval or data editing can
be quick, either locally or from a remote site. Since the data
is already in machine accessible form, computer analysis or manipulation of map data is easily accomplished. Finally, maps for
different purposes can be readily derived. Thus if one wants a
map of electric power transmission lines in a geographic area
spanning more than one map quadrangle and have the map also show
major highways and rivers but no terrain contours, it is a simple
matter for a computer to compile such a map from the stored data.
Upon compilation, the map can then be either displayed on a computer CRT terminal or delivered in hard-copy form. It can also
itself be stored in a data base to facilitate future access. That
is, with map data available in a computer data base, it is simple
and convenient to generate maps containing any combination of the
many different data categories - contour lines, highways, population centers, land usage markings, drainage lines, annotations,
political boundaries, etc. The maps may cover any specified area,
large or small, and can be displayed to any desired scale.
Furthermore, for data relating to large geographic areas on the
earth, the generation of maps utilizing a particular projections
system (e.g., x-y, Mercator, polar) is easily accomplished.

What are the problems of digital map generation and use?
The first problem is that of selecting a versatile data structure
which permits the map data to be stored compactly but in a way
that facilitates access for all intended purposes. Because maps
normally contain a large amount of data, this is a formidable
problem. The second concerns the development of algorithms for
analyzing and manipulating map data. Again because of the large
amount of data normally involved, considerations of algorithm
complexity take on particular importance. Next there are the
problems associated with displaying the data. The availability
of powerful computers and a great variety of new display devices
makes it possible to render maps in forms that would previously
not have been economically feasible.

Since maps are merely line drawings depicting spatially distributed data, all the many techniques that have been developed
in recent years for representing, manipulating and analyzing line
drawings can be applied to map data processing as well. In this

article we shall first take a close look at the problem of map
representation. Next we shall examine some of the available
techniques for map data analysis and manipulation, and finally
we shall describe some techniques for rendering map data in
different visual forms.

2. THE DIGITAL MAP

The kind of data that may appear on a map can take many different forms. It is possible to show, however, that the data can
be classified as belonging to five major categories and that any
map is simply a set of overlays of data sets drawn from these
categories. Four of the categories are shown in Fig. 1. These
are (a) regions (which may overlap), (b) contour lines (which
may never overlap), (c) drainage lines (which tend to form tree
structures terminating on the map boundary), and (d) area-sub-
division lines (which are used to represent political boundaries,
road systems, and the like). The fifth category, not shown here,
is that of annotation data, which is associated with the line

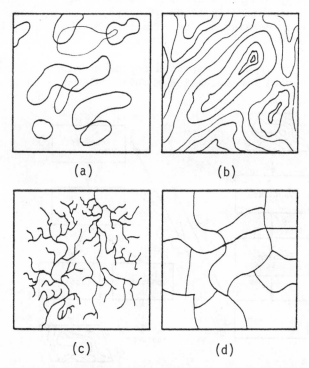

(a)   (b)

(c)   (d)

Fig. 1. Four different kinds of map data categories: (a) regions,
(b) contour lines, (c) drainage lines, and (d) area-
subdivision lines.

structure data of the other four categories. For purposes of
computer processing, however, it is advantageous to treat the
annotation data as being a distinct overlay data set rather than
directly associating it with the lines to which it pertains.
Such an arrangement then makes it convenient, for example, to
convert a map annotated in English to one annotated, say, in
French.

A map has been found to be such an effective means for depicting spatially-related two-dimensional data because of its inherent ability to preserve implicit spatial relationships. This
is not a priori true when map data is stored in a computer data
base. What is required then is to set up a data structure that
will preserve - and possibly even enhance - the spatial interrelationships. We are referring here to such pervasive and important
relationships as "north of", "close to", "inside of", "further
than", "larger than", "distant from", "joined to", "separated
from", "equally oriented as", and "similar to", as well as such
basic shape concepts as "round", "straight", "curved", "smooth",
"irregular", "narrow", and "convex". The problem here is how to
represent the map data in a computer so as to preserve this relationship, be efficient in terms of computer memory use, and
facilitate the processing of the data for any objective likely
to be encountered.

Some insight into the map data representation problem can
be gained from the "communication problem" illustrated in Fig. 2.

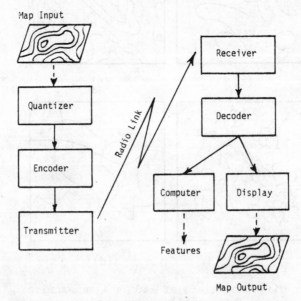

Fig. 2. Communication of Digitally Encoded Map Data

Suppose that a map is to be communicated via a radio link to a
distant location. The map must first be quantized, the quantized
data must be encoded into digital form and the encoded data then
used to modulate a radio signal which can be transmitted. At the
receiving station, the incoming signals are converted back into
digital data. At this point there are now two possibilities, depending on the purpose of the transmission. If the map is to be
processed (i.e., analyzed or manipulated) in a computer, it can
be entered at once and processing can commence. If, however, the
map is simply to be displayed, the digital data must be decoded
and "de-quantized" - with the latter involving interpolation and
digital-to-analog conversion - to obtain a display on either a
plotter or CRT. The result should contain all those features of
the original map that the sender deemed significant.

There are six criteria against which this communication process can be evaluated. These are precision, compactness, smoothness, facility for processing, ease of encoding and decoding, and
facility for display. If the objective is computer processing,
precision will be very important, as will be facility for processing. If the sole objective is display, smoothness is likely to
be paramount and facility for processing will be irrelevant.
Compactness and ease of encoding and decoding would be of importance in either case, with compactness possibly more important for
computer processing.

2.1 Quantization

Let us now look more closely at the quantization process.
Basically the process is one of analog-to-digital conversion,
that is, we are given a physical map (on paper or mylar) and we
desire a digital representation. The precision with which a map
can be represented is limited by the dynamic range of the original
analog medium and the ability to take measurements on it (e.g.,
1:1000, 1:10,000, etc.).

The process of quantization involves three factors: the _form_
of the quantization, that is, the rule according to which the
quantization takes place, the _size_ of the quanta, and the _approxi-_
mant used to represent the quanta. The choices made for these
factors will affect the performance relative to the aforementioned
six criteria.

We are at once offered two very different approaches to
follow, raster and lineal. In the raster approach we simply scan
over the map in TV line-by-line fashion and record the density or
luminance of each spot on the map - spatially resolving such spots
as finely as we desire and as the source medium (paper, mylar,
etc.) permits. In effect we convert the analog-medium map into a
(normally, rectangular) array of numbers, with each number giving

the "gray-value" of the particular map position to which it corresponds. Multicolor maps can be scanned with an appropriate multicolor scan head, yielding at least three numbers for each spot, one each for, say, red, green, and blue, into which any color can be resolved. A one-meter square map resolved to 0.1 mm will then yield 10^8 numbers if the map is monochrome, and at least three times as many numbers if polychrome. An 8-bit byte would normally suffice for each of the density values.

The raster representation preserves the spatial form of a map, and is a simple and easily handled form for communicating and then re-displaying a map. Some structuring of the map data can be realized by separating the data into distinct raster arrays, one for each data category, and then creating raster overlays to obtain particular data combinations. That is, one may create separate rasters for contour lines, for highways, for drainage lines, for annotation (separate ones for different languages!), for land usage, etc., and then assemble some subset of these to obtain a desired map.

The raster method of representation is not particular efficient in terms of its use of computer memory; though, there is, of course, no need actually to transmit each spot value obtained from the raster scan. Since there is always strong correlation among neighboring points in a map, extensive data compression can be used to reduce the amount that must actually be transmitted. The simplest scheme is to use run-length coding, where one indicates the particular density value at a point and the length over which this density remains unchanged. Further compression can be obtained by using also line-to-line coherence, or by using any of the more sophisticated image coding schemes that have been developed in recent years (1,2).

Since a map is a line drawing, in which information is conveyed by thin lines on a contrasting background, an alternative representation scheme is to encode the lines directly rather than to make a systematic raster scan over the entire map area. The lines in a map normally occupy only a small fraction of the total area, and hence the direct encoding of the lines should yield a representation that is both more compact and more directly expressive of the information actually conveyed by the map. A variety of such "lineal" representation schemes have been developed. Because of their importance we shall examine them in some detail.

Suppose that we are given a map and by means of an appropriate digitizer (e.g., data tablet, automatic line follower, etc.) describe the lines in the map in terms of a series of x,y pairs in a cartesian coordinate system. The precision will be limited by the smallest differences in x and y that can be reliably detected in defining adjacent points on the curve. If we indeed

use a cartesian coordinate system, then all points that can be
used for describing the curve must be nodes of an implied uniform
square grid, oriented parallel to the coordinate axes and with
spacing equal to the minimum detectable difference in the coordinate values. This is illustrated in Fig. 3, where the minimum
distances in x and y for distinguishing between two nodes are
labelled Δx and Δy, respectively. Since virtually all two-dimensional analog-to-digital conversion devices utilize a cartesian
coordinate system and possess uniform resolution, the uniform
square grid will almost always be the quantization form that is
used.

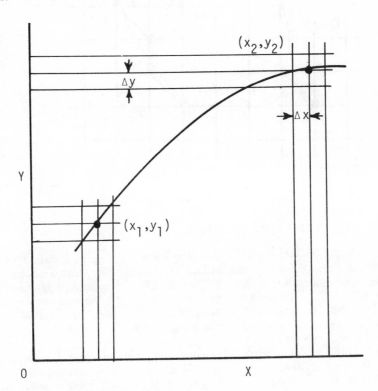

Fig. 3. Implicit uniform square map quantization grid

The minimum size is determined by the minimum resolvable
coordinate difference. Since this is the same as the grid spacing,
there is no useful information available about the curve between
two adjacent nodes. Hence we can join such adjacent nodes only
with the most primive approximant, namely, with a straight-line
segment. Such quantization is illustrated in Fig. 4; it is known
as the chain approximation scheme and has been widely used for
many years (3). Observe that the approximant consists of straight-
line segments of length T and T $\sqrt{2}$, where T is the grid spacing.

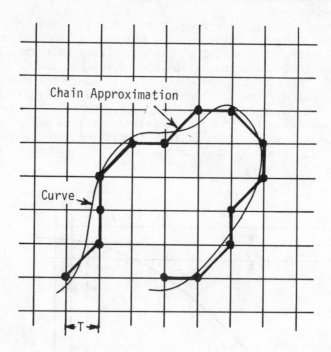

Fig. 4.   The chain approximation scheme for a curve

A contour map represented in this way is shown in Fig. 5.

It is well known that if a curve is to be approximated by a sequence of straight-line segments, the segments should be short where the curvature is great, and should be long where the curvature is small (4). This suggests that we might represent the curves in a map by means of line segments of varying length, selected so as to keep some error measure within a specified bound. (This error measure might be the maximum distance between the curve and the approximating line segment, the average distance, the mean-square distance, or the in-between area, etc.) The result will be what is commonly known as polygonal approximation. For this scheme, the quantization form is still a uniform square grid if a cartesian coordinate system is used and the spatial resolution is uniform over the entire map area. Alternatively, if the resolution varies in some non-uniform manner over the map, then the implicit grid will be similarly nonuniform (e.g., polar, logarithmic, curvilinear, etc.). The number of possible line-segment lengths will, however, always be finite because of the finite number of grid nodes in the map.

With polygonal approximation, we may connect any grid node to any other one. It can be shown that for an m x n image area,

Fig. 5. A chain-approximated contour map

where $m \leq n$, there can be at most $mn - m(m-1)/2 -1$ different-
length line segments (5). For a typical map the number of dif-
ferent-length line segments will in fact be smaller. Especially,
long line segments will occur only rarely, if at all. A line-
drawing coding scheme that provides for line segments that occur
rarely will tend to be inefficient. This suggests that we cur-
tail the number of permissible line-segment lengths.

In the chain approximation scheme of Fig. 4 we approximate a
curve by joining each node to one of its eight neighboring nodes.
This is unnecessarily restrictive. Suppose we view the "present"
node as being surrounded by a set of concentric squares of nodes
and permit the next node to be selected from any member of this
set, not only from the immediately surrounding square. As we add
additional squares to this set, the angular resolution improves
and we are able to progressively use longer line segments. The
requirement, of course, always is that the approximating line segment not differ from the curve by more than the allowed error.
This means that for sections of a curve exhibiting large radii of
curvature, we may use long line segments (say, from the 3rd or 4th

square) whereas in areas where the curvature is high, we will be
limited to using line segments to nodes in the 1st or 2nd square
(6,7).

The effect of using such a generalized chain representation
is illustrated in Fig. 6, where we show (a) a sample curve, (b)
the corresponding basic chain approximation, (c) a generalized
chain representation utilizing squares 1 and 2, and (d) a generalized chain representation utilizing squares 1 and 3.

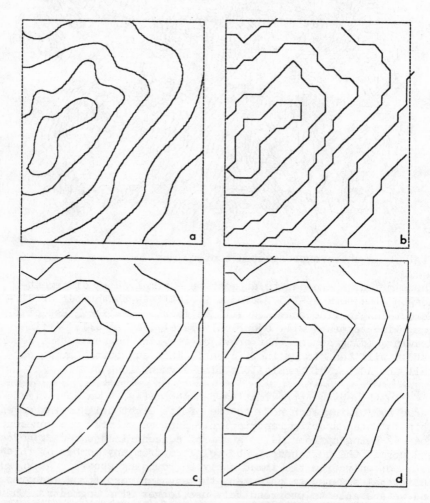

Fig. 6. Contour map section (a) in basic chain representation
         (b), in 1,2-generalized chain representation (c), and in
         1,3-generalized chain representation (d).

Fig. 7 shows the contour map of Fig. 5 re-quantized using only
square-5 segments. Note that the quantization <u>form</u> is the same
for all variations of a generalized chain representation - a

Fig. 7.   The contour map of Fig. 5 requantized using only line
          segments from a square-5 generalized chain code

uniform square grid. The size is simply the grid spacing. Only
the approximants change as different sets of line segments are
permitted for the approximation. It is important to note also
that the precision of the representation is solely a function of
the size of the quantization grid and is independent of the particular generalized chain configuration (6,7). Finally, polygonal
approximation can be seen to be only a special case of generalized
chain representation. For an m x n grid over the map area, $m \geq n$,
polygonal approximation is simply the generalized chain representation utilizing all squares 1 through m.

We have thus far considered only straight-line segment
approximants. This is the only correct choice when using the
basic chain representation scheme. However, when considering approximants for connecting nodes that are more distant than 2T, one
may well wish to examine the possibility of using curve segments.
One approach is simply to "generalize" the generalized chain

representation scheme to curve segments, and to regard a representation limited to straight-line segments as merely a special
case. The approximation process then consists of (1) selecting
the node to which the present node is to be joined, and (2)
choosing the curve segment to be used as approximant.

What should be the form of the curve segments? Of the many
possibilities, two are of particular interest: spline-fit curves
and standardized curves. With spline-fit curves nodes are selected to serve as the "knots" through which a spline is passed
with continuity in at least the first derivative (8). The resulting approximation will be smooth and precise. However, the
encoding effort is relatively high and the ease with which such
curves can be analyzed and manipulated in a computer is much less
than when a straight-line segment approximation is used.

Standardized curve segments are a direct extension of the
generalized chain scheme to include also a limited set of curve
segments (arcs) in addition to the set of straight-line segments.
The curve segments will be of a number of different lengths and
different degrees of curvature (5). Clearly with such fixed arcs
it will not be possible to assure continuity in the first derivative between connecting segments. Nevertheless, such an approximation should be smoother than any straight-line approximation,
and be simpler to process than a spline approximation. With
respect to compactness of storage there is not likely to be much
difference among any of these schemes (7).

2.2 Encoding

Thus far we have discussed only the problem of quantizing a
curve and representing it in terms of selected approximants. The
problem of encoding the resulting line structure, that is, of
actually representing it in some computer-processable form, is
a relatively much simpler task. As just explained, in all but
rare exceptions, the quantized form will inevitably be a uniform
square grid. Hence if we are restricted to using a straight-line
segment approximant, only the successive nodes need be identified;
if we use arcs, we must identify the type of arc as well. The
nodes can be identified by one of the following methods: (a)
giving full x,y-coordinate values, (b) giving the coordinate
differences for successive nodes, and (c) giving the changes in
the coordinate differences between successive nodes. It is clear
that (a) is the least efficient in terms of the number of bits
required for storage and is to be avoided. For (b) and (c) we
can improve the efficiency by taking note of the fact that different approximants are likely to occur with different probability
and assigning variable-length code words to the approximants,
with short code words assigned to frequently occurring approximants
and long code words to those that occur relatively rarely.

For the basic chain representation, the code used consists
of the octal digits 0-7, assigned in counterclockwise sense to
the 8 permissible line-segment directions connecting successive
nodes, as illustrated in Fig. 4 (using first-square nodes only).
Each line segment thus requires 3 bits. A generalized chain code
based on the use of the first two squares will have 24 permissible
line segments and will require 5 bits per segment. The code using
squares 1 and 3 will have 32 segments and will also require 5 bits
per segment. If the x and y changes were to be separately encoded, at least 4 bits would be required for the 8-direction code
and at least 6 for both the square-1,2 and square-1,3 codes.

The foregoing presumed that we assign a unique code word
(number) to each permissible line segment. A curve quantized into a chain of line segments can then be uniquely described by a
string of numbers, and such a representation implies a fixed
orientation (but not position) on the grid. For a curve to be
"well-quantized", the change in orientation from segment to segment will normally be small. This suggests the presence of strong
link-to-link coherence in any properly quantized chain, and this
coherence can be utilized for compressing the corresponding number
string. The most obvious and simple scheme is that of using first
differences. For the case of the common 8-direction code, this
leads to the following chain-difference code (1):

Angular difference	Code Word	No. of Bits
0°	0	2
+45°	1	2
-45°	2	2
+90°	31	4
-90°	32	4
+135°	331	6
-135°	332	6
Control	333	6

With the chain difference code we require fewer (2 instead of 3)
bits for those segments that occur most frequently, and more (4
or even 6 instead of 3) for those segments that occur rarely.
The net effect is usually a significant data compression - perhaps
by 30 percent. Similar difference schemes, with comparable savings,
can be realized for the generalized chain codes of any order (6,9).

2.3 Data Structures

Representation of maps in either raster or lineal forms,
with their numerous variations, exemplify the low-level data
structures for the map data. They are concerned solely with the
lines, curves and annotations on a map. But there are also the
even more important high-level data structures, which reflect the

syntax and semantics of the map in terms of the application for
which it is intended. The two data-structure levels are, of
course, closely linked, and, depending on the application, one
or the other low-level structure is more advantageous for a required high-level structure.

Generally speaking, the lineal form of low-level structure
is more likely to facilitate the high-level data accessing and
processing one can expect with a map. Thus, for example, a
contour map implicitly contains strong syntactic constraints
that are easily embedded (and their existence verified) with
lineal representation but are difficult to make explicit in a
raster representation (10,11,12). This includes such fundamental
properties as (1) that all contour lines must be closed (directly
or via the map boundary), (2) that contour lines may not cross
other contour lines, and (3) that in going from one contour line
to one of different elevation one must cross all contour lines
of intermediate elevation (10).

For higher-level data structuring one is interested in
easily discerning such spatial relationships as "contained in"
and "bordering on". Thus one may want quickly to find all the
cities of population over 20,000 in a particular political subdivision of a particular state for which the annual rainfall
exceeds 60 cm. Or one may want to know what cities of population
in excess of 30,000 are situated within 50 km of a road linking
two specified cities (12,13).

3. MANIPULATION AND ANALYSIS

Map data, in addition to being stored and retrieved,
frequently also needs to be manipulated and converted from one
form to another, or analyzed to extract specific features of
interest. Among the most basic such operations are the conversion from raster to lineal form and from lineal back to raster
form. Both these operations are relatively straight-forward,
though at times special difficulties occur and must be resolved
(14,15,16). Lineal-to-raster conversion is often troublesome
because of the difficulty of storing a full raster map in a computer's high-speed memory since the raster file can easily exceed
10^8 bytes in size. One normally handles this by dividing the
raster into strips (possibly as narrow as one line), making repeated scans over the lineal data file, and thus creating the
raster file on a strip-by-strip basis. Such a scheme of scan
conversion is also generally required when a lineal file is to
be displayed on a raster-type output device such as a raster CRT
or an electrostatic plotter.

Raster-to-lineal conversion can be carried out in a single

pass over the raster image by creating a moderately elaborate
data structure which keeps track of the various components of
the lineal map as they are extracted from the raster data (16,17).

Among some of the other "housekeeping" operations associated
with handling map data are "windowing", that is, extracting the
data associated with a sub-area (usually rectangular and with
sides parallel to the map boundaries) of the total map. This
operation is trivial for maps in raster format provided one of
the window boundaries parallels the raster scan lines. The operation is more complex for lineally structured maps, where now
each line segment (or arc) must be tested as to whether it lies
entirely inside, partially inside, or entirely outside the specified window. The algorithm for this is very similar to the well-
known clipping algorithm in computer graphics (18). One notes
the four lines bounding the designated window and determines
whether the end points of each map line segment are either both
above, both to the left of, both to the right of, or both below
the window-defining lines. In either of these four cases, the
entire line segment lies outside the window. If both end points
lie inside the window, the entire segment must be retained. If
one end point lies inside and one outside, the intersection with
one of the window-defining lines must be found, and the inside
part of the segment is then retained. Most difficult is the case
where both end points are outside the window but do not lie toward the outside of the same window boundary line. Such line
segments may or may not pass through the window. The matter is
resolved by finding the intersections of the segment with the
window-defining lines.

In most instances the line segments in a map are much smaller
than the window dimensions and it is a simple matter to determine
whether they are inside or outside the window. Only segments in
the vicinity of the window boundaries need be tested more carefully, as well as long line segments such as parallels or meridian
lines. Fig. 8 shows a detailed contour map containing many
thousands of contour lines. A window taken from the lower left-
hand corner of this map and enlarged 4-fold is shown in Fig. 9.

Other operations consist of deleting all but every fifth
contour line, of generating additional interpolated contours to
lie between given contour lines, or to determine the drainage
lines from contour line data. These operations are straightforward with lineally-organized data but cumbersome with raster-
organized data.

Manipulation of map data may also involve so-called "rectification", where the source data, possibly based on high-altitude
photography, must be correct to remove the effects of earth curvature. Strictly speaking, there is, of course, no way that an

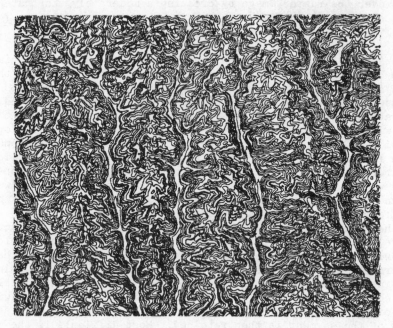

Fig. 8.   A section of a detailed contour map (Jewell Ridge
          Quadrangle, Virginia, USA)

area on a spheroid can be transformed without distortion into an
area in a plane. However, transformations are available that
preserve angles (so-called conformal transformations), that cause
a great circle to be represented by straight lines on a map
(gnomonic transformations), that preserve areas (i.e., closed
curves in the map have the same area as their corresponding
curves on the spheroid), and that preserve the angles relative to
the meridians and parallels (e.g., Mercator projection). All these
transformations are more easily and more rapidly obtainable for
lineal rather than raster-organized map data.

    Analysis computations, such as computation of area, distance
between points, determination as to whether a point is inside or
outside some specified contour - these can all be easily carried
out in lineal format (3). If the desired features are boundary-
oriented, the lineal form yields to especially efficient algorithms.
If the features are more area-oriented, another data structure
known as Merrill's parallel-scan code may be found advantageous
(12). In this representation, which can be regarded as a modified
or compressed raster form, the data is first generated in lineal
form and is then scanned using closely spaced parallel scan lines.
The intersections between the data lines and the scan lines are
recorded as x,y pairs, and then sorted according to increasing x

Fig. 9. A window, enlarged 4-fold, of the lower left-hand
corner of the contour map of Fig. 8.

value for each constant y-value. The storage efficiency of this
representation is much lower than for a lineal form such as the
8-direction chain code. It does, however, lead to particularly
fast algorithms for determining area, determining overlap of two
or more regions, and determining whether or not a specified point
lies within a given region.

     For the analysis of lineally-organized data using straight-
line segments, it is merely necessary to tabulate the x and y
components for each allowed line segment (of different length
or different orientations). The length of a curve is, of course,
simply the sum of the lengths of its constituent line segments.
The distance between two points is simply the square root of the
sum of the squares of the sums of the line segments' x and y
components of any curve joining the two points.

     The net area enclosed by a contour is given by

$$S = \sum_{i=1}^{n} a_{ix}\left(y_{i-1} + \frac{1}{2} a_{iy}\right)$$

where the a_{ix} and a_{iy} are, respectively, the x and y components
of line segment i, and the summation is over all the segments
forming the closed contour.

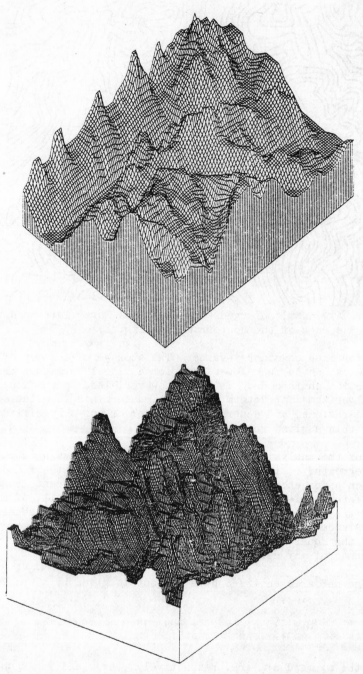

Fig. 10. Two perspective views of the area described by the contour map of Fig. 5.

A large variety of algorithms have been developed for processing lineally-represented map data and are described in the
literature (3,6,19). In general they tend to be of a complexity
that varies linearly with the amount of data and, if desired,
can be easily implemented on a pipeline type processor for rapid
computation.

One final "manipulation" of interest is that of rendering
map data in the form of orthographic or perspective projections.
A number of algorithms for this purpose have been developed by
researchers in the field of computer graphics (20 21). Most of
them are "approximate" to a greater or lesser degree but, nevertheless, yield good-quality "three-dimensional" representations.
Two such representations, representing two different views (and
resolutions) of the contour map of Fig. 5 are shown in Fig. 10.

ACKNOWLEDGMENT

This article describes research sponsored by the U.S. Air
Force through the Rome Air Development Center under Contract
F30602-78-C-0083. The information presented does not necessarily
reflect the position or policy of the U.S. Government and no
official endorsement should be inferred.

REFERENCES

1. A. Rosenfeld and A. C. Kak, Digital Picture Processing,
   Academic Press, New York 1976, Chapt. 5.
2. W. K. Pratt, Digital Image Processing, John Wiley & Sons, New
   York, 1978, Chapt's. 21-24.
3. H. Freeman, "Computer processing of line drawing images",
   Computing Surveys, vol. 6, no. 1, March 1974, pp. 57-97.
4. F. Attneave and M. D. Arnoult, "The quantitative study of
   shape and pattern perception", Psychol. Bulletin, vol. 53,
   1956, pp. 453-471.
5. H. Freeman and J. A. Saghri, "Comparative analysis of line-
   drawing modelling schemes", Comp. Graphics and Image Proc.,
   vol. 12, no. 3, March 1980, pp. 203-223.
6. H. Freeman, "Applications of the generalized chain coding
   scheme to map data processing", Proc. Conf. on Pattern Recognition and Image Processing, IEEE Computer Society, publ. no.
   78CH1318-5C, May 1978, pp. 220-226.
7. J. A. Saghri and H. Freeman. "Analysis of the precision of
   generalized chain codes for the representation of planar
   curves", IEEE Trans. on Pattern Analysis and Machine Intelligence, 1980, (to appear).
8. D. A. McClure, "Computation of approximately optimal compressed
   representations of discretized plane curves", Proc. Conf. on

Pattern Recognition and Image Processing, IEEE Computer
Society publ. no. 77CH1208-9C, June 1977, pp. 175-182.

9. D. Proffitt and D. Rosen, "Metrication errors and coding
efficiency of chain-encoding schemes for the representation
of lines and edges", Comp. Graphics and Image Proc., vol. 10,
1979, pp. 318-332.

10. H. Freeman and S. P. Morse, "On searching a contour map for a
given terrain elevation profile", Jour. Franklin Institute,
vol. 284, no. 1, July 1967, pp. 1-25.

11. S. P. Morse, "Computer storage of contour-map data", Proc.
23rd National Conf. of ACM, LasVegas, Nevada, August 1968.

12. R. D. Merrill, "Representation of contours and regions for
efficient computer search", Comm. ACM, vol. 16, no. 2, Feb.
1973, pp. 69-82.

13. L. D. Shapiro and R. M. Haralick, "A general spatial data
structure", Proc. Conf. on Pattern Recognition and Image
Processing, IEEE Computer Soc., publ. no. 78CH1318-5C, May
1978.

14. B. W. Jordan and R. C. Barrett, "A scan conversion algorithm
with reduced storage requirements", Comm. ACM, vol. 16, no.
11, November 1973, pp. 676-682.

15. R. C. Barrett and B. W. Jordan, "Scan conversion algorithms
a cell organized raster display", Comm. ACM, vol. 17, no. 3,
March 1974, pp. 157-163.

16. J. A. Saghri, Efficient encoding of line drawing data with
generalized chain codes, Tech. Rept. IPL-TR-79-003, Rensselaer
Polytechnic Institute, Troy, NY, August 1979.

17. A. Rosenfeld, "Algorithms for image/vector conversion",
Computer Graphics, vol. 12, no. 3, August 1978, pp. 135-139.

18. R. F. Sproull and I. E. Sutherland, "A clipping divider",
Proc. Fall Joint Computer Conf., 1968, AFIPS Press, pp. 765-
775.

19. H. Freeman, "Analysis of line drawings", in J. C. Simon and
A. Rosenfeld, Digital Image Processing and Analysis, Noordhoff,
Leyden, 1977, pp. 187-199.

20. B. Kubert, J. Szabo, and S. Giulieri, "The perspective
representation of functions of two variables", Jan. ACM, vol.
15, no. 2, April 1968, pp. 193-204.

21. T. J. Wright, "A two-space solution to the hidden-line
problem for plotting functions of two variables", IEEE Trans.
Computers, vol. C-22, no. 1, January 1973, pp. 28-33.

DIGITAL IMAGE PROCESSING OF REMOTELY SENSED IMAGERY

M. Goldberg

University of Ottawa

ABSTRACT

A survey of some digital image processing techniques which
are useful in the analysis of remotely sensed imagery, particularly Landsat, is presented. An overview of the various steps
involved in computer automated processing of Landsat imagery is
first given. A number of specific tasks, which employ digital
image processing methods are then discussed: on-board processing,
radiometric corrections, geometric transformation, image enhancement and multitemporal analysis.

A. INTRODUCTION

A.1 Digital Image Processing

An image can be considered as a two-dimensional representation of an object or a scene. In mathematical terms, an image
can be described by a function $f(x,y)$, where (x,y) are the spatial coordinates and $f(x,y)$ is the value of the image at the
point (x,y). The value $f(x,y)$ is called the grey level or radiance number at the point (x,y). In practical cases x and y
range over bounded intervals. An example of a continuous bounded
image is a black and white photograph.

For computer processing it is necessary to have the image
in discrete form. In mathematical terms, the image is represented
by a discrete set of real values $\{f(i,j): i = 1,2,\ldots,N;$
$j=1,2,\ldots.M\}$; that is, the value of the image is only given for
a finite number of points. The process of going from a continuous to a discrete form is called sampling. The individual points

383

J. C. Simon and R. M. Haralick (eds.), Digital Image Processing, 383–437.
Copyright © 1981 by D. Reidel Publishing Company.

(i,j) in the discrete image are called picture elements or "pixels". The f(i,j) are restricted to one of a finite set of radiance values by a process called quantization or digitization.

The function f(x,y), the recorded image, is related to the true scene, g(x',y'), by a transformation T,

$$f(x,y) = T[g(x',y')]$$

In general T is a complicated function of the sensors type, attitude and position, scene geometry, and the intervening medium which is usually the atmosphere (Hartl, 1976).

Digital image processing involves the manipulation of imagery in its sampled form, and has found applications in many different domains such as remote sensing, medicine, industrial inspection, astronomy, to name but a few. There are available a number of basic texts, which include Castleman (1979), Hall (1979), Pratt (1978), Gonzalez and Wintz (1977), Rosenfeld and Kak (1976), and Duda and Hart (1973). Examples of specific applications can be found in Simon and Rosenfeld (1977), Rosenfeld (1976) and Huang (1975).

A.2 Landsat Imagery

With the launch of Landsat 1 in July 1972, and with the subsequent launches of Landsat 2 in January 1975, and Landsat 3 in March 1978, remotely sensed imagery over wide areas of the world has become readily available. The multispectral scanner on board the satellite provides simultaneous images in four spectral bands in the visible and new infra-red regions to a ground resolution of 79 metres. A single frame covering an area 185 km by 185 km is sampled at the detectors at a rate of 3200 pixels per line and 2400 lines per frame. The spectral intensities are logarithmically compressed and quantized into 64 grey levels. An experimental 5th spectral band in the thermal infra-red region was added to Landsat 3 but is not discussed here. Landsat has been launched in a nearly polar sun-synchronous orbit of about 900 km, in order to give repetitive coverage: every 18 days Landsat passes over the same place on the earth. Further details on Landsat specifications are available in Richardson (1978) and Nasa (1976).

There are a myriad of applications of Landsat imagery to domains such as agriculture, geology, forestry, hydrology, to name but a few. Good sources of information include a collection of papers edited by Lintz and Simonett (1976); a review article by Haralick (1976); the journals, Remote Sensing of the Environment, Photogrammetric Engineering and Remote Sensing, and Canadian Journal of Remote Sensing; and various symposia, such as Canadian Symposium on Remote Sensing and International Symposium on Remote Sensing of Environment.

A.3   Automatic Processing of Landsat Imagery

The objective of computer processing of Landsat imagery is
the extraction of useful information. Techniques derived from
pattern recognition and image processing have been successfully
applied. Many review articles and a number of books have been
written on the subject, these include: Bernstein (1978), Swain
and Davis (1978), CNES (1978), Anuta (1977), Haralick (1976) and
(1973), Hartl (1976), Hajic and Simonett (1976), Fu (1976) and
Kazmierczak (1977). There is also a regular conference devoted
to this subject sponsored by LARS of Purdue University (1975,
1976, 1977, 1979).

The three properties of Landsat data which are important for
computer processing are the multispectral nature of the sensors,
the presentation of the data in image format, and the multitemporal coverage of the orbits. With the previous notation, $f(i,j)$,
the radiance value corresponding to pixel (i,j) now becomes a
vector value (x_1, x_2, x_3, x_4). The first component, x_1, is the

grey value of the first spectral band; x_2, is the grey level va-

lue of the second spectral band; and so on. The data transmitted
from Landsat can be considered for subsequent computer processing
as four parallel images (Figure 1), or alternatively, as an NxMx4
matrix. Figure 2 shows these four individual images for a typical
agricultural scene in Saskatchewan, Canada.

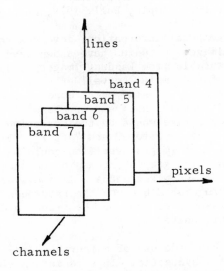

Figure 1.   Representation of a Landsat image as four 2-di-
            mensional parallel images, band 1 to band 4.

Figure 2 shows two of the three important sources of information in a Landsat image, the spectral and the spatial. The
third source, temporal, is illustrated in Figure 3, which represents the same forest scene in summer and winter.

From even a cursory inspection of the four images, the
striking resemblance between the four is clear. At the same time,
a closer inspection shows that there are differences; for example,
fields in bands one and two often merge but are clearly discernible in bands three or four. This information is contained in the
correlation, at the individual pixel level, between the 4 spectral
values. It gives rise to the notion of a spectral signature;
that is, four-dimensional spectral values can be associated with
the classes of interest, in this case crop types. The image is
then treated as a collection of spectral values and methods from
pattern recognition and multivariate analysis can be used to
analyze and automatically process the data (Fukunaga, 1972).

The spectral analysis just described ignores the spatial
coherence or correlation present in the image. A second examination of the images of Figure 2, shows that they exhibit certain
"regularity". As the images consist mainly of a collection of
fields, adjacent pixels will tend to be very similar. This spatial correlation can be exploited in processing. In general,
computer technique based upon spectral analysis are the simplest
to implement because the pixels can be treated independently of
one another. The exploitation of the spatial information requires
much more computer power as complex operations on groups of pixels
must be performed (see for example, Hall, 1979, chapter 7).

The analysis of temporal information, involves the use of
spatially registered images. Studies have shown that many temporal variation are measurable from Landsat imagery and can be
effectively used (see for example, Engvall et al., 1977). There
are two distinct objectives in using multitemporal images: detection of changes and improved analysis or classification. An
example of the first objective would be the detection of newly
cut forest areas from images "before" and "after". Figure 3
illustrates how a series of images over time could be used to
yield improved results. The greyish areas in the summer image
correspond to coniferous and mixed woods. In the winter, these
two groups can be distinguished because the hardwoods lose their
leaves and with the presence of snow on the ground are readily
differentiated from the conifers.

The last example has made use of a fourth source of information which can be called syntactic. This is information provided
by the user and which can then be employed in the computer processing. In the example above, the observation that hardwoods
lose their leaves in the winter could be used in syntatic infor-

Figure 2   The figures represent a 166 by 166 portion of a LANDSAT
           image over the Melfort Experimental Form in Saskatchewan
           in, respectivily, bands 4, 5, 6 and 7 of LANDSAT. Band
           4 a), is sensitive in the wave lenght region of .5 to .6
           micrometers, band 5 b), from .6 to .7 micrometers, band
           6 c), from .7 to .8 micrometers and band 7 d), from .8
           to 1.1 micrometers.

mation processing for separating hardwoods and conifers. Examples
of this approach can be found in Brayer et al. (1977) and Bajcsy
and Tavakoli (1975).

A.4 Computer Processing Model

     There are a number of steps involved in the computer proces-
sing of remotely sensed imagery. A simplified block diagram of
a general computer image processing and recognition system is
shown in Figure 4. The input/output subsystems interface to the

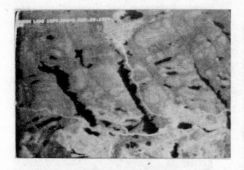

Figure 3   Example of a summer-winter change. In a), the summer image
           shows the black area corresponding to water, the grey to
           coniferous and mixed woods, and the light to clear cut
           and roads. In b), beside the added snow, the hardwoods
           have lost their leaves and can now be readily distinguished.

sensors on one side and the user on the other side. Often these
can be the most expensive components of the entire system. Preprocessing is concerned with those techniques related to computer
image processing; that is, to the manipulation of the images.
Segmentation and classification, relate to the image recognition
tasks; that is, to the extraction of information. The different
processing steps are shown in consecutive fashion in Figure 4.
In practice, however, there are feedback loops, and even inter-
changes in the order; for example, classification often preceeds
segmentation.

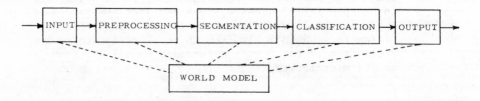

Figure 4   A block diagram of a general computer image
            processing and recognition system.

Figure 4.

Input:

      This step concerns the interface of the computer system to
the data acquisition system. It usually involves the problem of

data storage and may contain hardware for digitizing analogue
data such as maps. Hartl (1976) and Philipps and Swain (1978)
have treated this problem.

Output:

Once the data is processed, it may be presented to the user
in some visual form, a map, a picture, a list of statistics etc.
(Phillips and Swain, 1978). In other cases it may be integrated
directly into some geographical database (Simonett et al., 1978),
(Bryant and Zobrist, 1977), (Nagy and Wagle, 1979).

Preprocessing:

This step relates to operations which transform the images
so that subsequent analysis is facilitated and improved; examples
are compression, radiometric corrections, and geometric transformations. Image enhancement for subsequent visual analysis also
falls into this step.

Segmentation:

The first step in the image recognition process is often
the parsing of the image into a number of spatially contiguous
sets. For example, in the agricultural scene of Figure 2, these
would be ideally the individual fields. Conceptually, this step
corresponds to extracting the spatial information content. Good
general references to this subject include Rosenfeld and Kak,
(1976, chapter 7), Hall, (1979, chapter 7), Rosenfeld and Weszka
(1976), Riseman and Arbib (1977), and Zucker (1976).

Classification:

Once the "parts" of the image have been located they must
then be classified or labelled. In the remote sensing context,
this usually implies a classification by statistical methods based upon the spectral values. In many applications, segmentation
is not carried out, or only carried out after the classification.
In these cases, the classification proceeds on a pixel-by-pixel
basis, and corresponds to the extraction of the spectral information of the image. There are many excellent references to this
subject, Swain (1978), Haralick (1976), and Tou and Gonzalez
(1974).

World Model:

In Figure 4 dotted lines are drawn between all the blocks
and that one labelled as world model, in order to indicate their
nebulous connection. This last block corresponds to the use of
syntactic information in the broad sense. It is the individual

user who decides what processing is needed and defines the particular traits of the information of interest. Various aspects of
the interaction between the user and computer processing and recognition system are discussed in Swain and Davis (1978).

A.5 Outline

     This long introduction demonstrates that the computer pro-
cessing of Landsat imagery involves a number of complex steps
and uses different types of information. In the remaining portion
of this paper, attention is focussed only on some aspects of preprocessing and on the use of temporal information. The following
topics are covered in consecutive sections: on-board processing,
radiometric correction, geometric transformations, image enhancement, and a review of some techniques for temporal analysis.

B. On-Board Processing

B.1 Compression

     The first stage at which data can be processed is directly
on-board the satellite. One common objective of the processing
is to compress the data and so reduce the transmission requirements (Wintz et al., 1977). The newest Landsat satellite, D in
the series, contains a new thematic mapper in addition to the
earlier 4 channel multispectral scanner (MSS). This thematic
mapper is a 7 channel MSS, 3 bands in the visible, 3 in the infrared, and 1 thermal infrared, with a ground resolution of 30
metres and 256 grey levels per channel. The data output for this
scanner is 85 Megabits/second versus 15 Megabits/second for the
earlier 4 channel MSS. (Salmonson and Park, 1979). Future
satellites may have even higher data rates so that compression
of the data becomes a critical requirement.

     Facsimile devices and transmission of video signals are the
two areas of most active research in image compression. Excellent
overviews are presented in a collection of articles edited by
Pratt (1979), by Huang (1972), and in Pratt (1978, Part 6).

B.2 Compression, Spatial Coding

     In compression of images, use is made of the spatial corre-
lation; that is, the dependence between neighbouring pixels. By
suitably exploiting the statistical properties of the image, it
is possible to achieve compression without any distortion; that
is, the original image can be exactly regenerated. The basic
idea is to group the neighbouring pixels, calculate the frequency
of occurrences of the various groups,and then use a variable-
length code for these groups. The more frequently occurring
messages would have shorter code words. Huffmann (1952) gives a

procedure for generating the optimal codes. This method would
be rather difficult to implement for Landsat as the statistics
of the groups of pixels are not known beforehand.

Most compression techniques permit some degree of distortion
so that the original image can no longer be exactly regenerated.
Techniques can be classified depending upon whether the compression
takes place in the spatial domain of the image or the transform
domain.

Two effective spatial domain coding methods are delta modulation and predictive coding. In delta modulation, the differences between values of successive pixels is transmitted:

$$d(i,j) = f(i,j) - f(i,j+1).$$

The savings are obtained by coding $d(i,j)$ with a smaller number
of bits than in the original image. Distortion are therefore
introduced whenever there are large changes. Predictive coding
attempts to make use of the statistics of the pixels. From the
previous pixels values an estimate of the following value is made;

$$\hat{f}(i,j) = P(f(i,j-1), f(i,j-2)....).$$

This estimate is compared with the true value and the difference
is then transmitted:

$$\hat{d}(i,j) = f(i,j) - \hat{f}(i,j).$$

B.3 Compression, Transform Coding

In transform techniques, a two-dimensional transform is calculated for individual subsections of the image (Wintz, 1978).
Examples of transforms used are Fourier, Hadamard, and Karhunen-
Loève. The result of this transform is an array of coefficients
which are relatively decorrelated. More importantly, the energy
of the image is packed into a fairly small number of these coef-
ficents. By dropping the coefficients with low energy value,
compression is achieved with low distortion. At the same time,
as the coefficients are a function of all the pixels, the distortion is smeared over the entire image.

In the special case of multispectral images, transform techniques can also be applied to the spectral values considered as
vectors. Here use is made of the large spectral correlation
between the bands. For example, Ready and Wintz (1973) show
that by using the Karhunen-Loève transform 12 dimensional MSS
aerial images can be compressed, with little distortion, by
keeping only the first three coefficients. It is noted that the
Karhunen-Loève transform method is the optimal transform technique

with regards to the mean square error (MSE) criterion.

Values for the compression that can be achieved by various coding schemes are presented in Ready and Wintz (1973) for aerial multispectral scanner data, and in Pratt (1978, p. 729) for video. For statistical techniques a compression of from 1.5/1 to 3/1 can be expected; for predictive codes between 2/1 and 4/1; and for spatial transform codes up to 8/1 for MSS and 16/1 for video.

B.4   Compression, Clustering

A different approach to compression for video is to take advantage of the perceptual limitations imposed by human vision. Becuase of these limitations it becomes possible to introduce certain distortions into the image with negligible effect on perception (Sakrison, 1979).

A similar approach can be applied to Landsat. One of the main applications of Landsat imagery is for classification. Thus, any compression scheme which does not distort the resulting classification maps would certainly be useful. A compression technique based upon clustering or grouping a relatively small number of pixels at a time has been proposed by Hilbert (1975) and implemented by Lowitz (1978a, 1978b) with good results.

The basic steps are as follows:

Step 1:   Initialization. Arbitrarily choose 2^k clusters by specifying the cluster means.

Step 2:   Assign each pixel of the current line to the closest cluster by a distance to mean criterion.

Step 3:   Transmit the current dictionary of cluster means and the cluster labels for the pixels in the line.

Step 4:   Update the cluster means with the pixels of the line.

Step 5:   Return to Step 2 or end.

The compression ratio is defined as:

$$C.R. = \frac{\text{number of bits without compression}}{\text{number of bits with compression}}$$

$$= \frac{P \cdot B \cdot C}{k \cdot P + 2^k \cdot B \cdot C}$$

where,

    P = number of pixels per line
    B = number of grey level resolution bits
    C = number of spectral channels
    k = number of bits for cluster labels (2^k clusters)

As an example, for P=512, B=8, k=4, the following compression
ratios are obtained for different values of C.

C.R.(compression ratio)	C(number of channels)
1.87	1
3.73	2
7.46	4
11.2	6
14.9	8

It is clear that for images with one spectral channel the savings
are quite small, but that with increasing number of channels, the
compression possible is quite significant.

At the receiving station the images can easily be reconstructed. For each line a list of class labels and a cluster mean
dictionary is received so that reconstruction can proceed by a
simple table look-up procedure. Lowitz (1978b) shows an example
of MSS aerial images with compression ratio of 13 such that the
reconstruction quality is better than 30 db (peak to peak signal/
over RMS reconstruction noise). As an added bonus the dictiona-
ries of cluster means can be used for clustering and classifying
the entire image.

It is important to point out that large compression ratios
are possible for multispectral imagery because of the large spectral correlation in addition to the spatial one. A direct consequence of this spectral correlation between the bands is the good
representation that is possible with only a small number of clusters.

B.5 Control and Decision-Making

The on-board processing so far discussed has been concerned
with compression. New on-boarding processing applications related to decision making and control have been proposed. The impetus for these applications come from research on military guidance systems such as those used in the Cruise missile (Proceedings
of SPIE, 1979).

Wilson et al. (1979) have suggested the use of on-board processing to reduce the collection and transmission of unnecessary
data; for example, cloud covered scene. By detecting the presence

of clouds on board, the corresponding scenes are not imaged and
so result in reduced transmission requirements. More sophisticated processing could be used to monitor and track conditions
at only specified locations.

Aviv (1979) proposed a processing system, with extensive
on-board processing and storage capabilities which would result
in a substantial reduction in transmission requirements. The
idea is to store on-board the satellite signatures of the targets
of interest; for example, agriculture, geology, etc. Whenever
a scene with these pre-determined characteristics are encountered,
the corresponding image is transmitted. Real-time transmission
of the data to the appropriate interested user becomes feasible
and so can enhance the usefulness of the data.

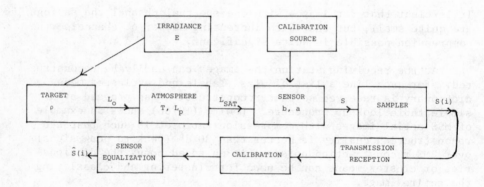

Figure 5: A block diagram depicting the various factors which
           enter into the final sampled output $\hat{S}(i)$.

C. Radiometric Corrections

C.1 Model

A block diagram depicting the various factors which enter
into the final digital value measured and transmitted by Landsat
is shown in Figure 5. The corresponding equations for one pixel
and sensor in simplified form are (Crane, 1971),

$$L_{sat} = \frac{\rho \, E \, T}{\pi} + L_P$$

where,

   L_{sat} = spectral radiance measured by the satellite

   ρ = reflectance of target

     E = downdwelling irradiance on the target
     T = atmospheric transmission
     L_p = path radiance; that is radiance scattered toward the
       satellite by the atmosphere

and (Vishnubhatla (1977))

$$S = b L_{sat} + a$$

$$S(i) = b L_{sat}(i) + a$$

$$\hat{S}(i) = f(s(i))$$

where,

     S   = analog voltage output of sensor
     b   = gain of sensor
     a   = offset or bias of sensor
     $S(i)$ = ith sample output
     $\hat{S}(i)$ = equalized sampled output

The relationship between the final digital value, $\hat{S}(i)$, and the intrinsic property of the target ρ, the reflectance, are seen to be quite complicated.

C.2 Atmospheric Effects

The irradiance, E, is a function not only of the sensor-sun-pixel geometry but also of the terrain and the atmosphere. Corrections for the first set of parameter can be made, for these are systematic and depend upon the illumination or sun angle and the view or look angle of the sensor. The second set of parameters are non-systematic and are difficult to correct.

The target is assumed to have a reflectance ρ. Only a fraction of the irradiance is therefore reflected to the satellite and this is given by,

$$L_o = \frac{\rho E}{\pi}$$

The reflected radiance L_o suffers a degradation in the atmosphere. The value measured by the sensor is, therefore, only linearly related to L_o (Ahern et al., 1979),

$$L_{sat} = T L_o + L_p.$$

The combined effect of T (transmission) and L_p (path radiance) is to reduce the contrast, as bright targets will seem darker, and dark targets lighter. Absolute reflectance values can be

obtained, by using appropriate computer models for the atmosphere and estimates of T and L_p (Turner and Spencer, 1972).

These estimates can be obtained from ground-based measurements
or from known standard targets in the scene such as oligotrophic
lakes (Ahern et al., 1979).

An alternate approach to absolute calibration is normalization by ratioing the measured radiances in adjacent spectral

bands, L_{sat}^i, L_{sat}^{i+1}, (Crane, 1971) (Haralick, 1976); for example,

$$x_i = \frac{L_{sat}^i}{L_{sat}^{i+1}}$$

$$x_i = \frac{L^i sat}{\Sigma L^i sat}$$

The first ratio is useful when $L_p \simeq 0$ for each channel and the

factor ET are approximately equal for adjacent channels; then,

$$x_i = \frac{\rho_i}{\rho_{i+1}}$$

The second yields a normalized coefficient between 0 and 1.

C.3 Sensor Corrections

At the sensor, the radiance, L_{sat}, is transformed into a

voltage which depends upon the gain and bias characteristics.
This voltage value, S, is sampled and quantized, which introduces
some additional noise. It is this digitized value, S(i), now
treated as a grey level value, which is transmitted.

In addition to image data, calibration data is periodically
transmitted. Two on-board calibration systems are available for
estimating the gain and bias of the sensor. One makes use of a
light source, and the second of the sun. The calibration data
is used for subsequent ground-based processing.

There are six independent sensors per spectral band, corresponding to six consecutive scan lines. As a consequence, non-
uniform sensor responses can give rise to a striping effect. By

equalizing the gain and bias of the 6 sensors, this effect can be reduced. Thus, for S(i), a new value $\hat{S}(i)$ is substituted. These corrections are easily implemented by using a table look-up procedure. A complete treatment of these correction and normalization procedures is given by Vishnubhatla (1977). (See also Horn and Woodham (1979)).

D. Geometric Transformations

D.1 Image Rectification, Statement of Problem

In most applications of remotely sensed imagery it is necessary to relate geographically the image and the pixels within the image to other geocoded data such as maps. Images acquired by Landsat, for example, exhibit significant distortions so that they cannot be used as maps. The process of geometrically transforming an image so that it can be used as a map is called image rectification. A convenient map projection for Landsat imagery is the universal transverse mercator (UTM) map projection, as this combines a cartesian coordinate system with a metric distance (Dept. of Army, 1958).

The problems involved in transforming an image into a geometrically corrected UTM map form are illustrated in Figure 6. The original recorded image is given in a coordinate system (x,y). The image must be transformed into the rectangular coordinate system (u,v) of the UTM map projection. The desired output are the corresponding pixel values at the lattice points in the (u,v) system. Distance functions in the (x,y) coordinate are very complicated and depend upon the geometric distoritions. The distance function in the (u,v) coordinate system is the usual Euclidean distance function. The first step is to determine a transformation which maps pixels (located at the lattice points in the (x,y) system) into points in the (u,v) system. These are shown as empty circles in Figure 6b and are not necessarily located at the (u,v) lattice point. The values at the lattice points in (u,v) can then be found by interpolation. Methods of calculating the transformation through the use of ground control points and of interpolation are described below.

D.2 Geometric Errors

The distortions in Landsat imagery induced by geometric errors are corrected in two passes. First of all, the effects of the known or systematic errors are compensated. Examples of systematic errors are earth rotation, mirror velocity, panoramic distortion, and map projection. These errors are well understood and their effects can be modelled and removed by an appropriate geometric transformation. A more complete discussion of the various distortions in Landsat is given by Bernstein (1976) and

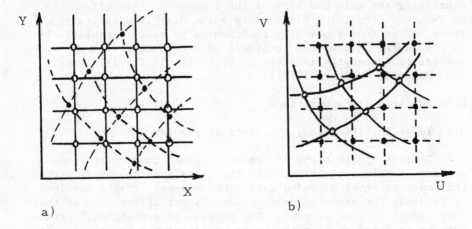

Figure 6   Geometric image transformation. Figure a) represents
           the original image in dark points in rectangular grid
           system in light points. Figure b) represents the
           geometrically corrected image in dark points with the
           original grid system in light points, which has been
           distorted. (Orhaug and Akersten, 1976).

Van Wie and Stein (1976). Horn and Bachman (1978) discuss methods which can compensate for the distortions induced by varying
surface slope. A digital terrain model of the area imaged is
required for these corrections.

     After the first pass, the remaining distortions are assumed
to be the result of random or at least unknown errors. These
are mainly due to changes in the position of the platform; roll,
pitch, yaw, and altitude. These changes are not known with sufficient accuracy as a function of time for compensation. What
is required is a way of measuring the effect of these errors
for each image.

     One method for estimating the random positional errors is
from the location of ground control points. A ground control
point is a physical feature detectable in the image and whose
precise location is known. Once all systematic distortions
have been corrected, the residual distortion should be due to
the random positional errors. Differences between the actual
and observed ground control point locations can then be used to
model these errors. The final result is then a function which
maps a pixel at (x,y) in the image into a point (u,v) in the new

map coordinate system. The functions can be bivariate polynomials of the following form, with N typically equal to 2 or 3,

$$v = \sum_{p=0}^{N} \sum_{q=0}^{N-p} a_{pq} \, x^p y^q$$

$$u = \sum_{p=0}^{N} \sum_{q=0}^{N-p} b_{pq} \, x^p y^q$$

This method is called polynomial spatial warping and requires no knowledge of the geometric errors. However, for each image a number of control points are required for a linear least squares estimate of the coefficient a_{pq}, b_{pq}. Details are given in Hall

(1979, pp. 186-188).

D.3 Ground Control Points

Twenty to thirty ground control points scattered over a Landsat image are usually sufficient to give good estimates of the transformation coefficients. Locating these points manually can be time consuming, so that automated methods which make use of matching techniques are employed.

The matching operation is illustrated in Figure 7. There are two subimages. S is the search image, an L x L array which consists of the pixels in the image to be mapped. W is the ground control point subimage or window, an M x M array of pixels which is stored in the database. The window W is permitted to be located anywhere within S, but must lie entirely within S. For each different location of W, a subimage of size M x M in S, S_M^{ij}, is specified, where (i,j) is the coordinate of the upper left corner. As seen in Figure 7, the window W can specify $(L - M \quad 1)^2$ different subimages. Each of these subimages must be compared with W. The subimage $S_m^{i'j'}$, which is most similar to W is then chosen and the location of the ground control point is found.

Two different functions are commonly used for matching W and S_M^{ij}. If $W(l, m)$, $S_M^{ij}(l, m)$ represents the (l, m)th element, then these are as follows:

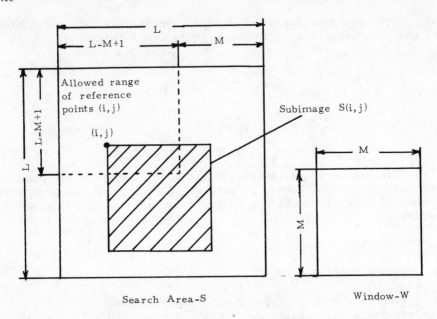

Figure 7   Search area and window for ground control point correlation.   (Bernstein and Silverman, 1971)

(i)   Normalized correlation coefficient:

$$R_N(i,j) = \frac{\sum\limits_{l=1}^{M} \sum\limits_{m=1}^{M} W(l, m)\, S_M^{ij}(l, m)}{[(\sum\limits_{l=1}^{M} \sum\limits_{m=1}^{M} (W(l, m)))^2 \; (\sum\limits_{l=1}^{M} \sum\limits_{m=1}^{M} (S_M^{ij}(l, m))^2)]^{1/2}}$$

(ii)   Absolute error function:

$$E(i, j) = \sum\limits_{l=1}^{M} \sum\limits_{m=1}^{M} |W(l, m) - S_m^{i,j}(l, m)|$$

The normalized correlation coefficient requires a considerable
amount of computer time to calculate. For large values of M (say
128) it is more effective to calculate the correlation in the
Fourier transform domain by making use of the fast Fourier transform (Anuta, 1970).   Barnea and Silverman (1972) have proposed
a very efficient scheme based upon the absolute error function.
Their technique, sequential similarity detection algorithm, results in reductions of two orders of magnitude in the number of
computations.

The principle of this algorithm is quite elegant. A threshold T is set and the error function is calculated sequentially for the pixels in the window by using some random numbering scheme. Thus, at each step in the algorithm a pixel in the window is chosen at random, the error with respect to the search image is calculated and accumulated. The calculation for a particular window stops when the threshold T is exceeded. Other windows are then tested. Only those search subimages which closely match the window require a complete calculation of the error. This can result in a considerable saving. After testing all search subimages, the one with the minimum error is chosen for the location of the ground control point. Barnea and Silverman (1972) have compared the number of basic operations for the three methods discussed. For example for a window size of 64 by 64 and a search image of 256 by 256:

(i)   Direct calculation of correlation coefficient: 6.9×10^8.

(ii) Calculation of correlation coefficient using the fast Fourier transform: 1×10^8.

(iii) Sequential Similarity Detection algorithm: 2.2×10^6.

A variation of this approach based upon a hierarchical search of the image at different resolutions is described in Hall, (1979, chapter 8). The first step is to reduce the number of pixels in both the window W and the search image S; that is, decrease the resolution. This can be effectively accomplished by averaging over disjoint sub-sections of the images (2 x 2, 3 x 3 ...) and retaining only the averaged value. The sequential similarity detection algorithm is then applied to these reduced images with a resulting decrease in processing time. The algorithm yields a number of promising locations which are then explored at a higher resolution. The sequential similarity detection algorithm is reapplied and the process continues until the best subimage at the highest resolution is found.

A significant increase in the accuracy of the correlation can be achieved by preprocessing the image. Ground control points usually lie on edges or intersection of edges; in other words, on the boundaries between different region. Nack (1975 and 1977) and Svedlow et al. (1978) describe a registration technique which first extracts the edges from the image and then the matching is performed upon the "edge images". Svedlow et al. (1978) used the following operator to construct the edge image:

$$I'(i,j) = \begin{aligned}&|I(i + 1, j) - I(i - 1, j)| \\ &+ |I(i, j - 1) - I(i, j + 1)| \\ &+ |I(i + 1, j-1) - I(i - 1, j + 1)| \\ &+ |I(i + 1, j+1) - I(i - 1, j - 1)|.\end{aligned}$$

Considerable storage and computational savings are possible if
the image is then thresholded into a binary image; that is,

$$I'(i,j) = \begin{cases} 0 & I'(i, j) > T \\ 1 & I'(i, j) < T. \end{cases}$$

A good choice of T is one in which 15-20 % of $I'(i, j)$ exceed
the threshold T. Swedlow et al. (1978) tested different techniques for finding the ground control points on several agricultural scenes. An excerpt of the results are presented in Figure
8. It is clear that use of the correlation coefficient on the
edge image yields the best results, at least for agricultural
scenes.

D.4 Interpolation and Resampling

     The remaining problem in forming the geometrically corrected
image is to evaluate the intensity values for the grid points in
the (u, v) coordinate system. By considering the inverse transformation in Figure 6a, it is seen that the black circles may
lie between four pixels of known intensity values. The value
at the point given by the black circle is a function of at least
the 4 nearest intensity values. Further mathematical analysis
shows that this value actually depends, to a much smaller extent,
on all values in the image. The function which gives the relation between the intensity value at the black point to the intensity values of the other pixels is called an interpolation function. The process is also called resampling as the image is
effectively being resampled at the new grid points. The simplest
and the most easily implemented interpolation function is the
nearest neighbour approach. The intensity value assumed at the
new point in the (u, v) grid is the intensity value of the pixel
in the (x, y) grid nearest to it. An example of nearest neighbour resampling is shown in Figure 9a, a portion of the image
has been blown up by a factor of 2 by using nearest neighbour
resampling. Effectively each pixel is repeated twice in each
direction. Notice the discontinuities in the field boundaries.

     In Figure 9b is now shown the same enlargement using the
following bilinear interpolation function:

$$X(u,v) = a_1' \, X(x,y) + a_2' \, X(x+1, \, y)$$

$$+ a_3' \, X(x,y+1) + a_4' \, X(x+1, \, y+1)$$

where

$$a_i' = 1 - \frac{a_i'}{\sum\limits_{j=1}^{4} a_j} \quad , \ i = 1,2,3,4,$$

Similarity Measure	Original Image	Magnitude of the Gradient	Thresholding the Magnitude of the Gradient
Correlation Coefficient	90%	100&	90%
Correlation Function	38%	74%	87%
Sum of absolute values of differences	69%	92%	87%

Figure 8   Percentage of acceptable registration attempts. It is clear that the use of the correlation coefficients on the gradient image yields the best results. (From Swedlow et al., 1978).

Figure 9   Interpolation and resampling. The result of a nearest neighbour resampling is shown in a). The bilinear interpolator result is shown in b) and the cubic convolution resampling result in c).

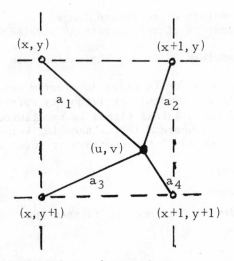

Figure 10   The bilinear interpolator. a_1, a_2, a_3 and a_4 represent the Euclidean distances from the desired point to the 4 neighbouring pixels.

where (u, v) lies between line x and x+1 and pixel y and y+1,
and a_1, a_2, a_3, a_4, are the respective Euclidean distances to

the 4 pixels in the (x, y) grid (see Figure 10). Notice the
smoother appearance of the image.

To quantify the effect of different interpolation functions,
the standard technique is to consider the entire process in the
Fourier transform domain. It is sufficient to consider the one
dimensional case, as in practice resampling is carried out first
line-by-line and then column-by-column. Resampling can then be
written as a convolution

$$g(x) = \int f_s(x) \ S \ (x - r) \ dr,$$

where, $f_s(x)$ is the original sampled function, $s(x)$ is the inter-

polation function, and $g(x)$ is the attempted recovery of the original continuous image. (Papoulis, 1962) (Jablon & Simon, 1978).

In Fourier transform domain convolution becomes multiplication, thus

$$G(w) = F_s(w) \ S(w)$$

where

G(w)   Fourier transform of recovered image
$F_s(w)$ Fourier transform of original image sampled

S(w)   Fourier transform of interpolation function.

The effect of the interpolation function is clearly seen from
this equation. It acts to reweight the frequency spectrum of
the sampled image. If the original signal is bandlimited to
frequencies less than π/T, where T is the sampling interval,
then no distortion is induced if the rectangular pulse is chosen
for S(w):

$$S(w) = \begin{cases} 1 & |w| < \pi/T \\ 0 & \text{elsewhere} \end{cases}$$

The corresponding interpolation function is then the sinc
function

$$\frac{\text{Sin } \frac{\Pi}{T} x}{\frac{\Pi}{T} x}$$

As this function has non-zero values for all t, the implication
of this result is that for **perfect** interpolation each new resampled value depends upon all the original samples.

As this interpolation function is not feasible, many approximations have been proposed (Shlien, 1979). But first, in Figures
11, 12 are shown the frequency spectrum of the nearest neighbour
and the linear interpolations. It is clearly seen that the
nearest neighbour interpolation passes more of the higher frequencies, which accounts for the discontinuities seen in Figure
9a. One possible approximation is to truncate the sinc function.
For example, if only 4 terms are permitted, then the frequency
spectrum is as shown in Figure 13. A different approximation
which yields good results is the so-called cubic convolution
(Riffman 1973):

$$S(x) = \begin{cases} 1 - 2x^2 + x^3 & 0 < |x| < 1 \\ 4 - 8|x| + 5x^2 - |x|^3 & 1 < x < 2 \\ 0 & 2 < x \end{cases}$$

This function also depends upon 4 terms, but as is shown in Figure 14, its frequency spectrum is more like the ideal than the
truncated sinc function. An example of the use of cubic convolution interpolation is shown in Figure 9c. The difference with
the bilinear is apparent upon a close inspection of the clover-
leaf. If more terms can be used, then a better performance is
to be expected. For example, if 16 terms are used, then the
truncated sinc Πx function has the frequency spectrum shown in
Figure 15. A better function to use in this case is the modified
sinc function shown in Figure 16:

$$S(x) = \frac{\sin x}{x} \left(1 - \frac{x^2}{64} \right)$$

As can be seen from the frequency response, it is very close to
the ideal interpolation function.

D.5 Example of Geometric Correction Process

As an example of the steps involved in geometric correction,
the digital image correction system (DICS) at the Canada Centre
for Remote Sensing is described. This is a dedicated computer
system built around a special purpose corrector subsystem which
carries out the spatial transformations and the necessary interpolations. It is controlled by a PDP 11/70 which is used for
finding the ground control point, calculating the coefficients
of the geometric transformation, and various other functions

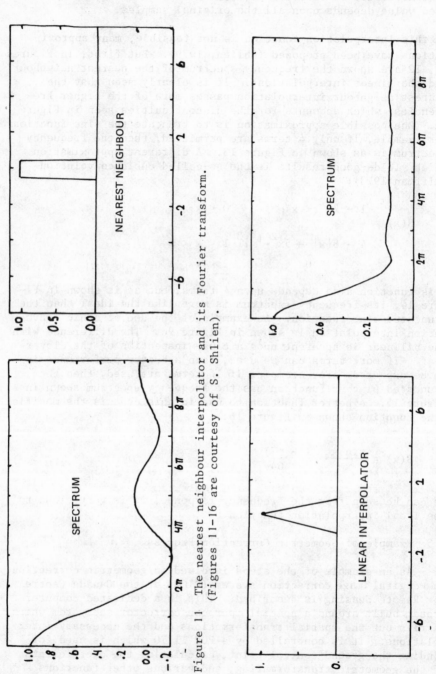

Figure 11 The nearest neighbour interpolator and its Fourier transform.
(Figures 11–16 are courtesy of S. Shlien).

Figure 12 The linear interpolator and its Fourier transforms.

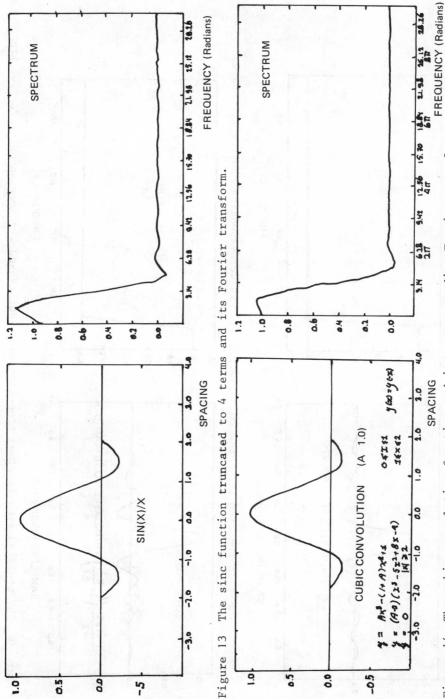

Figure 13 The sinc function truncated to 4 terms and its Fourier transform.

Figure 14 The cubic convolution function and its corresponding Fourier transform.

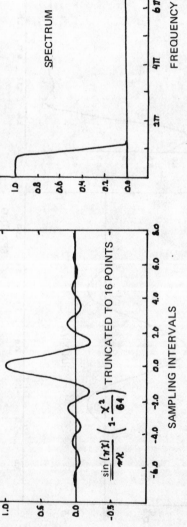

Figure 15   The sinc function truncated to 16 points and its corresponding Fourier transform.

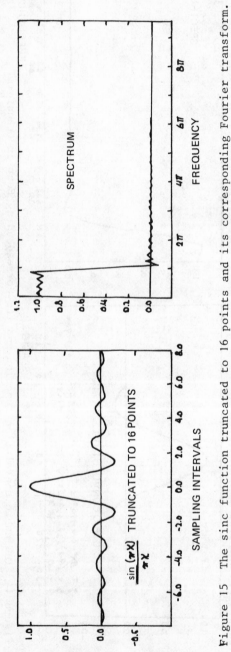

Figure 16   The modified sinc function truncated to 16 points and its corresponding Fourier transform.

such as radiometric correction (Butler et al. 1978, Guertin et al., 1979). The output is an image in UTM projection coordinates containing 1120 lines and between 400 to 1700 pixels. Each pixel corresponds to a 50 by 50 metre area and the positional accuracy is on the order of 0.5 pixels.

There are two distinct modes of operation of the system; rectification of the first Landsat image of a given area to a UTM projection coordinate system and registration of subsequent Landsat images to the first image. These are described separately.

The following steps are performed in rectification:

Step 1: Locate manually 20 ground control points choosen from the image on the map.

Step 2: For these 20 points, correct distortions due to the systematic geometric errors.

Step 3: Evaluate the coefficients of a second order polynomial transformation, from these ground control points. Note that this transformation now only corrects the remaining random errors.

Step 4: Apply correction for the systematic errors and the transformation for the remaining errors to the image.

Step 5: Interpolate the required intensity values by using the modified sinc function.

Steps one to three are performed by the computer, Steps 4 and 5 by the special purpose corrector.

Subsequent images are now registered with the rectified first image as reference. The steps are as follows:

Step 1: Align a small number (less than 5) of ground control points in the original reference image and the new image.

Step 2: Calculate a first order polynomial transformation between the two images from these ground control points.

Step 3: Select 60 32x32 pixel areas in the second image, and with the first order transformation, find corresponding 64x64 pixel search areas in the reference image.

Step 4: Evaluate the correlation function between the two corresponding areas. This is performed by using the Fourier transform technique.

Step 5:   Select ground control points of suitable quality; for
          example, high peak, not near edge. Correct for the
          systematic errors.

Step 6:   Evaluate coefficients of a second order polynomial
          transformation.

Step 7:   Apply transformation to the ground control points and
          eliminate those with large positional error. Return to
          step 6 if necessary.

Step 8:   Correct, transform, and interpolate the remaining pixels
          as in steps 4 and 5 for rectification.

Steps 1 to 6 are performed by the PDP 11/70, and Steps 7 and 8
by the hardware corrector. A Landsat image requires approximately
20 minutes of CPU time and one hour of operator time for registration.

E.   Image Enhancement

     The objective of image enhancement is to change the image
in such a fashion as to highlight the features of interest. In
most cases this implies improving the visual appearance of the
image. It is important to note that each observer may be interested in different features and so may have different criteria
of usefulness. There cannot be therefore, a single optimum enhancement procedure. General treatments of the image enhancement
process are found in Gonzalez and Wintz (1977, chapter 4) and
Pratt (1978, chapter 12).

E.1 Spectral Enhancement Techniques

     Contrast enhancement is an operation which is applied to a
black and white image to augment visibility. In many images the
entire dynamic range of the sensor is not fully utilized. For
example, in Landsat imagery, scenes of water areas occupy only a
fraction of the available dynamic range, and therefore have low
contrast. This is best observed from the one dimensional histogram of band 4 (see Figure 17). Only a small number of grey
level values actually occur. A redistribution of these values,
to fill the entire dynamic range can yield considerable improvement. There are a number of ways of carrying out this redistribution. These techniques go under the name, histogram modification. Hummel (1975) presents a good review of these techniques.

Equal interval quantization

     In this technique the grey level values are linearly stretched
to fill the available dynamic range. Thus, if X_{min} and X_{max}

are, respectively, the minimum and maximum intensity, the new
intensity values are given by

$$X_{new} = \frac{K \, X_{old}}{X_{max} - X_{min}} \, ,$$

where X_{old} is the original grey level, and X_{new} is the new grey
level, and K is some constant. The transformation is best implemented by the use of a look-up table which relates the old
and new values. An example of such a transformation is shown
in Figure 18.

Logarithmic transformation

This is an example of a non-linear quantization of the intensity values:

$$X_{new} = K \log X_{old}.$$

The effect of this transformation is to compress the high grey
level values and expand the low ones. An example of such a
transformation is given in Figure 19.

Equal probability quantization

The objective of this transformation is to yield a flat
histogram; that is, the grey level values are uniformly distributed. This quantization scheme can be shown to maximize the
expected information content or entropy.

Density slicing

This is a generalization of equal interval quantization.
N threshold values are chosen, usually at non-uniform intervals.
The grey levels are now mapped into one of N+1 values. An example of such a transformation is shown in Figure 20. Note the
formation of contours in the transformed image. These contains
may not always be significant. As an example see Figure 21
which was formed by setting different threshold values.

Other methods

Dunne et al. (1971) apply a logarithmic transformation
about the mean grey level value. This results in a flatter and
broader histogram. Frei (1977) describes a transformation which
takes into account the properties of the human visual system.
Carroll and Robinson (1977) describe a logarithmic transformation

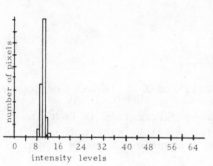

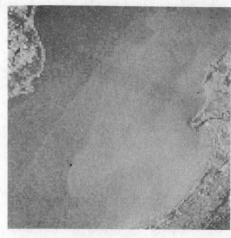

Figure 17   The histogram in b) shows that the image a) does not
            cover entirely the dynamic range of the display device,
            resulting in a low contrast picture.

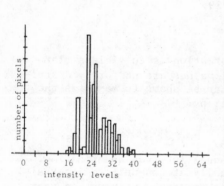

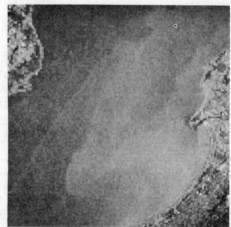

Figure 18   The histogram of Figure 17 has been linearily expanded
            as shown in a), thus covering the entire dynamic range
            of the display device. The resulting picture b) has
            therefore more contrast.

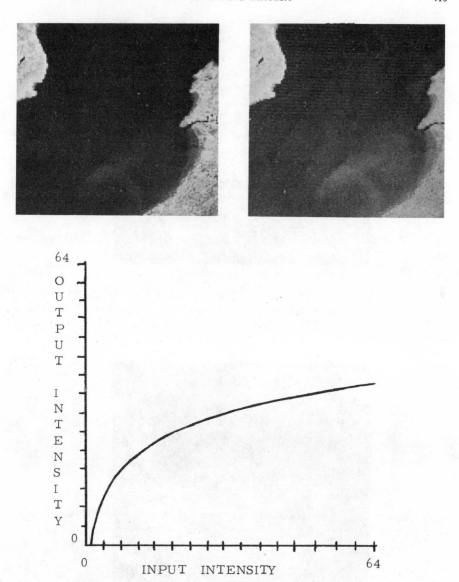

Figure 19 The histogram of the image in a) has been passed through
          a non-linear transformation, namely a logarithmic
          compression function b). The result is shown in c).
          It has for effect to compress the high grey levels and
          to expand the low grey levels. As a result, more detail
          can be seen in the dark areas, namely in the water in
          this figure.

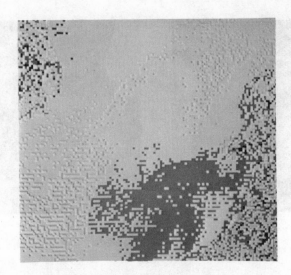

Figure 20   Density slicing has been applied to Figure 19). By
            setting 4 threshold values, the image has been sliced
            into 5 intensities.

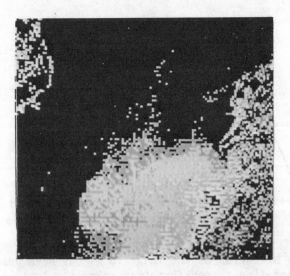

Figure 21   The same density slicing operation as in Figure 20
            has been applied, but the threshold values have been
            changed. Note that the boundaries between each level
            have changed position. Therefore the boundaries
            obtained by density slicing may not always be relevant.

in terms of the homomorphic transformation (Stockham, 1972).
Gonzalez and Wintz (1977, p. 127) describe a method in which the
histogram of the transformed grey level values is specified. This
method is useful for interactive analysis and is a generalization
of equal probability quantization.

E.2 Spatial Enhancement Techniques

The transformations described so far act on each pixel independently of its neighbourhood; that is, no use has been made
of spatial information. In this section we consider two examples
of transformation which operate on a number of pixels at a time
to give one new pixel value: image smoothing and image sharpening.
These transformations require considerably more computer time
than the previous techniques.

Image smoothing is employed when the image appears to have
many random isolated pixels in error. A simple technique is to
replace the grey level value by an average of the neighbouring
value. For example, if X_{ij} is the intensity of the pixel in line

i and column j, then

$$X'_{ij} = \sum_{r=-m}^{m} \sum_{s=-n}^{m} W_{rs} X_{(i+r)(j+s)}$$

This operation tends to smooth out noisy pixels, but unfortunately
blurs the image as well. Here W_{rs} is a weighting function and

n, m determine the window size of the neighbourhood. A simple
example is n=1, m=1, all W_{rs}=1/9. It is important to note that

the averaging operation can often be carried out recursively and
is independent of the number of pixels in the window. This can
be seen by first performing the running average (recursively)
along the row, and then performing the running average (recursively) along the column. Only two adds, two subtracts and one
divide are required for each pixel. A different smoothing algorithm, which does not cause as much blurring, is the median filter. Here, the pixel value under consideration is replaced by
the median value of the neighbourhood; that is, all the values
are ordered and the middle value is chosen. Narendra (1978)
describes such a filter, and its implementation by a recursive
method. An example is shown in Figure 22.

The opposite operation to image smoothing is image sharpening; also called edge sharpening, as it enhances edges. This
operation is performed by carrying out some derivative operator
on the image. One example is the Roberts gradient, which is the

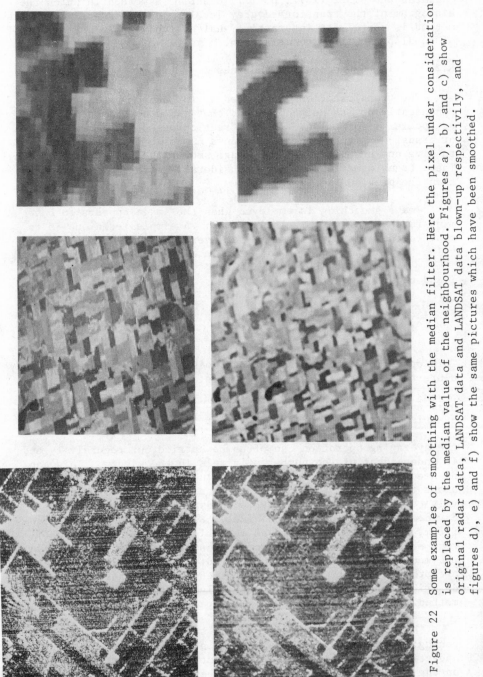

Figure 22   Some examples of smoothing with the median filter. Here the pixel under consideration is replaced by the median value of the neighbourhood. Figures a), b) and c) show original radar data, LANDSAT data and LANDSAT data blown-up respectivily, and figures d), e) and f) show the same pictures which have been smoothed.

sum of the magnitude of differences between diagonally adjacent pixels:

$$X'_{i,j} = \left|X_{i,j} - X_{i+1,\ j+1}\right| + \left|X_{i,j+1} - X_{i+1,j}\right|$$

An example is shown in Figure 23. Note that field boundaries are enhanced.

Spatial enhancements are examples of filtering operations which can also be implemented in the two dimensional spatial frequency domain by Fourier transform techniques. Image smoothing corresponds to low-pass filtering and image sharpening corresponds to high-pass filtering. In the first case only low spatial frequencies are left; that is, areas which are slowly changing. In the second case, the high frequencies are emphasized; that is, only rapidly changing areas, such as edges.

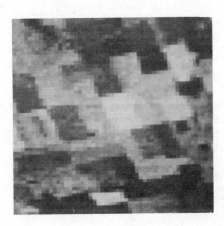 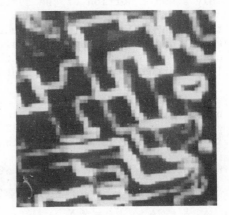

Figure 23 The Roberts gradiant operator has been applied to a), yielding the result in b) where it is seen that the boundaries have been greately enhanced.

E.3 Colour Display

The use of a colour display device increases significantly the potential for visual analysis. One simple, but effective, technique is to map a black and white image into a pseudo-colour image. For example, in density slicing each interval could be mapped into a different colour and so enhance the contrast. In

the following, the problem of displaying multiband imagery on
a colour television monitor is discussed. Pearson (1975) is
a good source of background information on colour display techniques.

Landsat imagery contains 4 channels of imagery. A colour
television monitor can only accept three streams of image data;
as a result some compromise must be made. The simple approach
is to choose three bands and map each directly into one of the
three primary colours, red, green, and blue. For Landsat, the
usual product is band 4 into blue, band 5 into green, and band
6 or band 7 into red. This yields the traditional false colour
photograph which simulates infrared colour film. In this case,
vegetation is represented by shades of red and brown.

This procedure has two basic deficiencies: one related
to the data and the other to the response of the human visual
system. At the data level, only 3 bands can be displayed at
anytime. Furthermore, the data is highly correlated so that
only a small portion of the available colour space is used.
As an example, yellow is seldom seen in Landsat imagery. At
the human visual level, the three primary colours used, red,
green, and blue (R, B, G) are also correlated; that is, the
eye does not respond independently to these three colours.
Thus, further correlation is introduced to the displayed data.

A standard technique to eliminate correlation between the
bands is the use of the Karhunen-Loève (or principal components) expansion (Fukunaga, 1972, p. 227), (Taylor, 1974). If
$X = (x_1, \ldots x_i)$ is an n-dimensional random vector, then X can

be represented by an expansion of the form:

$$X = \sum_{i=1}^{n} y_i \emptyset_i = \emptyset\, Y$$

$$Y = (y_1, \ldots y_n)\, T$$

$$\emptyset = (\emptyset_1, \ldots \emptyset_n)$$

Here Y is the new random vector, and the $\emptyset_i$ are the new basis

vectors. If the $\emptyset_i$ are chosen to be the eigenvectors of the

covariance matrix of X, Σ_x; that is

$$\Sigma_X \emptyset_i = \lambda_i \emptyset_i \quad,$$

then the components y_i are mutually uncorrelated and the covariance matrix of Y is

$$\Sigma_y = \begin{bmatrix} \lambda_1 & 0 \\ & \\ 0 & \lambda_n \end{bmatrix}$$

The λ_i are the eigenvalues and turn out to be the variances of the y_i. The usual procedure is to reorder the indices so that

$$\lambda_1 > \lambda_2 > \ldots > \lambda_n.$$

The larger the variance λ_i, the greater the variability in the corresponding component, the greater the information content. The transformation for the two dimensional case is shown in Figure 24.

The steps involved are summarized:

(i)   The individual bands are first logarithmically enhanced,
(ii) The covariance matrix is calculated from a sample of pixels,
(iii) The eigenvector and eigenvalues are then calculated and ordered.

For typical Landsat scenes, Taylor (1974) found that the first component accounts for 90% of the variance, the second for 79% of the remainder, and the third for 74% of the remainder. Thus very little information is lost by dropping the fourth component. Figure 25 shows individually the four principal components. The decrease in information content is dramatic.

To reduce the correlation introduced by the human visual system, the 3 principal components must be mapped into three uncorrelated dimensions in the colour space. Taylor (1974) has chosen these as, (see Figure 26),

(i)   brightness = R + G + B,

(ii) red-green = R - G,

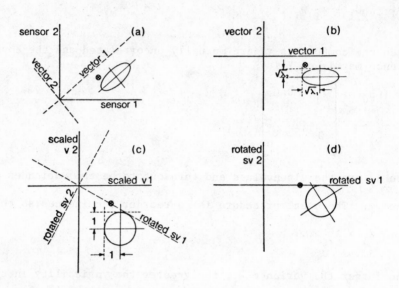

Figure 24   Stages in the construction of the linear transformation
            from sensor data to independent vector form.
            (a) The data from two sensors in correlated, and
            therefore is distributed in a roughly elliptical form
            with the main axes diagonal. A single isolated data
            point is shown. The eigenvector solution of the
            covariance matrix determines vectors parallel to the
            main axes of the ellipse. (b) The data distribution
            rotated into the eigenvector representation. The
            description in this space is not correlated, as is
            shown by the fact that the axes of the ellipse are
            parallel to the coordinate axes. The standard deviations
            are unequal, being the square roots of the eigenvalues.
            (c) Each dimension is scaled by the square root of the
            eigenvalues, so that the standard deviation of the
            distribution is now unity in every direction. There
            are no longer any main axes to the data distribution,
            and the space can be rotated freely without introducing
            correlation into the description. (d) An arbitrary
            rotation which places the isolated data along an axis.
            Its representation in this space is now (k, 0).
            (From Taylor, M.M., 1974).

Figure 25   Example of a principle components expansion. For
typical LANDSAT scenes, at this one, Taylor (1974)
found that the first component a) accounts for 96% of
the variance, the second b) for 79% of the remainder,
the third c) for 74% of the remainder, and thus very
little information is lost by dropping the fourth
component d). The decrease in information is seen to
be a very dramatic one.

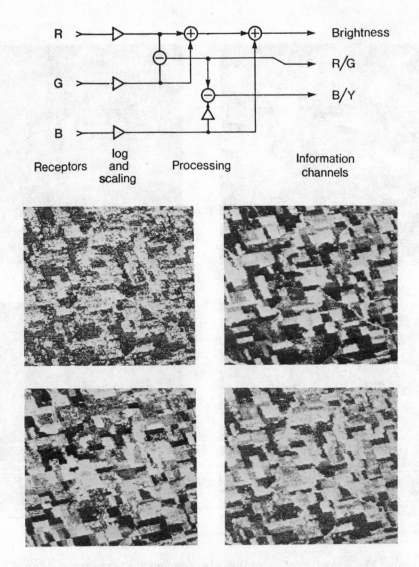

Figure 26   a) The three principal components are mapped into
three uncorrelated dimensions in the colour space
(Taylor, M.M., 1974). This reduces, the correlation
introduced by the human visual system. An example
shows the b) Br component, c) the R/G component and
d) the B/Y component. Figure e) shows in black and
white what would appear in colour on a display screen.
Note that since the brightness of each component is
now equal, the use of this transformation is only valid
for colour evaluation.

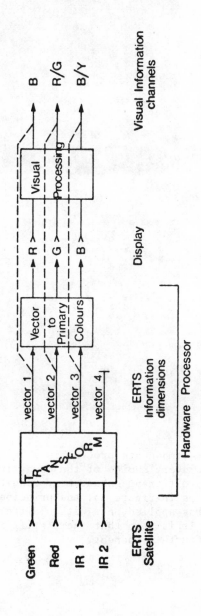

Figure 27 Figure is a schematic description of the principal components colour display. The four data streams are transformed into four streams of uncorrelated data (vector 1-4), three of which are to be mapped onto the three dimensions of colour space. The algorithm to convert from the vector representation to colour primaries is supposed to be the inverse of the processing applied by the visual system to produce the three dimensions from the sensed primaries of the display. (From Taylor, M.M., 1974).

Figure 28 Since the principal components process is a self-
optimizing one, for every choice of the covaraince
matrix, an entirely different enhancement results.
For the same scene as in figure 25, the principal
component process was applied, but with a different
covariance matrix. It is seen that there still is
information in the fourth component.

(iii) blue/yellow = (R - G) - B.

Before being sent to the guns the three components must be contrast enhanced.   The complete process of principal component enhancement is summarized in Figure 27.

The principal component process is a self-optimizing one; that is, for every different choice of the covariance matrix an entirely different enhancement results.   This point is illustrated in Figure 28.   Different samples were used to calculate the covariance matrix, so that the enhancement, as shown by the display of the four components is very different. The interpretation of the enhanced images is therefore difficult.   Nevertheless, they clearly accentuate very subtle differences in the image.

A different solution to the display problem has been described by Goetz et al. (1975).   They suggest the use of brightness, hue and saturation as the three dimensions of the visual system.   The image data itself is not transformed, and only three bands are chosen for display.   This method has the advantage of being more invariant to different images.

F.   Multitemporal Images

F.1   Introduction

Temporal analysis, the exploitation of spatially registered imagery of different times, adds a new dimension of information.   Studies have shown that temporal variations are clearly evident in Landsat imagery and can be effectively used.   There are two different objectives in using multitemporal images: change detection and improving the classification maps.   A good overview of temporal analysis techniques for Landsat imagery is given in Anuta and Bauer (1973).

F.2   Change Detection

The objective of change detection methods is to localize or highlight those regions which have undergone some change. This would be of value in the many monitoring applications of Landsat, such as snow coverage, floods, urban use, to name but a few.   Change is very subjective; it is possible that the spectral values change from one image to the next, but there be no change in the user defined classes of interest.   For example, in Figure 3 are shown the same forest area containing hardwoods and softwoods in summer and winter.   Obviously, there has been a change.   Besides the added snow, the hardwoods have lost their leaves and can now be more readily distinguished.   At a different level, the classes of user interest, hardwood and softwood, have <u>not</u> changed. They are still present but in a different form.

Different methods have been proposed for change detection and can be conveniently categorized as follows:

(i)   Statistical tests on groups of pixels
(ii)   Differencing individual pixels
(iii) Derived indices or features
(iv)   Artificial intelligence approach.

Examples of each method are given.

In the first method, groups of pixels (for example, 64 by 64 pixel areas) are compared en bloc by some statistical method. Lillestrand (1972), for example, computes a spectral correlation function

$$R(I_1, I_2) = \text{fraction of pixels with intensity value } I, \text{ in first image and } I_2 \text{ in second.}$$

Most of the values lie along some straight line and would correspond to areas of no change. Areas of change, on the other hand, would correspond to the components of $R(I_1, I_2)$ at

some distance from this line. Eghbali (1979) employed the Kolmogorov-Smirnov (K-S) test to compare 16 by 16 pixel areas for two Landsat images. The hypothesis is that the distribution of intensity values in the two images is the same. The K-S test is used to accept or reject this hypothesis and thus indicate whether some change has occurred.

Conceptually, the simplest method for detecting changes, is to subtract, on a pixel-by-pixel basis one image from the next and form a difference image:

$$f(i, j) = f_1(i, j) - f_2(i, j)$$

This new image $f(i, j)$ can be enhanced and then visually inspected (Frei et al., 1979). Areas of change will tend to have high grey scale values. A different approach is to statistically classify the difference image. In this way, different classes of change can be identified (Anuta and Bauer, 1973). If the images have been previously individually classified, then change detection can be accomplished by simply comparing the class assignments of the individual pixels. This method has the advantage of not requiring normalization and also can be used where there are large seasonal variations. Aaronson et al. (1977) show that this approach can be employed to differentiate stressed from non-stressed corn. With a similar technique, Swain (1977) demonstrated, by comparing the classifications from two different times, urban encroachment on agricultural land can be detected.

The third approach to change detection is to derive an index which can indicate certain changes. For example, Williams et al. (1979) suggest the use of various vegetative indices which can indicate and quantify changes due to defoliation in forested areas.

An example of the artificial intelligence approach to monitoring for change is given by Tenenbaum et al. (1979). In this approach the computer has stored a map of the area to be monitored and incoming imagery are compared with the map. The exact nature of the comparison depends upon the application. For example, in reservoir or shore line monitoring, only the boundary between water and land must be found and compared for changes with the map.

F.2 Multitemporal Classification

The goal here is to incorporate time information as one supplemental element in classifying imagery. Often, from a single observation in time it is not possible to distinguish two classes of interest, an example is fields of rye from wheat. If these crops have different growth patterns, then this supplemental information can help separate the two. Seeley et al. (1978) review various models for the growth of wheat.

A number of different approaches to multitemporal classification have been proposed and include the following:

Concatenate the Images:

In this method, all the observations are combined into a higher dimensional space. For example, in the Lacie project (Heydorn et al., 1978) Landsat imagery from four different dates are concatenated to form a 16-dimensional vector for each pixel, which is then classified by statistical methods.

Index Approach:

Here some index or feature is derived on a pixel basis for each observation. Thus, for each pixel a set or vector of values is formed. Engvall et al. (1977) classify the pixels

by considering a trajectory, in some two-dimensional space,
defined by this set of values. A similar technique is described by Misra and Wheeler (1978).

Other Methods:

Hlavka et al. (1979) and Kaneko (1978) attempt to combine the
spectral observations at different times with a growth model
for wheat. Swain and Hauska (1977) describe a decision tree
(or multilayered) classifier which can handle multitemporal
observations. Swain (1978b) proposes a Bayesian approach using
a cascaded classifier. Goldberg and Kourtz (1977) propose a
technique based upon comparing pixel classifications in conjunction with a reliability measure.

G. Conclusions

     A survey of some digital image processing techniques used
in the analysis of Landsat imagery has been presented. The
emphasis has been on general techniques which are normally
classified as preprocessing; that is, normalize or improve
the image. The remaining techniques which fall under the category of image recognition have only been superficially mentioned. Finally, a short review of multitemporal methods for
Landsat imagery analysis is given. Much research remains to
be performed in this area.

REFERENCES

(1) Aaronson, A.C., Buchman, P.E., Wescott, T., and Fries, R.E.
     (1977) "A Landsat Agricultural Monitoring Program", 1977
     Machine Processing of Remotely Sensed Data 4th Annual Sym-
     posium, Laboratory for Applications of Remote Sensing,
     Purdue University, West-Lafayette, Indiana, June 21-23,
     1977, pp. 44-49.

(2) Ahern, F.J., Teillet, P.M., and Goodenough, D.G. (1977)
     "Transformation of Atmospheric and Solar Illumination
     Conditions on the CCRS Image Analysis System", Proc. of
     Machine Processing of Remotely Sensed Data, West Lafayette,
     Indiana. pp. 34-51.

(3) Anuta, P.E. (1970) "Spatial Registration of Multispectral
     and Multitemporal Digital Imagery Using Fast Fourier
     Transform Techniques", IEEE Transactions on Geoscience
     Electronics, Vol. GE-8, no. 4, pp. 353-368.

(4) Anuta, P.E., and Bauer, M. (1973) "An Analysis of Temporal
     Data for Crop Species Classification and Urban Change De-
     tection", LARS Information Note 110873, Laboratory for
     Applications of Remote Sensing, Purdue Univ., West Lafayette,
     Indiana.

(5) Anuta, P.E., (1977) "Computer-Assisted Analysis Techniques
     for Remote Sensing Data Interpretation", Geophysics, Vol.
     42, no. 3, pp. 468-481.

(6) Aviv, D.G. (1979) "New On-Board Data Processing Approach
     to Achieve Large Compaction", Proc. SPIE Real-Time Signal
     Processing II, Vol. 180, pp. 48-55.

(7) Bajcsy, R., and Tavakoli, M. (1975) "Image Filtering -
     A Context Dependent Process", IEEE Trans. on Circuits
     and Systems, Vol. CAS-22, pp. 463-474.

(8) Barnea, D.I., and Silverman, H.F. (1972) "A Class of
     Algorithms for Fast Digital Image Registration", IEEE
     Transactions on Computers, Vol. C-21, pp. 179-186.

(9) Bernstein, R. and Silverman, H. (1971) "Digital Tech-
     niques for Earth Resources Image Data Processing", Pro-
     ceedings of the American Institute of Aeronotics and
     Astronotics, 8th Annual Meeting and Technology Display,
     Washington, D.C., Vol. C21, No. 2, AIAA 71-978, New-York,
     N.Y.

(10) Bernstein, R. (1976) "Digital Image Processing of Earth
      Observation Sensor Data", IBM Journal of Research and
      Development, Vol. 20, no. 1, pp. 40-57.

(11) Bernstein, R. (1978) Editor, Digital Image Processing
      for Remote Sensing, IEEE Press.

(12) Brayer, J.M., Swain, P.H., and Fu, K.S. (1977) "Modeling
      of Earth Resources Satellite Data" in Syntactic Pattern
      Recognition, Applications, Springer-Verlag, pp. 215-242.

(13) Bryant, N.A. and Zobrist, A.L. (1977) "IBIS: A Geogra-
      phic Information System Based on Digital Image Process-
      ing and Image Raster Data Type", IEEE Transactions on
      Geoscience Electronics, Vol. GE-15, no. 3, pp.152-159.

(14) Butlin, T.J., Guertin, F.E., and Vishnubhatla, S.S.
      (1978) "The CCRS Digital Image Correction System",
      Proc. Fifth Canadian Symp. on Remote Sensing, Victoria,
      B.C., pp. 271-284.

(15) Caroll, S. and Robinson, J.F. (1977) "Homomorphic Pro-
      cessing of Landsat Data", Canadian Journal of Remote
      Sensing, Vol. 3, no. 1.

(16) Castleman, K.R. (1979) Digital Image Processing, Pren-
      tice-Hall.

(17) CNES (1978) Mathematical and Physical Principles of
      Remote Sensing, Centre National D'Etudes Spatiales,
      Toulouse, France.

(18) Crane, R.B. (1971) "Preprocessing Techniques to Reduce
      Atmospheric and Sensor Variability in Multispectral
      Scanner Data", Proc. 7th International Symposium on
      Remote Sensing of Environment, 17-21 May 1971, University
      of Michigan, Ann Arbor, Michigan, 1345-1355.

(19) Department of the Army (1958) "Universal Transverse Mer-
      cator Grid", Department of the Army Technical Manual
      TMS-241-8.

(20) Duda, R.O. and Hart, P.E. (1973) Pattern Classification
      and Scene Analysis Wiley.

(21) Dunne, J.A. et al. "Maximum Discrimiability Versions of
      the Near-Encounter Mariner Pictures", Journal Geophysics
      Research, Vol. 76, p. 438. (1971)

(22) Eghbali, H.J., (1979) "K-S Test for Detecting Changes from Landsat Imagery", IEEE Trans. Systems, Man, Cybernetics, Vol. SMC-9, pp. 17-23.

(23) Engvall, J.L., Tubbs, J.D., and Holmes, Q.A. (1977) "Pattern Recognition of Landsat Data Based Upon Temporal Trend Analysis", Remote Sensing of Environment, Vol. 6, pp. 303-314.

(24) Frei, W. (1977) "Image Enhancement by Histogram Hyperbolization", Computer Graphics and Image Processing 6, pp. 286-294.

(25) Frei, W., Shibata, T., and Huth, G.C. (1979) "Environmental Change Detection in Digitally Registered Aerial Photographs", Proc. SPIE Applications of Digital Image Processing III, Vol. 207, pp. 26-31.

(26) Fu, K.S. "Pattern Recognition in Remote Sensing of the Earth's Resources", IEEE Transactions on Geoscience Electronics, Vol. GE-14, no. 1, pp. 10-18. (1976)

(27) Fukunaga, K. (1972) Introduction to Statistical Pattern Recognition, Academic Press.

(28) Goetz, A.F.H., Billingsley, F.C., Gillespie, A.R., Abrams, M.J., Squires, R.L., Schoemaker, E.M., Lucchitta, I., and Elston, D.P. (1975) "Application of ERTS Images and Image Processing to Regional Geologie Problems and Geologic Mapping in Northern Arizona", NASA Technical Report 32-1597, Jet Propulsion Laboratory, Pasadena, California.

(29) Goldberg, M. and Kourtz, P. (1977) "The Use of Landsat Imagery for Forestry Mapping: Description of a Proposed Operational System", Proceedings of the Workshop on Picture Data Description and Management, April 21-22, 1977, Chicago, Illinois.

(30) Gonzalez, R.C. and Wintz, P.A. (1977) Digital Image Processing, Addison-Wesley.

(31) Guertin, F.E., Butlin, T.J., and Jones, R.C. (1979) "La Correction Géométrique Des Images Landsat au Centre Canadien de Télédétection", Canadian Journal of Remote Sensing, Vol. 5, no. 2, pp. 118-127.

(32) Hajic, E.J. and Simonett, D.S. (1976) "Comparison of
      Qualitative and Quantitative Image Analysis", Remote
      Sensing of Environment, Edited by Lintz, J. Jr. and
      Simonett, D.S., Addison-Wesley.

(33) Hall, E.L. (1979) Computer Image Processing and Recogni-
      tion, Academic Press.

(34) Haralick, R.M. (1973) "Glossary and Index to Remotely
      Sensed Image Pattern Recognition Concepts", Pattern
      Recognition, Vol. 5, pp. 391-403.

(35) Haralick, R.M. (1976) "Automatic Remote Sensor Image
      Processing", Digital Picture Analysis, Edited by Rosen-
      feld, A., Springer-Verlag, pp. 5-63.

(36) Hartl, P. (1976) "Digital Picture Processing", Remote
      Sensing for Environmental Sciences, Erwin Schande editor,
      tor, Springer-Verlag, pp. 304-349.

(37) Heydorn, R.P. et al. (1978) "Classification and Mensu-
      ration of LACIE Segments", The LACIE Symposium, NASA
      Johnson Space Center, pp. 73-86.

(38) Hilbert, E.E. (1975) "Joint Pattern Recognition/Data
      Compression Concept for ERTS Multispectral Image",
      Proc.SPIE, Vol. 66.

(39) Hlavka, C.A. et al. (1979) "Multitemporal Classification
      of Winter Wheat Using a Growth State Model", Proc.
      Machine Processing of Remotely Sensed Data, West Lafa-
      yette, Indiana, pp. 105-115.

(40) Horn, B.K.P., and Bachman, B.L. (1978) "Using Synthetic
      Images to Register Real Images with Surface Models",
      Communications of the ACM, Vol. 21, no. 11, pp.914-924.

(41) Horn, B.K.P., and Woodham, R.J., (1979) "Destriping
      Landsat MSS Images by Histogram Modification", Computer
      Graphics and Image Processing, Vol. 10, pp. 69-83.

(42) Huang, T.S. (1972) "Bandwidth Compression of Optical
      Images", in Progress in Optics, Volume X, Ed. Wolf, E.,
      North-Holland Publishing, pp. 1-44.

(43) Huang, T.S. (1975) Ed., Picture Processing and Digital
      Filtering, Springer-Verlag.

(44) Huffman, D.A., (1952) "A Method for the Construction of
      Minimum Redundancy Codes", Proc. IRE, Vol. 40, pp.
      1098-1101.

(45) Hummel, R.A. (1975) "Histogram Modification Techniques",
      Computer Graphics and Image Processing 4, pp. 209-224.

(46) Kaneko, T. (1978) "Crop Classifications Using Time Fea-
      tures Computed from Multitemporal Multispectral Data"
      Proc. Fourth Int. Joint Conf. on Pattern Recognition,
      Kyoto, Japan, pp. 943-945.

(47) Kazmierczak, H. (1977)"Processing of Image Data for Remote
      Sensing Application", in Digital Image Processing and Ana-
      lysis, Ed.Simon, J.C., & Rosenfeld, A., Noordhoff, pp.317-337.

(48) Lars (1975, 1976, 1977, 1979) Symposium on Machine Pro-
      cessing of Remotely Sensed Data, Purdue University,
      West Lafayette, Indiana, IEEE.

(49) Lillestrand, R.L. (1972) "Techniques for Change Detec-
      tion", IEEE Trans. Computers, Vol. C-21, pp. 654-659.

(50) Lintz, J. Jr., and Simonett, D.S. (1976) Editors,
      Remote Sensing of Environment, Addison-Wesley.

(51) Lowitz, G.E. (1978a) "Image Data Reduction", Proc. Int.
      Conf. on Spacecraft On-Board Data Management, Nice,
      France, pp. 291-298.

(52) Lowitz, G.E. (1978b) "Compression des Données Images
      Par Reconnaissance des Formes et Clustering", Proc.
      Int. Conf. on Earth Observation from Space and Manage-
      ment of Planetary Resources, Toulouse, France, pp.243-
      250.

(53) Misra, P.N., and Wheeler, S.G. (1978) "Crop Classifica-
      tion with Landsat Multispectral Scanner Data", Pattern
      Recognition, Vol. 10, pp. 1-13.

(54) Nack, M.L. (1975) "Temporal Registration of Multispec-
      tral Digital Satellite Images Using their Edge Images",
      A AS/AFAA Astrodynamics Specialist Conference, Nassau,
      Bahamas, July 28-30, 1975, paper no. AAS 75-104.

(55) Nack, M.L. (1977) "Rectification and Registration of
      Digital Images and the Effect of Cloud Detection",
      1977 Machine Processing of Remotely Sensed Data Symposium,
      Laboratory for Applications of Remote Sensing, Purdue
      University, West-Lafayette, Indiana, pp. 12-23.

(56) Nagy, G., and Wagle, S. (1979) "Geographic Data Pro-
      cessing", Computing Surveys, Vol. 11, no. 2, pp. 139-
      181.

(57) Narendra, P.M. (1978) "A Separable Median Filter for
      Image Noise Smoothing", IEEE Proceedings on Pattern
      Recognition and Image Processing, Chicago, pp.137-141.

(58) Nasa (1976) Landsat Data Users Handbook, Document no.
      76SD4258, Goddard Space Flight Centre, Greenbelt, Md.

(59) Orhaug, T., and Akersten, S.I. (1976) "Digital Pro-
      cessing of Multispectral Data", FOA Rapport C 30075-E1,
      Forsvarets forskningsanstalt, Huvudavdelning 3, 104 50
      Stockholm 80.

(60) Papoulis, A. (1962) The Fourier Integral and its Appli-
      cations, McGraw-Hill Book Company, Inc.

(61) Pearson, D.E. (1975) Transmission and Display of Pic-
      torial Information, John Wiley and Sons.

(62) Philipps, T.L., and Swain, P.H. (1978) "Data-Processing
      Methods and Systems", in Remote Sensing: The Quantita-
      tive Approach, Ed. Swain, P.H., and Davis, S.M., pp.
      188-226.

(63) Pratt, W.K. (1978) Digital Image Processing, Wiley-
      Interscience.

(64) Pratt, W.K. (1979) Editor, Image Transmission Techniques,
      Academic Press.

(65) Ready, P.J. and Wintz, P.A. (1973) "Information Extrac-
      tion, SNR Improvement, and Data Compression in Multi-
      spectral Imagery", IEEE Transactions on Communication,
      Vol. COM-21, No. 10, pp. 1123-1131.

(66) Richardson, B.F., Jr. (1978) Editor, Introduction to
      Remote Sensing of the Environment, Kendall Hunt.

(67) Riffman, S.S. (1973), "Digital Rectification of ERTS
      Multispectral Imagery", Symp. on Significant Results
      Obtained from the Earth Resources Technology Satellite,
      NASA Sp 327, Goddard Space Flight Center, pp. 1131-1142.

(68) Riseman, E.M., and Arib, M.A. (1977) "Computational
      Techniques in the Visual Segmentation of Static Scenes",
      Computer Graphics and Image Processing 6, pp. 221-276.

(69) Rosenfeld, A. (1976) Editor, Digital Picture Analysis,
      Springer-Verlag.

(70) Rosenfeld, A. and Kak, A.C. (1976) Digital Picture
      Processing, Academic Press.

(71) Rosenfeld, A., and Weszka, J.S. (1976) "Picture Recogni-
      tion" in Digital Pattern Recognition, Ed. Fu, K.S.,
      Springer-Verlag, pp. 135-166.

(72) Sakrison, D.J. (1979)"Image Coding Applications of Vision
      Models", in Image Transmission Techniques, Ed., Pratt,
      W.K. Academic Press, pp. 21-71.

(73) Salmonson, V.V., and Park, A.B. (1979) "An Overview of
      the Landsat-D Project with Emphasis on the Flight Seg-
      ment", Proc. Machine Processing of Remotely Sensed Data
      LARS, West Lafayette, Indiana, pp. 2-11.

(74) Seeley, M.W. et al. (1978) "Prediction of Wheat Pheno-
      logical Development: A State-of-the-Art Review", The
      LACIE Symposium, NASA Johnson Space Center, pp. 981-990.

(75) Simon, J.C. and Rosenfeld, A. (1977) Editors, Digital
      Image Processing and Analysis, Noordhoff, Leyden.

(76) Simonett, D.S., Smith, T.R., Tobler, W., Marks, D.G.,
      Frew, J.E., and Dozier, J.C. (1978) "Geobase Information
      System Impacts on Space Image Formats", SBRSU Techni-
      cal Report 3, Santa Barbara Remote Sensing Unit, Santa
      Barbara, California.

(77) SPIE (Society of Photo-Optical Instrumentation Engineers),
      (1979) Digital Processing of Aerial Images, Vol. 186 of
      Proceedings, Alabama.

(78) Stockham, T.G., Jr. (1972) "Image Processing in the Con-
      tent of a Visual Model", Proceedings of the IEEE, July
      1972, pp. 828-842.

(79) Svedlow, M., McGillem, C.D., and Anuta, P.E. (1978)
      "Image Registration: Similarity Measure and Prepro-
      cessing Method Comparisons", IEEE Transactions on
      Aerospace and Electro mc Systems, Vol. AES-14, no. 1,
      pp. 141-149.

(80) Swain, P.H. (1977) "Advancements in Machine-Assisted Analysis of Multispectral Data for Land Use Applications", 1977 Machine Processing of Remotely Sensed Data 4th Annual Symposium, Laboratory for Applications of Remote Sensing, Purdue University, West-Lafayette, Indiana, June 21-23, 1977, pp. 336-343.

(81) Swain, P.H., and Hauska, H. (1977) "The Decision Tree Classifier: Design and Potential", IEEE Transactions on Geoscience Electronics, Vol. GE-15, no. 3, pp. 142-147.

(82) Swain, P.H. (1978a) "Fundamentals of Pattern Recognition in Remote Sensing" in Remote Sensing: The Quantitative Approach, Ed. Swain, P.H., and Davis, S.M. McGraw-Hill, pp. 136-185.

(83) Swain, P.H. (1978b) "Bayesian Classification in a Time-Varying Environment", IEEE Trans. on Systems, Man, and Cybernetics, Vol. SMC-8, pp. 879-883.

(84) Swain, P.H., and Davis, S.M. (1978), Editors, Remote Sensing: The Quantitative Approach, McGraw-Hill.

(85) Taylor, M.M. (1974) "Principal Components Colour Display of ERTS Imagery", Proc. 2nd Canadian Symposium on Remote Sensing, Guelph, Ontario, April 29-May 1, 1974, pp. 295-312.

(86) Tenenbaum, J.M. et al. (1979) "Map-Guided Interpretation of Remotely-Sensed Imagery", Proc. Pattern Recognition Image Processing, Chicago, Ill., pp. 610-615.

(87) Tou, J.T., Gonzalez, R.C. (1974) Pattern Recognition Principles, Addison-Wesley.

(88) Turner, R.E., and Spencer, M.M. (1972) "Atmospheric Models for Correction of Spacecraft Data", Proc. of Eight Int. Symp. on Remote Sensing of Environment, Michigan, pp. 895-934.

(89) Van Wie, P., and Stein, M. (1977) "A Landsat Digital Image Rectification System", IEEE Transactions on Geoscience Electronics, Vol. GE-15, no. 3, pp. 130-137.

(90) Vishnubhatla, S.S. (1977) "Radiometric Correction of Landsat I and Landsat II MSS Data", CCRS Technical Note 77-1.

(91) Williams, D.L., Stauffer, M.L., and Leung, K.C. (1979) "A Forester's Look at the Application of Image Manipulation Techniques for Multitemporal Landsat Data", Proc. Machine Processing of Remotely Sensed Data, West Lafayette, Indiana, pp. 368-374.

(92) Wilson, R.G., Sivertson, W.E., Jr., and Bullock, G.F. (1979) "Adaptive Remote Sensing Technology for Feature Recognition and Tracking" Proc. Pattern Recognition and Image Processing, Chicago, Ill., pp. 623-629.

(93) Wintz, P.A. et al. (1977) "Satellite On-Board Processing for Earth Resources Data", in Digital Image Processing and Analysis, Ed., Simon, J.C. and Rosenfeld, A., Noordhoff, Leyden, pp. 269-293.

(94) Zucker, S.W. (1976) "Region Growing: Childhood and Adolescence", Computer Graphics and Image Processing 5, pp. 382-399.

(95) Shlien, S. (1979) "Geometric Correction Registration and Resampling of Landsat Imagery", Canadian Journal of Remote Sensing 5, pp. 74-89.

(96) Jablon, J. and Simon, J.C. (1978), Application des Modèles Numériques en Physique, Birkhauser Verlag, Basel

A DECISION THEORY AND SCENE ANALYSIS BASED APPROACH
TO REMOTE SENSING

Ezio Catanzariti

Istituto di Fisica Teorica
Mostra d'Oltremare Pad. 19
80125 - Napoli, Italy

ABSTRACT

Most automatic interpretation systems for remote sensed images have used techniques based on the statistical Pattern Recognition classification model. This approach is inadequate for
fully exploiting and using the large amount of spatial and semantic information present in the image. On the other hand, some
computational models have been developed in Scene Analysis that
try to capture the complexity of the vision process. We suggest
that integration of the main features of the classification approach with these Scene Analysis models can provide adequately
complex computational structures for Remote Sensing tasks. Some
classification algorithms based on one of these models are briefly discussed.

1. BACKGROUND

1.1. The Basic Approach

Classification systems for remote sensed images must satisfy
several equally important but often conflicting requirements,
e.g., accuracy of classification, computational efficiency, generality of the system in terms of its applicability to different
image domains, etc.

The most used systems are based largely on traditional Pattern Recognition techniques. Bayesian decision theory provides
a complete, hence very powerful, formal framework within which
such techniques have been developed.

J. C. Simon and R. M. Haralick (eds.), Digital Image Processing, 439–447.
Copyright © 1981 by D. Reidel Publishing Company.

The classification can be supervised or unsupervised, depending on the availability of ground truth data. In both cases
statistical properties of the classes of interest are taken into
account in order to arrive at a classification scheme. In this
model the data samples are represented as points in an n-space.
Each coordinate in such a space represents a measurable feature
of the data sample. Each class is assumed to have associated
with it a multivariate probability distribution in n-space. Once
the form of the distribution is known, the probability of membership in each class is computed for each pixel in the scene. The
pixel can now be assigned to a class according to a criterion
that minimizes the probability of misclassification (Bayes criterion). Usually the Bayes criterion is taken in the form of the
maximum likelihood criterion: a pixel is assigned to the class
that gives it the highest probability of membership (point by
point classification).

The attraction of the classification model is its computational simplicity, due to certain 'heuristic' circumstances, such
as the validity of the normal density assumption for the multispectral data (1). Moreover, it is a domain indipendent theory,
and hence very suitable for general Remote Sensing applications.

The point of view underlying this approach is that each pixel can be classified context free on the ground of its spectral
signature only. Unfortunately, such a simple hypothesis has been
proved inadequate for taking into account real world classification problems such as noise, heavily textured areas, spectrally
overlapping classes, and scenes with class statistics not stationary over the image.

1.2. Other Attempts

The awareness (2) of the fundamental weakness of the p. by
p. classification scheme has led to many attempts to introduce
contextual information into the classification process.

Some use of spatial context is frequently made by post-processing procedures which relabel single pixels or groups of pixels
if they appear to be too isolated (3) (4). Techniques like these
are concerned only tangentially with classification accuracy;
rather they are most effective in improving the readability of a
picture by eliminating 'salt and pepper' noise from classified
images.

At other times some spatial features (like texture) are computed for groups of pixels and added as additional dimensions to
the feature space. Classification is then performed using a p.
by p. classifier which operates in this augmented feature space
(5). Improvements are reported in classification accuracy over

plane (i.e., using only spectral features) p. by p. classification. These methods represent already successful attempts at
generalizing the classification model.

In (6), a two-step classification procedure is presented.
In the first step the whole scene is recursively partitioned in
rectangular blocks such that each block is 'likely' to contain
pixels from a single class, and each class of interest is approximated by a union of blocks. In the second step the blocks are
classified using supervised and unsupervised classification.
Following this technique, classification of a pixel in the scene
is a result of the spectral properties of its neighbours as well
as its own. In (7) analogous methods are reported which partition the image in blocks of statistically similar pixels before
classifying them. Each block is then considered a statistical
sample, and a sample classifier, rather then a p. by p. classifier, is used to classify each sample.

1.3. Scene Analysis

Most of the above methods represent attempts to force spatial
features into the p. by p. classification scheme. The essence of
this scheme, however, remains basically unchanged, as do its
deficiencies.

A generally accepted formalization of the problem of digital
images classification, analogous to the classification model, but
able to take into account contextual information, has yet to
emerge from the work that has been done to date in Pattern Recognition, even if the need for such formalization is felt and work
is being done in this direction (8).

The situation changes drastically when we turn our attention
to the Artificial Intelligence Scene Analysis domain of research.
The aim is not simply to partition the picture into regions and
label them (classification), but to solve even more complex tasks
as well, such as grouping labelled regions into meaningful objects,
finding relationships among objects, etc.; in other words, to
construct, as complete as possible, a structural symbolic description from images of the world.

For this kind of problem the classification model is by itself inadequate. But, more important, a statistical or, in general, a purely mathematical approach to computational vision is
not pursued in Scene Analysis. The main point is that any ab-
stractly formulated approach has certain inherent limitations.
Many important things have to be taken into account and modeled,
such as our a priori expectations about the scene-domain (9), the
underlying physical structure of the scene and the image formation
process (10) or principles underlying human perception (11).

Great effort is therefore spent on working out suitable models
able to use knowledge from different sources in carrying out the
interpretation process.

The computational counterpart of such theoretical effort
generally consists of computer programs characterized by the use
of very deep knowledge in segmenting and interpreting images belonging to restricted classes of scenes such as polyhedrals or
simple natural scenes. These programs are therefore very heavy
in terms of their computational load and strongly dependent on
characteristics of their particular domain. Furthermore, their
image domains are typically simpler than those, say, for a satellite image. For this reason, in spite of interesting results on
remote sensed images (12) (13), these programs hold little attraction for the R.S. field of application.

Neverthless, they cannot be left out of consideration because
they can provide the necessary computational structures for incorporating contextual information into the classification process.

We feel the above briefly stated considerations form valid
motivations for conceiving approaches that would combine the best
features of the classification paradigm with some of the Scene
Analysis ideas on vision.

In (14) Bayesian decision theory is already used in connection with region merging techniques for interpreting an outdoor
scene.

More generally, we suggest that computational models from
Scene Analysis be adapted so as to provide a structured environment into which various types and levels of information (other
than the spectral type and pixel level) can be exploited by using
decision theory as a general mechanism to collect this information
and include it in such an environment.

In the second part of this paper some classification algorithms based on this approach are briefly discussed. Results
obtained with these algorithms on LANDSAT scenes have been documented elsewhere (15) (16). We are further experimenting on this
direction. Our aim in presenting them here is two-fold:
i) to put them into a Scene Analysis perspective, so as to bring
to light the unifying conceptual framework which encompasses them;
ii) to discuss the results from a R.S. point of view, by showing
in practical cases how the problem of meeting the various R.S.
related requirements has been approached and how further work on
the same area might be pursued.

2. SOME EXAMPLES

2.1. Interpretation-directed Segmentation

Most of the above described classification techniques can be seen as applications of a Scene Analysis paradigm for vision, which has only recently been called into question. According to this paradigm, classification would involve two main phases which must be sequentially performed: segmentation and classification. Before the objects in the picture could be assigned meaning they would have to be explicitly located. In the segmentation phase of the picture recognition process, the given image would be partitioned into a certain number of subsets on the ground of a simple equivalence relation induced by the similarity of the gray scale. Each subset would then be given a class label on the basis of its likeliness to belong to one of the classes of interest (in straight p. by p. classification the picture does not necessarily need to be segmented first).

A different approach to picture segmentation, viz. region merging, was first taken by Brice and Fenema (17). They recognize that a segmentation into regions on the basis of the above criterion is generally insufficient to reach a simple interpretation within the problem domain. Therefore these authors start with 'atomic regions' (regions of constant gray values) and use more global information to merge them into larger coherent regions. It is only at the end of the segmentation process that they assign an interpretation to the region.

Feldman and Yakimovsky (14) use Brice and Fenema's ideas as their starting point. Their work goes beyond Brice and Fenema's, however, in that they provide direct incorporation of specific problem information into the segmentation process. Their system allows the interpretation of the initial regions to control the subsequent region growing (i.e., segmentation) process. This algorithm is a computational example of an alternative paradigm for vision recently formulated in A.I. as part of the rather ambitious A.I. attempt to build up a theory of perceptual processes. According to this paradigm (18), it takes semantic information (interpretation) for segmentation to be performed sensibly. Segmentation is inseparable from interpretation and vice versa; to be meaningful segmentation needs to be guided in some way in the context of a real world model, but such a model cannot be elaborated without having first interpreted the picture in some way. Thus, the whole vision process becomes an alternation of segmentation and interpretation (the island-driving model of perception).

The following algorithm, reported in (15), is an example of how similar concepts can be successfully adapted to a R.S. task.

An initial segmentation is first obtained by making use of a slightly modified maximum likelihood classifier: for each pixel, the probability of membership in each of the classes of interest is computed. If the largest probability, say p_i, is greater than a given threshold, then the pixel is labelled as class i (strong pixel). Otherwise, the pixel is labelled as class(i,j), where j is the class corresponding to the second largest probability (ambiguous pixel).

A three-step region merging program is then applied that will get rid of the ambiguous regions, merging each of them into an adjacent unambiguous region, if the following hold:
i) The ambiguous region has received the highest probability of membership from the same class as the strong one.
ii) The newly formed region remains a strong region.
iii) The 'weak' common boundary between the two regions is above a certain threshold.
These criteria are applied in a less stringent fashion in the following two steps; in particular, the second class label of the ambiguous region is taken into account.

Results show that a 10% improvement in classification accuracy (as compared to the p. by p. classification) is obtained at the expense of an increase in computing time (about four times more, for details see (15)).

The computational success of this algorithm is strongly based on the concept, motivated by the island-driving paradigm, of allowing those regions to influence the region merging process, whose interpretation is unambiguous.

This same concept can also be used with a view toward increasing the correctness of the classification without making the computational complexity to rise at a rate not commensurate with it. An immediate analysis show that:
I) The initial segmentation seems too limitative. In fact, an ambiguous pixel, hence an ambiguous region, will end up being assigned one or the other of the two interpretations which have been considered the most likely ones from the very beginning of the classification process. This criticism has been taken into account in devising the algorithms discussed in the next section.
II) A minimum of spatial information is used in the region growing process. Regions are chosen for merging mainly on the basis of local spectral similarities along the boundaries. One can try to improve classification results by collecting (and using) statistics on more general domain related information. For example:
i) Compute certain region properties, e.g. the average signature vectors for the bands used, shape, size, a quick computable texture. Introduce a 'region classifier' into the algorithm under examination to evaluate the results of different mergings.

ii) Use a statistically evaluated 'weakest boundary first' criterion for merging. This is a global criterion which uses statistics
on boundary properties as well (14).
We are currently studying certain procedures based on the above
ideas.

2.2. A Pyramid Image Structure

Kelly (19) first describes an approach based on the idea of
planning. He considers an image of a human face and extracts a
smaller picture from it. The idea is that this reduced picture
exhibits only the gross features (large edges) of the face without the surrounding noise of the fine features. These features
are detected and used to refine the segmentation of the original
picture, i.e., to find all the edges in the picture.

Several authors have extended Kelly's idea on planning.
Different kinds of structures (20) (21) (22) have been developed
which are capable of handling hierarchical picture processing.
This same idea underlies certain multistage techniques which make
sequential use of pictures at different scales in the sampling of
remotely sensed data (23). Information gathered at any stage is
used to direct the selection of samples at the successive lower
stage.

This method of segmenting a picture through different levels
seems just another application of the island-driving model of vi-
şion we have discussed above. In fact, the segmentation obtained
at each level gives a context to (and guides) segmentation at the
successive lower level. In the algorithm we next describe, one
of such structures, Tanimoto's pyramid (20), is used for a R.S.
task.

The pyramid is built up through the top level. Level 1 of
the pyramid is the original image. The spectral signature of a
pixel at level L+1 in the pyramid is obtained by successively averaging the signatures of a square cell of four adjacent pixels
at level L. The top level is application dependent. For more
details including an 'homogeneity test' performed on boundary
pixels, see (16).

A labelling procedure is begun at the top level (i.e., a modified maximum likelihood classifier is applied). Each pixel must
be classified if the probability of belonging to one of the classes
of interest is above a threshold (a strong pixel); otherwise, it
is to be labelled as ambiguous.

Each pixel is then expanded into a block of four pixels which
retain the same label. In this way, ambiguous pixels will be sent
to the lower level to be reconsidered by the labelling algorithm.

This process is continued until eventually the lowest (highest resolution) level is reached. At this point one proceeds to classify all the remaining unclassified pixels.

Results (16) vary according to the settings of the program parameters. Notice that one of these parameters, the number of pyramid levels, form part of the a priori knowledge used to influence the segmentation process. In fact, it can be optimally chosen by having an estimate of the average sizes of the classes of interest. If this number is carefully chosen, considerable improvements result in classification accuracy over p. by p. classification, while computing time remains the same or is shortened. This improvement is due to a reduction in the total number of pixel classifications. In fact, a pixel labelled as strong at level L implies a reduction factor of $L^{(L-1)}-1$ over the p. by p. classifier, which is usually more than enough to compensate the time needed to build the pyramid.

Following this technique, another R.S. main requirement is met; the readability of the segmented image as expressed by the number of regions: a very clean output obtains as a side effect of the pyramid structure.

2.3. Region Merging and Splitting

The computational simplicity of the pyramid classifier consists in the fact that a pixel at level L represents a square region of $L^{(L-1)}$ pixels in the original image; by classifying a pixel at a certain level, one classifies a semantically uniform area at that level of resolution. This feature has been further explored by devising techniques that could make use of more refined semantics in performing the segmentation.

The following step is performed at each level in the pyramid after the labelling algorithm has been applied:
The eight neighbourhoods of each pixel are scanned. A pixel already lebelled as strong that does not have a 'sufficient' number of neighbors belonging to its own class, is considered to be possibly misclassified and is sent to the next lower level to be classified again (a split). An ambiguous pixel which has a 'large' number of strong neighbours all belonging to the same class has its own classification influenced by theirs and is therefore merged into that class.

Results (16) show that, applying this algorithm, classification accuracy can be still further improved without significantly affecting computing time.

We noticed before that in these simple pixel manipulation techniques already experimented, some use of non trivial semantics

is implied. A few experiments performed with the algorithm described in (15) show that classification performance does not improve if one gives as input to that program the image segmented
by using the pyramid technique. This is not surprising as they
use similar semantics. Future experiments will be concerned with
the use of higher level segmentation schemes in conjunction with
the pyramid structure.

REFERENCES

(1)   Crane,R.B., Malila,W.A., and Richardson,W.: 1972, IEEE Trans.
      Geo. Electron. GE-10(4), pp.158-165
(2)   Landgrebe,D.A.: 1976, Symp. on Machine Proc. of Re. Sensed
      Data, Purdue Univ., Indiana
(3)   Goldberg,M., Goodenough,M., and Shlien,S.: 1975, Third Can.
      Symp. on Re. Sensing, Edmonton, Alberta
(4)   Kan,E.P.: 1976, Fall Conv. of ASP, Seattle, Wash.
(5)   Haralick,R.M., Shanmugan,K., and Dinstein,I: 1973, IEEE
      Trans. on Syst., Man and Cyber. SMC-3, pp.610-621
(6)   Robertson,T.V.: 1973, Machine Proc. of Re. Sensed Data,
      Purdue Univ., Indiana
(7)   Landgrebe,D.A.: 1980, Comput. Graph. Image Proc. 12,pp.165-175
(8)   Haralick,R.M.: 1980, NATO ASI Digital Image Proc. and
      Analysis, Bonas, France
(9)   Parma,C.C., Hanson,A.R., and Riseman,E.M.: 1980, NATO ASI
      Digital Image Proc. and Analysis, Bonas, France
(10) Tenenbaum,J.M., Fishler,M.A., and Barrow,H.G.: 1980, Comput.
      Graph. Image Proc. 12, pp.407-425
(11) Zucker,S.W.: 1980, NATO ASI Digital Image Proc. and Analysis,
      Bonas, France
(12) Bajcsy,R, and Tavakoly,M.: 1976, IEEE Trans. on Syst., Man
      and Cyber. SMC-6, pp.623-637
(13) Tenenbaum,J.M., Barrow,H.G., Bolles,R.C., Fishler,M.A., and
      Wolf,H.C.: 1979, Pat. Rec. Image Proc. (Conf. Proceed.),
      Chicago, pp.610-615
(14) Feldman,J.A., and Yakimovsky,Y.: 1974, Art. Intel. 5,pp.349-371
(15) Starr,D., and Mackworth,A.K.: 1978, Can. J. of Re. Sensing
      4-2, pp.101-107
(16) Catanzariti,E., and Mackworth,A.K.: 1978, Fifth Can. Symp.
      on Re. Sensing, Victoria, B.C.
(17) Brice,C.R., and Fenema,C.L.: 1970, Art. Intel. 1, pp.205-226
(18) Mackworth,A.K.: 1978, Comput. Vision Syst.,Academic P.,pp.53-60
(19) Kelly,M.D.: 1971, Machine Intel. 6, pp.397-409
(20) Tanimoto,S.L., and Pavlidis,T.: 1975, Comput. Graph. Image
      Proc. 4-2, pp.104-119
(21) Klinger,A., and Dyer,C.R.: 1976, Comput. Graph. Image Proc.
      5-1, pp.68-105
(22) Uhr,L.: 1978, Comput. Vision Syst., Academic P., pp.363-377
(23) Nichols,J.D., Gialdini,M. and Jakkola,S.: 1973, Third Earth
      Res. Tech. Sat. Symp., NASA SP-351

EXPERIMENTS IN SCHEMA-DRIVEN INTERPRETATION OF A NATURAL SCENE*

Cesare C. Parma, Allen R. Hanson and Edward M. Riseman

Computer and Information Science Department
University of Massachusetts
Amherst, Massachusetts 01003

ABSTRACT

The system under development, VISIONS, is an investigation into general issues in the construction of computer vision systems. The goal is to provide an analysis of color images of outdoor scenes, from segmentation (or partitioning) of an image through the final stages of symbolic interpretation of that image. The output of the system is intended to be a symbolic representation of the three-dimensional world depicted in the two-dimensional image, including the naming of objects, their placement in three-dimensional space, and the ability to predict from this representation the rough appearance of the scene from other points of view. Research in segmentation and interpretation has been separated into the development of two major subsystems with quite different methodologies and considerations.

The focus of this paper is upon the interpretation system. The primary emphasis will be on the development of strategies by which several knowledge sources (KSs) can be integrated using expected knowledge stored in structures called 3D and 2D schemas, each of which may be general or specific to the scene under consideration. A series of increasingly more difficult experiments is outlined as an experimental methodology for developing schema-driven (e.g., top-down) control mechanisms; each succeeding experiment will assume a set of weaker

_____
*This research was supported by the Office of Naval Research under Grant N00014-75-C-0459, and the National Science Foundation under Grant MCS79-18209.

449

J. C. Simon and R. M. Haralick (eds.), Digital Image Processing, 449–509.
Copyright © 1981 by D. Reidel Publishing Company.

constraints, representing image interpretation tasks where a
decreasing amount of knowledge of the situation is available.
Experimental results show current capabilities of a number of
KSs and the effectiveness of a specific 2D schema in the
interpretation of a scene.

In Memoriam

Cesare Parma, 1947-1979

On August 30, 1979 Cesare Parma, a graduate student in the COINS
Department, was struck and killed by lightning during a sudden
thunderstorm in Amherst, Massachusetts. Many of the results on
schema-driven image interpretation in Section V of this paper
were due to the hard work and creativity of Cesare. All the
members of the VISIONS group benefited greatly from the blend of
his strong intellect and the natural warmth of his personality.
We are deeply saddened by this loss, and this paper is dedicated
to the memory of this fine individual.

Acknowledgements

We wish to acknowledge the COINS graduate students, members of
the VISIONS group, who have contributed to this work. This
includes Frank Glazer, Ralf Kohler, John Lowrance, Paul Nagin,
Terry Weymouth, Tom Williams, and Bryant York. The details of
their work have appeared or will appear in a variety of reports
and doctoral dissertations.

I. KNOWLEDGE DIRECTED PROCESSING

The system being developed is called VISIONS and is designed to
provide an analysis of color images of outdoor scenes, from
segmentation through symbolic interpretation.   The   VISIONS
system is decomposed into two major subsystems: a "low-level"
system which processes the large numeric arrays of sensory data,
and then feeds the "high-level" interpretation processes, which
construct a description of the world portrayed in the scene.
The output of the system is to be a symbolic model of the
three-dimensional world depicted in the two-dimensional image,
including   the   names   of   objects,   their   placement   in
three-dimensional space, and the ability to predict from this
model the rough appearance of the scene from other points of
view.

The original design of the VISIONS system was heavily influenced
by a commitment to knowledge-directed interpretation, and this
commitment has been maintained. The emphasis of this paper is
on the form of knowledge structures, called schemas, and on the
control structures necessary to coordinate a variety of complex
processes, which are referred to as knowledge sources, or KSs
[LES77]. A knowledge source is a process which specializes in
the formation of an hypothesis about an interpretation of the
image, based upon a particular type of available visual cue and
partially processed sensory data. For example, the perspective
KS might infer the physical size of an object depicted by some
region in the image, and the object size KS might order, in
terms of a confidence measure, the plausible object identities
based upon that size. There are serious problems to be faced in
the general application of these processes in an integrated
fashion.   In our system schemas are the means by which we deal
with the problems of control of the KSs.   A schema is a
knowledge structure about a particular visual concept, say a
road scene, with procedural components for properly invoking a
subset of KSs in a coordinated manner.

The effectiveness of many AI systems appears to be derived from
either the constraints available via prior knowledge, or the
restrictions of a specific task domain, or a combination of
both.   The natural language understanding system of Schank
[SCH75] is heavily directed in a top-down manner by knowledge
structures   called   scripts;   recently,   they have proven
sufficient for extracting summary descriptions of a large number
of actual wire service news stories [SCH79]. The HARPY speech
understanding system [LOW76], one of the most effective speech
systems to date, embeds a grammar and vocabulary in a network of
expected utterances. The system operates top-down by matching
paths in the network (which represent possible sentences)
against the utterance. One can view this system in terms of a

schema for each sentence and the representation of this information in a storage efficient form.

There are various special-purpose vision systems whose effectiveness may be traced directly to the utilization of domain-dependent simplications, for example blood-cell analysis, assembly line parts inspection, etc. It is our belief that the use of schemas ([ARB77] or frames [MIN75], or scripts [SCH77]) provides a bridge between general-purpose and special-purpose systems [BAL78, HAN78c, NEV78]. The development of an individual schema and the verification that it is applicable may be as tractable as the development of a particular strategy in a special-purpose system. Knowledge of the front view of a particular house to some degree should be usable in a manner similar to knowledge of the structure of a complex machine part on a conveyor belt.

Our initial efforts have been directed at the construction of a system with sufficient flexibility and generality to explore a variety of issues without requiring substantial systems modifications as the research evolved. As to be expected, the price of such efforts at generality is slower development of the system than we desired, slower than would have been possible with a less flexible special-purpose system. Because of the magnitude of the problem, our research methodology has been to focus on modular components of the system under the constraints of a general system design.

We wish to make it clear that we do not believe that computer vision ought to be primarily a top-down process. Many important mechanisms of human vision appear to be constructive processes which transform sensory data without recourse to semantics [BAR78, HOR77, MAR76, MAR78]. The research effort proposed here, however, attempts to direct the application of some of these processes under the guidance of knowledge-oriented constraints. It will be interesting to see the degree to which this approach can be made general.

There is a very large body of literature that is relevant to the development of effective computer vision systems. In fact it spans the fields of computer science, electrical engineering, cognitive psychology, mathematics, art, etc. with topics that include the physics of light and surfaces, shadows and highlights, image segmentation, color, texture, two- and three-dimensional shape, perspective, occlusion, motion, stereopsis, representation of knowledge, inference, and more. It is not feasible to review this literature here, but a recent book Computer Vision Systems [HAN78a], edited by the authors, documents the state-of-the-art in many of these areas. Here we choose just to mention a few of the many efforts in image understanding systems and leave reference of others for the more detailed sections of the paper.

There have been interesting and somewhat successful attempts to integrate the segmentation and interpretation processes [YAK73, FEL74, BAR76, TEN77]. There are a variety of image interpretation systems where the analysis does not employ three-dimensional representations and processes [SAK76, SHI78, BAL78, LEV78, BAJ76, UHR78, DUD77, RUB77, LOW76, MAC78, HAV78]. Interpretation systems using three-dimensional representations can be applied to a wider class of imagery but are correspondingly far more complex. Consequently, much of the work in 3D scene description (interpretation) has primarily been restricted to polyhedral models of objects [ROB65, WAL75], although there has been interesting work on generalized cylinders as a representation for curved surfaces [NEV77, AGI72, MAR77]. Finally, there is related work in speech understanding that has influenced our research, in particular the Hearsay system [ERM75, LESS77] whose general structure has been followed in our own research.

II. AN EXPERIMENTAL METHODOLOGY FOR SCHEMA-DRIVEN INTERPRETATION

II.1. Strategies for Controlling the Interpretation Process

In a system as complex as VISIONS, there exists a wide range of plausible strategies for guiding the interpretation process. In the past we have raised two important issues of control in our system: the basis upon which KSs are to be invoked and the means by which alternative hypotheses provided by KSs are to be used. Our system was organized to deal with the selection of appropriate KSs and a search space of interpretations by employing a hierarchical modular control strategy [HAN78b, WIL77]. This computational mechanism allows user-defined strategies to be constructed hierarchically out of modular components.

This approach required considerable machinery for dealing with issues of search, and some of these issues drew our attention away from the central issues of vision. The top-down approach that is suggested by schemas bypasses problems of recovering from errors and the inherent combinatorics of a search space of alternatives, at least until we more fully understand the reliability, robustness, and redundancy of our KSs when used in this manner.

We separate the schema-driven operation of our system into distinct tasks:

a) Top-Down Interpretation of Images Via Schemas – this
   involves the utilization of a relevant schema as a
   top-down plan for interpretation; it requires
   coordinated application of the KSs, guided by the schema,
   to various portions of the image.

and b) Bottom-Up Instantiation of Schemas – this is the process
   of selecting a schema that is relevant to the
   interpretation of the image; in effect, it is the
   problem of finding cues in the image and paths of
   inference through long term memory which imply a
   prototypical context which ought to be used [PAR80,
   DUD76, LOW80].

The top-down approach does not imply a complete avoidance of
bottom-up issues. Schema instantiation and the application of a
general schema to specific images, for example, will require the
use of bottom-up processes. In this case, however, the purposes
and goals of the bottom-up processes are more specific and
well-defined.

II.2. Top-Down Interpretation via Schemas

In the following sections we outline a highly structured
approach to the development of general top-down image
interpretation. A key problem is to develop effective ways to
employ schemas after they are somehow accessed. In some of the
experimental stages that we will outline, the goal is to
interpret an image using either a specific or a general scene
schema from either a known or unknown perspective viewpoint.
Thus, the relevant scene schema is assumed to be known, but the
specificity of the information varies. Before describing our
experimental methodology, let us note the difference between
specific vs. general schemas, 3D vs. 2D schemas, and known vs.
unknown perspective viewpoints.

specific schema – a schema capturing a particular instance of
   a given type of scene or object, e.g., a particular
   house, a familiar section of road, or a specific car such
   as your own;

general (prototypical) schema – a schema representing a
   standard or prototypical model of a scene or object, such
   as a house scene, road scene, or car scene, but not any
   specific house, road, or car scene;

3D schema – the 3D description of a scene or object in a
   local coordinate system; this involves the
   representation of surfaces and volumes, and the
   relationships between them;

2D schema - the 2D appearance of a 3D schema relative to a
    viewer-centered coordinate system; this is the way a 3D
    schema would appear from a particular point of view;

unknown perspective viewpoint - in this case a known schema
    (general or specific) can only be used as a 3D schema,
    since the relationship between its local coordinate
    system and the viewer's coordinate system is unknown.

known perspective viewpoint - if the relationship between the
    coordinate systems of the schema and viewer is known,
    then the 3D schema can be used to generate a plan for the
    scene in terms of a 2D schema.

Under this categorization, a general 3D schema is a structure
describing default features of objects and general relationships
between sets of objects which are expected to hold across a
schema class [MIN75]. A specific 3D schema is a general schema
in which features and relationships have been assigned (more)
precise values and in which features and relationships unique to
the particular environment have been added.   In fact top-down
interpretation of, let us say, a road scene using a general 3D
schema would then involve the construction of a specific 3D
schema of that road scene.

A specific 2D schema is a transformation of the corresponding
specific 3D schema, given an assumed view angle.   The
transformation according to view angle is necessary in order to
match the specific 3D schema to the image. Similarly, a general
2D schema represents a transformation of the general 3D schema
given a view angle; in this case, the general 3D features and
relationships are mapped into general 2D features and
relationships.

This paper will focus upon schema-controlled strategies for
employing the KSs.

We propose to explore a variety of strategies by means of a set
of carefully defined experiments of increasing difficulty and
generality. By controlling the amount and type of a priori
information, different portions of the system can be exercised
and different strategies to use the information can be
developed.   The remainder of this section of the paper will
outline experimental stages of system development, and later
sections will provide experimental results for the first of
these stages. There is also a continuing effort on our part on
the instantiation of the relevant schema [LOW80], and a longer

version of the current paper [PAR80] contains approaches and results on schema instantiation.

II.3. Experimental Stages in Schema-Driven Interpretation

Stage 1: The specific scene schema is known;
         the viewpoint is known.

In Stage 1 experiments, the system is, in effect, told what it will see. It must merely match its highly constrained expectations to what appears in the particular scene. In these experiments, a specific 2D schema is directly available. The research focus is on the structure of the schema, the control structure for driving the KSs directly from the schema, and on mechanisms for consistently integrating the hypotheses returned by the KSs into the schema. This experiment is an exercise of all the components of the system and its success is fairly well ensured. Since the specific 3D schema is available and the point of view is known, a 2D schema can be generated which closely matches the appearance of the 2D image. The 2D schema provides a powerful plan for directing various KSs in processing the image and interpreting the scene. Some of the results cited later in this paper are a partial exercise of this capability. Those results, we emphasize, should be viewed as exercises in demonstrating the integration of the system.

Stage 2: The general scene schema is known;
         the viewpoint is known.

Stage 2 tests the system's ability to interpret a scene using a prototypical schema instead of the specific schema. Thus, the general knowledge of road scenes would be used to interpret an image of some particular road scene. The spatial constraints are more general and any given object in the schema may or may not appear. Since the viewpoint of the general schema is known (e.g., looking down the road), the general 3D schema can be used to generate a general 2D schema which then provides a list of key region, line, and vertex features, as well as rough spatial locations and spatial relationships between features that might appear. Strategies are needed that have flexibility in locking onto any relevant characteristics which are extracted from the 2D image. The processed sensory data must be used by the schema in constructing the description of the particular road scene. While certain relationships are expected, for example converging lines of the sides of the road, their existence and location in the image can only be determined by application of some of the KSs.

Stage 3: The specific scene schema is known;
         the viewpoint is unknown.

Stage 3 exercises a different processing capability of the
system: the ability to manipulate 3D representations in the
selection of the probable view angle. It must rotate and
translate a 3D description of a particular scene in order to
generate a 2D view which matches the scene. The problem is
simplified from the general case because the specific 3D schema
is made available. Therefore, if the proper viewpoint can be
determined, a very good match is ensured (c.f. results of Stage
1 in Section V.). Here, important information about the
viewpoint may be provided by the orientation of line segments,
the 2D shape of regions, and spatial relationships between
regions in the image. In addition we can attach information
about standard viewpoints to the 3D schema.

> Stage 4: The general scene schema is known;
>   the viewpoint is unknown.

Stage 4 is an integration of the techniques developed in the
first three. The focus here is on the use of bottom-up
information to constrain the general relationships found in the
general schema and to obtain the most likely view angle. It is
a non-trivial extension of Stages 2 and 3 because even the
proper viewpoint still leaves a potentially large degree of
variability in the matching and interpretation process. Success
here will be dependent upon the quality of the KS's developed
during the first three stages and the effectiveness of the
control strategies developed in the last two stages.

III. THE SEGMENTATION ALGORITHMS OF THE LOW-LEVEL SYSTEM

The VISIONS research group has maintained a long-standing
research effort in low-level image analysis. Our goal has been
to produce a system which can initially provide a segmentation
to drive the image interpretation process, and which later can
receive semantic feedback to direct low-level processing in the
refinement of that segmentation. We cannot discuss the full
range of our segmentation efforts; they are documented in a
series of reports and papers [NAG79, KOH79, HAN78b, PRA79,
PRA80, OVE79, HAN80a]. Here, we limit our discussion to a brief
description of two algorithms, an edge relaxation algorithm and
a histogram-guided region relaxation algorithm. Both the edge
relaxation process and the region formation process are
undergoing continuous development.

All algorithms are implemented in a simulation of a parallel
hierarchical machine architecture, called a "processing cone",
for processing images [HAN74, HAN80b]. The cone is related to
similar structures proposed by [UHR74, TAN78, TAN80, ROS79a,b].
The segmentation processes basically involve two complementary

relaxation labelling processes [ROS76, ZUC77, DAV76] for partitioning images into regions and boundaries, either of which can be preceded by a sophisticated smoothing algorithm [OVE79] in a preprocessing pass on the image. The boundary formation process responds to local changes in the data, while the region formation process is sensitive to global similarities in the data. An earlier version of the region algorithm has provided the data upon which the interpretation processes in this paper are applied.

III.1. Edge Relaxation and Boundary Continuity

The edge/boundary analysis utilizes a representation of local discontinuities in some visual feature (e.g., intensity or color) as a collection of horizontal and vertical edges located between individual pixels. The iterative edge relaxation processes then allow contextual interactions to organize collections of edges into boundary segments [PRA79, HAN78b]. Figure 1 provides sample results of this process.

III.2. Histogram-Guided Region Relaxation

Region analysis is based on cluster detection in the histogram of some visual feature [HAN78b, NAG79]. Prominent peaks in the probability density function of a feature or in the joint density function of a pair of features indicate the most frequently occurring (or co-occurring) values in the feature space. The region formation process therefore utilizes global histogram cluster labels, defined by the peaks, with pixels. These peaks also allow likelihoods of cluster labels (computed as a function of the spatial location of the peaks relative to the spatial location of each individual pixel in feature space) to be associated with each pixel. Interactions between the label sets of pixels in local neighborhoods are then used to organize connected sets of pixels into regions (i.e., connected sets of pixels all with high probability of the same label constitute a region). Figure 2 outlines results of applying the histogram-guided region relaxation algorithm.

Results of an earlier version of the region relaxation algorithm appear in Figure 3. These results form the basis of experiments in the remainder of the paper. Because of previous limited computational resources on our old computer facilities (PDP-15 with 96K bytes core), the segmentation was obtained from an image with a resolution of 128x128 pixels. This image was derived from a 256x256 quarter of a 512x512 array, which was then further reduced by averaging to 128x128. The current processing is on a VAX 11/780 with 3 megabytes core, and processing of images with higher spatial resolution is now typical.

(a)

Figure 1. Boundary segmentation via edge relaxation. (a) Intensity image of a 128×128 portion of a suburban house scene. (b) Closeup of a portion of roof trim and a sequence showing the effect of iterative updating of edge likelihoods via constraints of boundary continuity. (c) Initial edge probabilities. (d) Edge probabilities after 2 iterations. (e) Edge probabilities after 20 iterations.

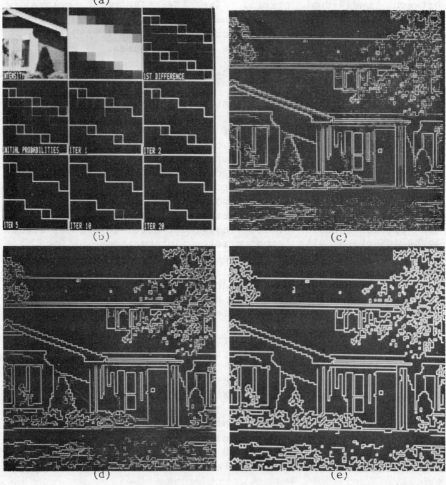

(a)

(b)

Figure 2. Region segmentation via relaxation histogram cluster labels. (a) Initial intensity image of a 128×128 portion of a house scene derived by averaging from an image of higher resolution (previous limitations on computational resources dictated this limitation). (b) Resultant segmentation superimposed on intensity image. Note that there is a difference in aspect ratio in this image due to differences in the displays used to generate the picture.

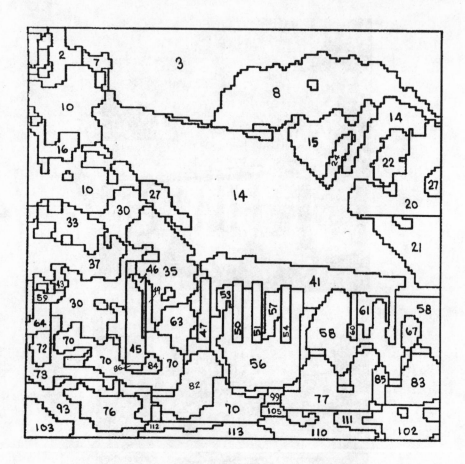

<u>Figure 3</u>.   Segmentation data used in experiments. These results
of region formation via relaxation on cluster labels were
produced by an earlier version of the algorithm which produced
the results in Figure 2.   The region segmentation has been
converted to a region boundary representation and region labels
are shown.   They form the basis of the experiments described in
later sections.* Note that only large regions or regions
mentioned in paper are numbered, but <u>all</u> regions have a unique
label.

*The integration of the edge and region segmentations is the
focus of the current Ph.D. research of Ralf Kohler.

IV. SUMMARY DESCRIPTION OF THE KNOWLEDGE SOURCES AND INITIAL EXPERIMENTS

This section provides a general overview of the knowledge sources in the VISIONS interpretation system. Knowledge sources are the means by which hypotheses are generated and verified. In some cases, the KSs have been developed only to the point where the results are reasonable. The advantage of this approach is that it allows a minimally complete system to be configured and run. The input/output and functionality of each KS is clearly specified and can be improved as time and resources permit.

A set of seven modular KSs and several representations will be briefly reviewed. While we cannot discuss each of these in detail in the limited space of this paper, a short discussion of each KS and, wherever possible, a simple example of local results is provided. However, these local results must be viewed in the context of the evolving design of the whole system [HAN78b,c].

A base-level system has been implemented and is operational to the point where interesting experiments, such as the ones described in the following sections, are being performed. In building this base-level system, an attempt was made to provide sufficient generality of processes and representation -- function and structure -- to allow us to work on different types of scenes, to easily add knowledge in both active and passive form, and to define and execute different types of interpretation strategies.

The reader should note that the results cited in this section were obtained from a version of the system running on the University Computing Center's CYBER-74 time-sharing system. The system is implemented in GRASPER [LOW78], a high level graph processing language built in ALISP [KON75]. The system has been transferred to the COINS Department VAX 11/780 and integration with the VISIONS low-level system is in progress.

Table 1 provides an overview of the set of KSs currently available and briefly discusses the representations employed in various parts of the system. Cross references to more detailed discussions and/or results are included.

IV.1. Long-Term Memory (LTM) and Short-Term Memory (STM)

General knowledge about the physical world (or the task domain

TABLE I.	SUMMARY OF KNOWLEDGE SOURCES AND REPRESENTATIONS
Name	Brief Statement of Function or Purpose
Low-Level Segmentation System	The goal of the low-level system is the segmentation of an image into visual primitives (regions, boundary segments, and vertices), and the extraction of a range of features to be used by the various knowledge sources (KSs) of tne interpretation system.
RSV Structure	RSV is a symbolic layered graph structure of regions, line segments, and vertices containing the segmentation results and feature descriptors. This data structure is stored in short-term memory (STM; see below) and represents the processed visual data upon which the interpretation is based.
LTM (Long-Term Memory)	LTM is a hierarchical representation of general (i.e., non-image specific) world knowledge organized into natural levels of abstraction: schemas (stereotypical scenarios), objects, volumes, surfaces, regions, line segments, and vertices.
STM (Short-Term Memory)	STM is a hierarchical structure of the same form as LTM and used for constructing an interpretation by means of the knowledge sources. An interpretation is then the collection of instantiated nodes in STM. The RSV structure is the bottom three levels -- all other levels are initially empty.
Inference Net KS	It is a network of a priori probabilities of nodes and conditional probabilities between nodes; it is defined on the arcs and nodes in LTM, and are the means by which implications of local hypotheses may be propagated upward and downward through the layered structure. Any hypothesis generated by a knowledge source can then be used to generate further hypotheses.
2D Shape KS	This KS allows symbolic classification of the shape of regions. The confidence that a given image region has a particular primitive 2D shape will be returned. The results allow paths for surface & volume hypotheses via LTM.

Name	Brief Statement of Function or Purpose
Spectral Attribute Matcher KS	It hypothesizes object identities of a region on the basis of a comparison between region attributes (color and texture) and statistics of these features attached to the object nodes in LTM. It is designed primarily for objects for which these attributes are reasonably invariant across images (currently sky, bush, grass, tree, road).
Perspective KS	The goal is the hypothesis of surface orientation, size, and/or distance in order to produce a partial volume/surface plan of the scene. The current version focusses on relationships between elevation, height, range, and width of surfaces given a camera model and a set of assumptions regarding surface orientation.
Horizon KS	It uses the horizon schema (the most general outdoor schema which relates sky, ground, and horizon) and the camera model to fix the location of the horizon. It is used to filter other hypotheses on the basis of their relationship to the horizon.
Object Size KS	This module is designed to generate object hypotheses on the basis of the image size of a region. It compares the computed physical (i.e., real world) size of a surface, determined by the perspective module, to the physical size of objects in LTM.
3D Schema	The 3D schema captures stereotypical visual events by organizing subsets of information in LTM into higher order complexes of expected scenarios (e.g., a road scene schema). It may be either specific (a particular known scene) or general. The representation is stored in a local coordinate system and contains control information for top-down interpretation. A projection of a 3D schema produces a 2D schema.
2D Schema	A 2D schema is a projection of a 3D schema from a given point of view. The projection carries along control strategy information and features of the projection (e.g., surface orientation, relative to viewpoint, etc.). It is used to direct top-down interpretation of the image.

of interest) is stored in "long-term memory" (LTM). An image will be "understood" in terms of the concepts and relations found in LTM. This knowledge is hierarchically organized into levels which represent a natural abstraction of world knowledge (Figure 4).

Nodes in LTM represent visual primitives with which the system can construct an interpretation, while the arcs represent relations (primarily AND/OR relations) which exist between the primitives. Inter-level arcs represent the paths by which primitives at one level may be related to primitives at levels above and below. These arcs represent paths for hypothesis formation (possible inferences) within LTM; they are used in various ways by other knowledge sources during the interpretation process. Figure 5 depicts a representative fragment of the network.

The interpretation of an image is viewed as a set of instantiations of the nodes in LTM. These instantiations constitute short-term memory (STM) and are shown on the left side of Figure 4. This representation of knowledge, as well as its relationship to the inference net, is the subject of ongoing research by J. Lowrance [LOW80], a graduate student in our research group. Both STM and LTM are implemented as a layered graph in GRASPER [LOW78], a graph processing language extension to LISP which follows the general approach of [FRI69, PRA71].

IV.2. 2D Shape

The 2D shape of a region may be an important cue to the identity of an object, or to attributes of a visible surface (such as the 3D orientation of the surface). Many simple relationships between the physical world and its 2D image projection are captured in LTM. For example, the 3D shape of simple volumes (e.g. cylinders and rectangular solids), as well as the 2D shapes of 3D surfaces (e.g., the rectangular surface of a window), are related to standard 2D shapes (e.g. rectangles, trapezoids, circles and ellipses). Therefore, in order to gain access to paths by which 3D hypotheses may be formed, symbolic attributes of shape, where they are relevant, must be associated with regions.

First, we outline the strategy for labelling geometric shapes formed by straight lines. Figure 6 is a portion of LTM which captures an informal definition of several shapes in terms of the straight line segments forming them. The shape classification is hierarchical; that is quadrilaterals are a superclass of both trapezoids and parallelograms, the latter being a superclass of rectangles and rhombi, etc. The definitions of shapes involve increasingly restrictive

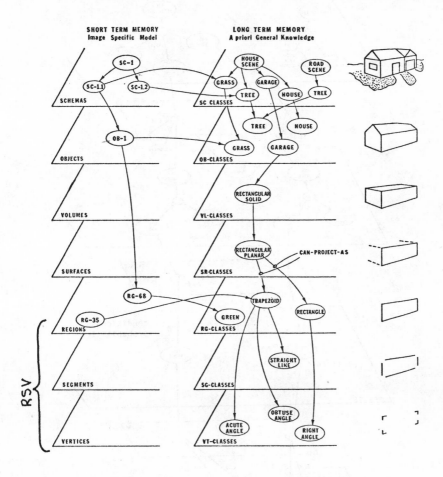

<u>Figure 4</u>. Hierarchical decomposition of long-term memory (LTM) and its relationship to short-term memory (STM). LTM contains the stored knowledge to which the system has access. An interpretation of an image is viewed as a set of instantiations in STM of nodes in LTM.

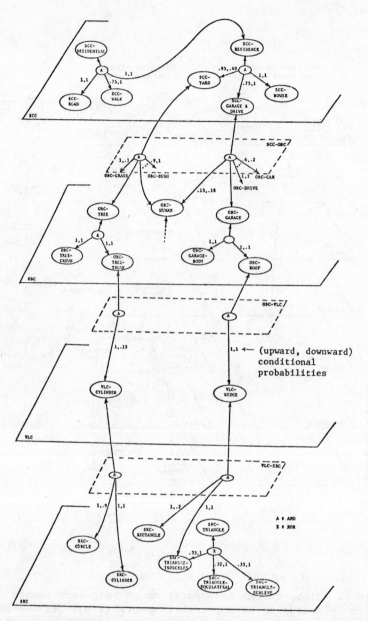

<u>Figure 5.</u> Detailed structure of a fragment of LTM. Relation nodes (AND and XOR) are circled, while names of primitives are prefixed by the level on which they reside. The pairs (x,y) on solid lines are the upward and downward condition probabilities, respectively, for use by the inference net [PAR80, LOW80]. Not shown are the priors for the nodes, attribute lists, control information, etc.

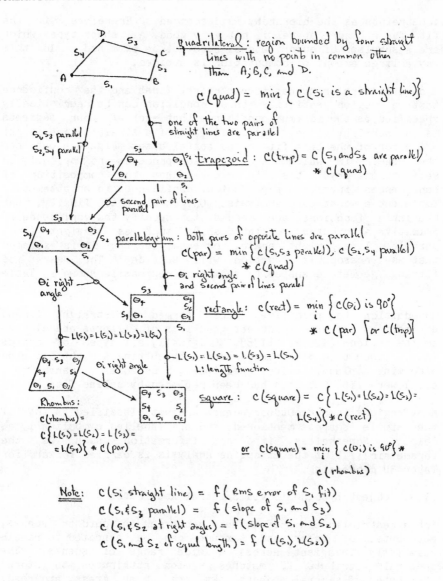

quadrilateral: region bounded by four straight lines with no points in common other than $A, B, C,$ and D.

$$c(quad) = \min_i \{ c(S_i \text{ is a straight line}) \}$$

one of the two pairs of straight lines are parallel

trapezoid: $c(trap) = c(S_1 \text{ and } S_3 \text{ are parallel}) * c(quad)$

second pair of lines parallel

parallelogram: both pairs of opposite lines are parallel

$$c(par) = \min \{ c(S_1, S_3 \text{ parallel}), c(S_2, S_4 \text{ parallel}) \} * c(quad)$$

θ_i right angle and second pair of lines parallel

rectangle: $c(rect) = \min_i \{ c(\theta_i) \text{ is } 90° \} * c(par) \; [\text{or } c(trap)]$

$L(S_1) = L(S_2) = L(S_3) = L(S_4)$
L: length function

square: $c(square) = c \{ L(S_1) = L(S_2) = L(S_3) = L(S_4) \} * c(rect)$

or $c(square) = \min_i \{ c(\theta_i) \text{ is } 90° \} * c(rhombus)$

Rhombus:

$c(rhombus) = c \{ L(S_1) = L(S_2) = L(S_3) = L(S_4) \} * c(par)$

Note: $c(S_i \text{ straight line}) = f(\text{rms error of } S_i \text{ fit})$
$c(S_1 \& S_3 \text{ parallel}) = f(\text{slope of } S_1 \text{ and } S_3)$
$c(S_1 \& S_2 \text{ at right angles}) = f(\text{slope of } S_1 \text{ and } S_2)$
$c(S_1 \text{ and } S_2 \text{ of equal length}) = f(L(S_1), L(S_2))$

<u>Figure 6.</u>   Hierarchical definition of a portion of the shape types currently in LTM. c(x) represents the heuristic confidence measure of shape type x. The classification of shapes loosely follows the classification based on affine symmetry [NEW65].

constraints as the hierarchy is descended.   Therefore, if the
fit for a quadrilateral is not very good, all shape types which
are a subclass of quadrilateral need not be examined.   In this
way a large amount of computation is avoided.

A quadrilateral requires four straight lines and the confidence
that a region satisfies this condition can be heuristically
specified as the minimum confidence that each of four segments
is a straight line. The confidence of a straight line is the
RMS error of the best fit to the actual data. Figure 6 outlines
the manner in which the computation proceeds and hopefully is
self-explanatory. It should be noted that the composition of
confidences involves a product of confidences in an attempt to
implement a worst-case analysis. One should note, finally, that
heuristic functions are needed to specify the confidence of
primitive attributes or relationships such as straight line,
parallel line, right angle, or equal length; it is expected
that any reasonable function will suffice.   The result of
fitting geometric shapes to segmented regions is shown in Table
II.

In addition to primitive shapes formed by straight lines,
quadratics are used to detect good fits of ellipses and circles
to the regions [AGI72, SHI78]. Originally all types of conics
(i.e., the type of curves produced by cutting a right circular
cone with a plane, including ellipse , hyperbola , parabola ,
etc.) were fit, but this has been replaced by spline fits.

Most regions in our outdoor scenes are not classified as any of
the simple shapes mentioned and are labelled symbolically as
'blob'. Nonetheless, important information such as the
parametric fit of the 2D spline analysis is carried forward for
later 3D processing.

IV.3.   Object Hypothesis via Spectral Attributes

For a restricted class of objects occurring in outdoor scenes,
attributes of color and texture can be expected to remain
relatively invariant across a wide range of scenes.   The
spectral attribute KS matches region attributes to stored
attributes of several objects (sky, tree, bush, grass, and road)
and returns a measure of the degree of match, ranging from -100
(no match) to +100 (excellent match).   The stored attributes
were obtained by measuring 60 features across samples of each
object extracted from a data base of 25 images.   A piecewise
linear decision function which reflects the expected variability
of each feature of an object is then formed.   The matching
process extracts an identical set of features from the region
(or union of regions) to be identified, and uses the decision
function to generate a degree of match for each object.   This

Table II. Summary of 2D shape fits to selected regions of
Figure 3.

Region	Shape	Probability	Aspect Ratio
RC--0047	Rectangle	.937	6.33
	Trapezoid	.96	——
RG–0050	Rectangle	.99	6.33
RG–0051	Rectangle	.99	6.33
RG–0054	Rectangle	.99	6.00
RG–0060	Rectangle	.80	7.5
	Trapezoid	.85	——
RG–0045	Rectangle	.80	6.25
	Trapezoid	.85	——
RG–0049	Rectangle	.85	10.33
	Trapezoid	.90	——
RG–0086	Rectangle	.99	3.00

research is part of the Ph.D dissertation of T. Williams; more detail appears in [HAN78b, WIL80].

The attribute matcher can only be used to hypothesize the presence of certain "target" objects based upon the expected invariance of their color and texture attributes. There are many objects such as cars, shirts, and most other man-made objects which vary in their spectral characteristics. This KS, however, will return a confidence value for any region, regardless of whether the region represents a target object or not. Therefore, we require mechanisms for filtering these hypotheses.

Figure 7 illustrates the results obtained by applying the spectral attribute matcher KS to the 21 largest regions of our example. Of the 21 regions, 14 were target regions (3, 8, 10, 30, 20, 79, 15, 37, 82, 96, 90, 83, 110, and 93) and 7 were non-target regions (14, 58, 41, 56, 35, 70, and 21). Of the 14 targets, 8 were correctly identified on the basis of a maximum confidence decision. If bush and tree are collapsed into a single object (which is not unreasonable given the similarity of spectral attributes), then 11 of the 14 are correctly identified. Of the remaining three target errors, the correct hypotheses had the second highest confidence in two cases (regions 15 and 96); region 8 represents a mixture of sky and small tree limbs and the correct hypothesis is debatable.

Of the 7 non-target regions, 5 of the regions (58, 41, 56, 35, and 70) represent portions of the white house wall and all were hypothesized as sky. In the absence of any additional information, such hypotheses are reasonable and cannot be eliminated. The remaining two regions are both roof (regions 14 and 21) and both were hypothesized as grass, probably due to similarity of values for several crude texture measures. Both of these hypotheses, and three of the previous five, can be filtered if the location of the horizon is known and the ground is assumed to be flat. Regions 14 and 21 cannot be grass and be located above the horizon, while regions 58, 56, and 70 are either below or straddle the horizon and hence cannot be sky. This will be discussed again in the "horizon" KS (Section IV.5).

The statistics on the remaining 93 regions are approximately the same, although if the size of the region falls below a minimum size, reliable texture measures cannot be extracted and performance falls off. A number of the regions have negative confidence values for all target objects and no hypotheses are generated for these regions.

Region	Area		Confidence Measure (maximum confidences circled)					Hypothesized Identity (max. conf.)	Actual Identity (visual)	Correct Hypothesis?	Correct hypothesis? (Bush/Tree one object)	Comments	Correct Hypotheses Filtered by Horizon KS
	# pixels	% of picture	Bush	Grass	Road	Sky	Tree						
14	3101	18.9	-10	32	-55	-17	-16	Grass	House Roof	No	No	above assumed horizon	no hyp.
3	1939	11.8	-62	-48	41	74	-84	Sky	Sky	Yes	Yes		Yes
8	971	5.9	-20	-6	31	47	-46	Sky	Tree	?	?	mixture: tree without leaves and sky	?
10	793	5.9	-33	-40	-88	-53	46	Tree	Tree	Yes	Yes		Yes
58	606	3.7	-52	-41	-49	20	-21	Sky	House Wall	No	No	white house wall in sunlight; region straddles assumed horizon	no hyp.
41	560	3.4	-8	-23	-70	-38	76	Tree	House Wall	No	No	white house wall eaves, & gutter in shadow matches tree on brightness, texture	No
30	518	3.2	-54	-41	-53	-41	14	Tree	Tree	Yes	Yes		Yes
56	486	3.0	-60	-51	-62	23	-23	Sky	House Wall	No	No	white house wall; region straddles assumed horizon	no hyp.
20	427	2.6	10	37	-88	-50	54	Tree	Tree	Yes	Yes		Yes
35	410	2.5	-2	13	-61	32	-16	Sky	House Wall	No	No	white house wall with shadow of tree	No
79	373	2.3	45	0	-92	-71	46	Tree	Bush	No	Yes	confidence for bush is almost as large as tree	Yes
70	354	2.2	-2	7	-61	29	-25	Sky	House Wall	No	No	white house wall; part of region above horizon	?
15	330	2.2	-27	20	-56	-32	2	Grass	Tree	No	No	region above horizon; tree next most	Yes
21	310	1.9	-8	40	31	-35	-16	Grass	Roof	No	No	region above horizon; likely knocks out road also	no hyp.
37	308	1.9	-33	-30	-92	-53	44	Tree	Tree	Yes	Yes		Yes
82	238	1.5	2	0	-77	-53	-2	Bush	Bush	Yes	Yes		Yes
96	217	1.3	4	6	-53	-53	-10	Grass	Bush	No	No	bush next most likely	No
90	202	1.2	29	9	-85	-71	0	Bush	Bush	Yes	Yes		Yes
83	198	1.2	39	0	-94	-71	44	Tree	Bush	No	Yes	bush next most likely	Yes
110	196	1.2	-48	32	25	-35	-46	Grass	Grass	Yes	Yes		Yes
93	196	1.2	29	-20	-89	-74	35	Tree	Bush	No	Yes	bush next most likely	Yes

Figure 7. Results obtained by applying the spectral attribute matcher KS to the 21 largest regions (ordered by decreasing size) of Figure 3. The results obtained by filtering the hypotheses by the horizon KS (see Section V.11) are also shown; if there are no positive confidences after filtering, no hypothesis is generated.

IV.4.   3D Schemas

There is a great deal of expected structure in our visual
environment and it seems evident that such expectations are
important in processing visual information. One of the
functions of the 3D schema is the organization of subsets of
information in LTM into higher order complexes of stereotypical
situations in such a way that the spatial relationships between
objects, volumes, and surfaces which might occupy or define that
space are made explicit. The 3D schema would allow rotation and
translation of the prototypical scene so that its appearance
from any point of view can be generated. Thus, the processing
of a 3D schema allows the generation of potentially relevant 2D
schemas.

A system for separately defining arbitrary surface patches,
combining patches into volumes, combining volumes into objects,
building specific 3D schemas, and rotating schemas subject to
the assumption of a given point of view is partially developed.
A 3D representation is being developed by B. York in his
Ph.D. dissertation [YOR80] employing Coons surface patches whose
four sides are delimited by cubic splines [COO74, YOR79].
However, these components have not yet been fully integrated
into our system.

The results given in Section V demonstrate top-down
interpretation of an image. In order to do this it was assumed
that a specific 3D schema was available, that it could be
rotated given an assumed point of view, projected onto a 2D
image plane, and then hidden lines removed. While those 3D
facilities were not available then, and 2D schema information
was supplied directly, they are now available.

Our current version of specific 3D and 2D schema have for each
schema region a centroid of the expected central location and a
radius representing the decreasing likelihood that the schema
region appears at that location. Thus, one can think of a
spherical or circular probability cloud denoting expected
spatial position. This crude representation of location allows
selection of regions in the image for matching against schema
objects; furthermore, alternative region selections can be
ordered by degree of location match. Figure 8 depicts wire
frame and surface representations of a model of the house image.
The 3D schema we have described attempts to capture approximate
relative spatial information of the entities appearing in Figure
8. There are still interesting problems remaining that are
associated with the generation of 2D schema from 3D schema. For
example, the likelihood that a 2D schema region is visible will
be related to the likelihood that another schema region will
occlude it. Many issues related to the generation of specific

Figure 8.   Wire frame and surface representations of a model of
the house image seen from two points of view. The current 3D
house scene schema is actually an abstract representation of the
approximate relative spatial locations of the entities in these
images. The components (volumes, surfaces, straight line
segments) are actually represented by a position in space and a
radius associated with a decreasing likelihood of the component
appearing at that location.

2D schema from specific 3D schema are under examination.

IV.5. Perspective

The perspective knowledge source concentrates on the ways in which the general relationships governing perspective transformations can be used to extract or explain information concerning surface orientation, distance, and size [DUD73, HAR78]. A region (or the union of a group of regions) represents the projection of a 3D surface onto the 2D viewing plane. The problem then is to recover some of the 3D attributes of that surface from the segmented image. Figure 9 is a simple sketch depicting the relationship of the distance and height in the physical world and their associated parameters in the image.

The current version of the perspective KS focusses on the relationship between the following variables:

a) elevation - vertical distance above the ground plane,

b) height - vertical distance from visible bottom edge to visible top edge of surface,

c) range - horizontal distance from viewing location to a distinguished point on the surface,

and d) width - horizontal distance from the visible left edge to the visible right edge of the surface.

The interrelationship of these variables depends on the orientation of the surface in three-space. For simplicity, we assume the orientation is either vertical (i.e., perpendicular to the ground plane, such as a tree) or horizontal (i.e., in the ground plane,such as a road). While these assumptions may appear to be unnecessarily restrictive, they are sufficient to cover many surfaces of interest.

The four variables described above are interrelated. Given the assumption of ground planarity and a camera model (angle of inclination to ground plane, focal length, and height above ground plane), knowledge of any one implies knowledge of the remaining three, although the form of the relationship depends on whether the orientation of the surface is assumed vertical or horizontal. We are continuing to explore ways to use perspective under weaker assumptions in our current research.

In general, there are usually several unknown quantities to be determined and depending on the assumptions made one can solve for different variables. Applying the perspective KS to selected regions of Figure 3, it is easily determined that the

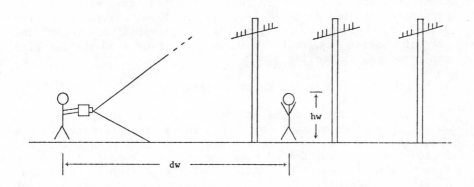

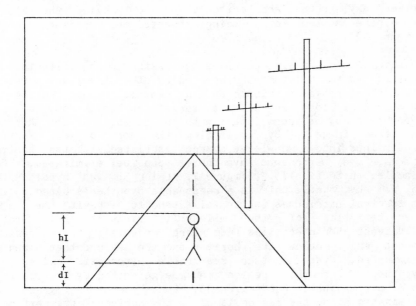

<u>Figure 9</u>.   Perspective - ground plane, vanishing points, projective geometry.

range of region 79, for example, is about 37 meters and its
height is 1.61 meters; this required assumptions of ground
planarity, and that the surface projected as region 79 is
perpendicular and attached (i.e., zero elevation) to the ground
plane. More extensive results from the perspective KS, and the
use of these results for the development of a 3D spatial plan
are presented in Section V.

IV.6. Horizon Schema and Horizon Filter KS

It should be clear that the effects of perspective and distance
on the projection of surfaces in the image are determined by the
observers position, the camera model (of height, pan, tilt, and
focal length), and the orientation of the ground plane. These
factors also determine the position of the horizon in the image,
if it is visible. The horizon schema is perhaps the simplest
and most general of the schemas present in the system. The
function of the horizon schema is to define the relationship
between sky, ground, and horizon , and to provide the global
coordinate system for placing objects and schemas in space
(Figure 10).

The horizon schema also provides the basis for a filtering KS
applied to the hypotheses generated by other knowledge sources.
Since the spectral attribute KS, for example, has no notion of
the spatial location of its target objects, some of its
hypotheses may be inconsistent with the location of the horizon
in the image. By collapsing the more obvious spatial
constraints into a knowledge source associated with the horizon
schema, many erroneous hypotheses can be eliminated. For
example, in Section IV.3, Figure 7, region 58 was hypothesized
to be sky. While this is a reasonable hypothesis based solely
on spectral attributes (white walls tend to "inherit" the color
characteristics of the ambient illumination or reflected
illuminant characteristics from nearby objects), "sky" regions
cannot exist below the horizon and the sky hypothesis can be
eliminated. Since no other reasonable hypothesis exists, no
hypothesis for this region can be generated by the spectral
attribute matcher. For region 15, the hypothesis "grass" is
eliminated since the region is above the horizon; the next most
likely hypothesis (tree) cannot be eliminated and becomes the
final hypothesis.

The results from the horizon filter KS applied to the output of
the spectral attribute matcher KS are shown in Figure 7. Note
that the assumption of ground planarity is built into the
current version of the horizon schema, and that the real world
in many instances presents us with more complex situations.

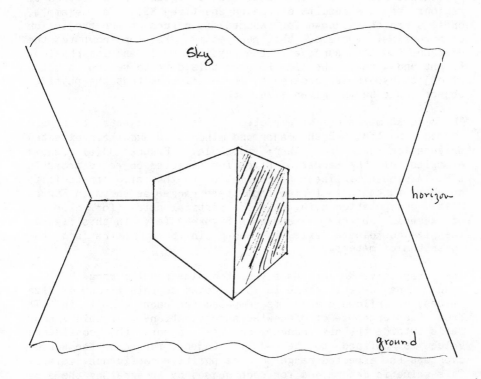

Figure 10. Illustrative diagram of the Horizon KS and its uses
as a hypothesis filter. Hypotheses which violate the spatial
constraints imposed by the horizon can be eliminated. Regions
which extend below the horizon cannot be labelled sky, while
regions which extend above the horizon cannot be in the ground
plane (e.g., road, grass). Therefore, the cross-hatched region
cannot be labelled sky, grass, or road.

IV.7.   Object Size KS

The object size KS is responsible for generating object
hypotheses based on the size of a region (or collection of
regions) and the results of the perspective KS.   For example,
once a region is known (or assumed) to represent the projection
of a vertical surface, the perspective KS can compute the
distance to the surface in the physical world and its physical
height and width.   The size KS uses this data to return a list
of object hypotheses ordered by the confidence that the physical
object could be the given size.

The size KS makes use of expected sizes of objects that are
stored in LTM.   Both major and minor axes and their expected
orientation are used where possible.   Figure 11(a) shows
examples of the ranges of sizes for selected object classes in
LTM.   A piecewise linear approximation to the size probability
density function is formed from these ranges as shown in Figure
11(b).   Computation using only the vertical axis (for clarity)
of several objects is shown in Figure 11(c);   in this figure,
the size coordinate axis is shown in both meters and the
logarithm of meters.

The perspective KS returns a computed size and the range of the
size;   the default range is $\pm 5\%$. Based on this window of size
values, a confidence value is computed for each object in LTM
from the ensemble of piecewise approximations.   If this window
falls outside the size range for the object, the confidence
value is defined to be -100.   Objects for which the window
overlaps the expected range produce positive confidence values.
This value is determined for each object by integrating the area
under the curve (for that object) within the error window, and
then normalizing by the largest value produced for any object
(times 100 so that the largest will have a confidence value of
100).

Applying the size KS to RG-50 (a window shutter) of Figure 3
results in the hypotheses: tree trunk with confidence 100,
shutter with confidence 35, and all others are negative.

V.   RESULTS OF INTERPRETATION WITH A SPECIFIC 2D SCHEMA

V.1.   Introduction

One of the purposes of 3D schemas is to generate the appearance
of prototype scenes from any point of view.   For example the 3D
schema of a road scene can be rotated and projected to produce
the image of a road scene as it would be expected to appear to
an observer looking down the road.   A particular projection of a

Object	Horizontal Axis				Vertical Axis			
	Smallest Possible	Smallest Probable	Largest Probable	Largest Possible	Smallest Possible	Smallest Probable	Largest Probable	Largest Possible
Building-Door	.6	.84	1.20	1.40	1.40	2.0	2.40	3.40
Building-Shutter	.25	.30	.60	1.0	.5	.71	2.0	2.80
Building-Side	3.40	4.80	11.30	32.0	1.70	2.40	16.0	27.0
Building-Window	.30	.60	1.20	2.40	.25	.35	2.40	3.40
Bush	.60	.84	1.70	4.80	.60	1.0	2.0	2.80
Car	2.0	3.40	4.80	6.70	.60	1.0	1.40	2.0
House	5.70	9.50	16.0	27.0	3.40	4.80	9.50	13.50
Human	.30	.42	.60	.84	1.0	1.40	2.0	2.40
Roof	2.0	4.80	16.0	27.0	2.0	2.40	4.80	6.70
Tree	1.70	3.40	6.70	13.50	2.0	3.40	13.50	32.0
Tree Crown	1.70	3.40	6.70	13.50	.71	1.0	9.50	19.0
Tree Trunk	.25	.25	.60	1.70	.71	1.0	5.70	23.0
Utility Pole	.25	.30	.42	.60	2.40	3.40	9.50	13.50

(a)

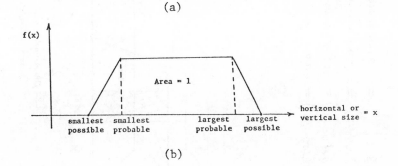

(b)

Figure 11. Object Size KS. (a) Typical size ranges for horizontal and vertical axes of some objects in LTM; all sizes are given in meters. (b) Approximation to a probability density function formed from the values in LTM. (c) Object confidences given a size are based on the probability of the size falling in a default ± 5% window (exaggerated for clarity), although the actual window can be set by the perspective KS on the basis of an error analysis of the computed size. These probabilities are scaled up, making the highest equal to 100. For expository purposes, only the vertical axis computation is shown; in actual applications, both horizontal and vertical extents are incorporated, resulting in a more constrained set of hypotheses.

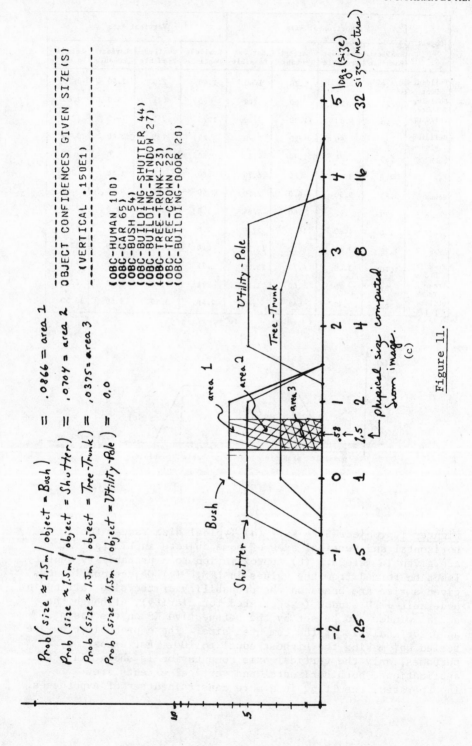

Figure 11.

3D schema is referred to as a 2D schema and will be very useful in directing top-down analysis of the image. It is best thought of as a plan (a set of constraints) for interpretation of the image.

A 2D schema of a specific house scene viewed from a front diagonal perspective is implied by the illustration in Figure 12. This 2D schema is not a projection of a general house scene, but rather of a particular house scene. The general house schema would need to specify the expected variability of the general house scene.

The current representation of the 2D schema involves a set of information for each region including location of centroid, area, 2D symbolic shape with an aspect ratio of major to minor axis, color and texture features, location and properties of boundaries, object identity, 3D surface orientation and 3D size. To perform these experiments the 2D schema was generated manually, and current work will make it possible to drive a 3D schema representation and automatically form the projection, estimate likelihood of occluding schema surfaces, and fill out the required attributes from the LTM knowledge base.

It should be clearly understood by now that the current spatial representation of a 2D schema is not a direct copy of the model drawn in Figure 12(b,c), but instead approximates the location of this information.

The position of a 2D schema region is defined by two parameters, the position of its centroid and its area. The squiggly boundaries in the 2D schema of Figure 12 are for display purposes. Actually, the positions of the schema regions are not known except to the degree that constraints are implied when there is a distinctive shape, such as rectangle with a particular aspect ratio. On the other hand, there are sometimes boundaries with known characteristics (e.g. long and straight) appearing in expected positions such as those bounding the roof in our house scene schema. Lines whose shape are known are drawn without squiggles where they are roughly expected to appear in the image.

Top-down control of the KSs in the interpretation of an image is relevant in the case where expectations about a given scene are available. The experiments in this section are intended to depict the case where the system is attempting to interpret a known scene via a specific 2D schema, i.e., from a known point of view (Stage 1 from Section II.3). We also assume that the camera model (focal length of lens, tilt angle, pan angle, and height above ground) is known. Results will be presented of the control by 2D schemas of the 2D shape KS, spectral attribute KS,

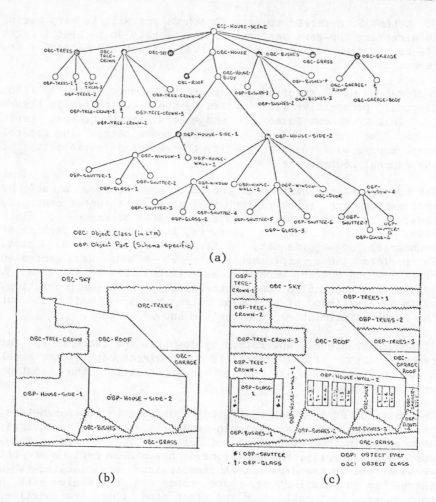

Figure 12. 2D schema for a specific house scene. (a) Hierarchical structure of schema components. (b) The schema regions represented by the dark nodes in the hierarchy. (c) Schema regions associated with tip nodes in the hierarchy; this is the schema used in all the experiments of Section VI. The squiggly boundaries in the schema are for aesthetic purposes. Currently the position of schema regions is defined by parameters of color and symbolic shape, and any subset of these four parameters may be used by a matching function applied to image regions. Straight lines (without squiggles) represent boundaries whose shape and rough position is known, and can also be used to direct matches to nearby straight line segments in the image.

fits of straight line segments, the perspective KS, and the size KS.

The 2D schema KS directs matching of schema regions to image regions and some schema line segments to image line segments. The matching process employs a weighted evaluation function on features of symbolic 2D shape, size, color, and position between regions in the image and in the schema. We will not go over the details of the heuristic match function here, although we note that any non-empty subset of the features can be used for matching. Note that it is necessary, in general, to expand or contract the schema in order to correlate schema size with image size. This is a function of distance and camera parameters and would have to be part of schema processing if it is to be robust in its application.

Matches can also be defined to operate between a schema region and groupings of several image regions, or a schema line segment to a group of image line segments, but then a search is necessary to discover the best groupings. The search for good matches can be directed by a variety of strategies. We will present simple results of a few.

V.2. Semantically Directed Merging via 2D Schema

The first experiment will demonstrate the matching color and texture attributes in order to improve a fragmented segmentation. It involves the interaction of the spectral attribute KS and the 2D schema in an attempt to merge many adjacent regions whose object identities are the same. The strategy attached to the specific 2D schema for applying KSs to perform semantic merging is outlined in Figure 13. It first involves calls to the spectral attribute matcher to get a list of object types which it can match. The 2D schema contains information on the areas of the image in which these objects (tree, bush, sky, and grass) are expected to appear. Thus, it distinguishes the areas expected to be target objects from the areas of non-target objects. The 2D schema then accesses the region segmentation to select candidate image regions for matching. Each schema region which is expected to contain one of the four types of objects above will be used to direct semantic merging via the attribute match KS. In these areas adjacent regions will be merged if their identities are verified by the attribute KS to be consistent with the schema. Thus, the attribute KS can be viewed as verifying the 2D schema plan.

Figure 14 shows that semantic merging allows most of the fragmentation in the tree to be merged, and separate grass and bush regions to be linked as well. The image is greatly cleaned up and more representative of the semantics of the scene. The

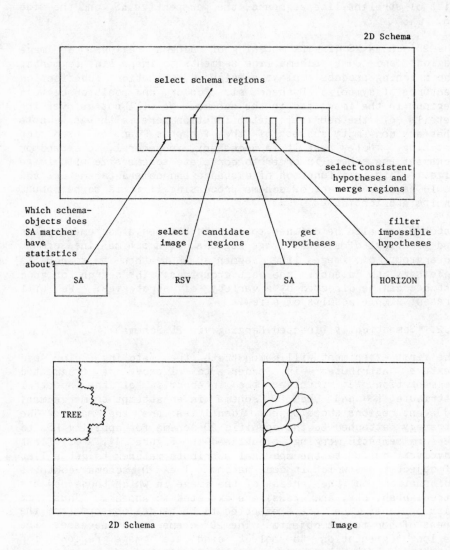

Figure 13. Semantic merging strategy. The 2D schema determines from the spectral attribute matcher which schema objects it can classify, then selects schema regions which are the expected locations of those objects, then determines all image regions in those vicinities, checks which objects are implied by those attributes, filters object categories which are inconsistent with the horizon model, and then merges regions with identical labels that are consistent with the schema.

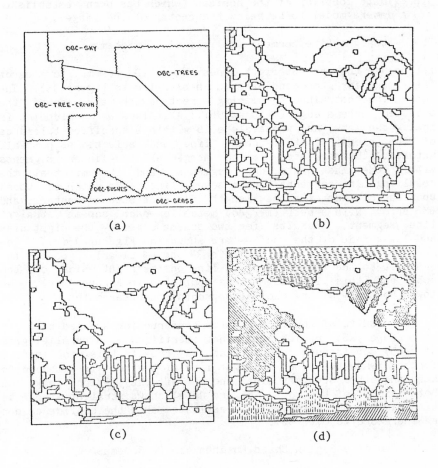

Figure 14. Merged regions via top-down guidance from 2D schema
and bottom-up results from spectral attribute matcher. Adjacent
regions are merged if the attribute match KS produces identities
which are consistent with the 2D schema and horizon KS. (a) 2D
schema showing area in image where tree, bush, sky, and grass
are expected to appear.
(b) Original segmentation.
(c) Semantically merged image.
(d) Same as (c). Cross-hatched
semantically merged image.

tree   grass

bush sky

results might be further improved by applying the horizon KS to filter the object hypotheses that are inconsistent with the approximate location of the horizon (which has been established via a camera model to be below the center of the image).

V.3. Straight Line Segment Analysis via 2D Schema

Let us use the long straight lines in our 2D schema to search the image for good candidate matches (refer to Figure 15). The search is constrained by placing a rectangular mask around the selected schema edge (Figure 15b). All lines whose midpoint is inside the mask, and whose slope is within a specified tolerance of the slope of the schema line are selected as possible matches. The next step is to merge all co-linear segments within the mask into new segments, and then match all the resulting line segments to the schema line. The match is based upon attributes of slope, length, distance between centers, and RMS error, with a best (merged) match for each schema straight line segment. Results for two schema edges -- the right side and lower side of the roof -- are shown in Figure 15(c). The merging of image line segments 34 and 94 clearly produces the best fit to the schema straight line on the right side of the roof and this line segment is completed in Figure 15(d). The lower boundary of the roof also produces a clear match.

If the results of the line segment construction are fed back to the shape KS, region 14 is now identified as a parallelogram with 65% confidence. Note that we expect this to improve further, when the lower straight line of the roof is extended to meet the other straight lines and cut off the region leaks on both sides of the roof. Figure 15(e) shows straight line fits with minimum RMS error. It is estimated that the confidence can be increased to over 90%.

V.4. Symbolic Region Shape Matches via 2D Schemas

Certain regions in the schema and the image have symbolic attributes of simple geometric types such as rectangle, trapezoid, ellipse, etc. The shape attributes of schema regions, where they are relevant, are pre-defined (or else will be generated during 3D schema projection). The shape attributes of image regions can be determined by the 2D shape KS.

Let us examine the strategies depicted in Figure 16 for 2D shape matching. The schema requires access to the results of the 2D shape KS and the list of schema regions with distinctive geometrical shape. The shape matching function then can use shape and position to determine a degree of fit. There are three types of matching capabilities of schema and image regions using any subset of the four features:

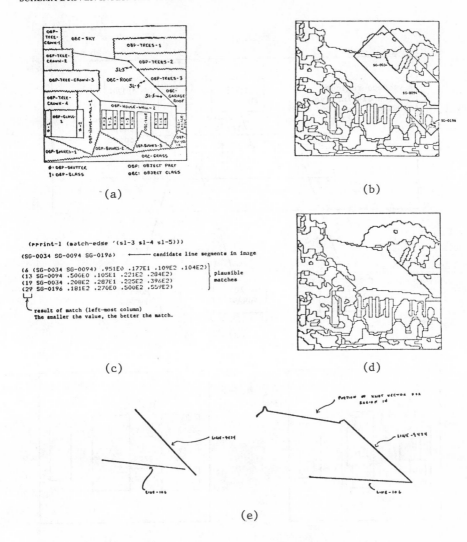

(a)

(b)

(c)

(d)

(e)

<u>Figure 15.</u>   Results from schema-directed straight line segment
analysis.   (a) High-level schema used to direct merging of
segments.   (b) Original segmentation showing mask used to locate
candidate line segments.   (c) The candidates for matching
against schema segments SL-3, SL-4, SL-5 are SG-34, SG-94,
SG-134, SG-224, as found in the masked area of (c).   The results
of matching combinations of these segments are shown in the left-
most column.   Clearly, segments SG-34 and SG-94 form the closest
match.   (d) Insertion of roof boundary segment as a result of
schema match of SL-3, 4 and 5 to segments SG-34 and 94.   (e) If
straight line fits are used to improve the roof boundary, the
confidence of a parallelogram can be increased sharply.

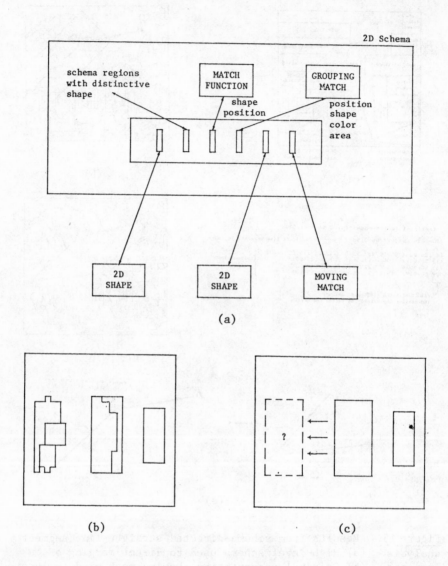

Figure 16. 2D shape matching. The 2D schema calls the 2D shape KS to extract regions with primitive geometrical shapes and then matches them with schema regions by symbolic shape label and position. The match can be applied to individual or groups of image regions. Additionally, features of color and expected area can be used. Image regions which are matched can be used as a template to be moved over the image.

    a) location of centroid,
    b) symbolic shape,
    c) intensity/color, and
    d) area.
The matching function can be applied:
    a) directly between a schema region and an image region,
    b) between a schema region and a group of adjacent image
       regions, and
    c) between a template (possibly derived from a previous
       match of image regions) and groups of image regions.
Let us now examine the results summarized in Figure 17.

First, consider an attempted match -- without the use of postion
information -- of all schema regions and image regions which
have distinctive geometric shapes. This will show that the 2D
schema can be robust without an exact spatial plan for the 2D
image. The 2D shape KS was run on all regions within the
expected house area, and those image regions which have a high
confidence of a primitive shape type can be further processed.
Each of these regions is used to match against schema regions
with similar shape based on attributes of size, shape/aspect
ratio, and color. The result of matching regions 45 and 50 with
the best five schema regions is tabulated in Figure 17(c). They
are found to match reasonably well with the various shutter
regions in the schema and poorly with other schema regions. The
left shutter has fragmented in the original segmentation and
region 45 is closer to a trapezoid than a rectangle.
Consequently, it has a poorer match with rectangular shutters
than region 50. It should be noted that the evaluation function
is scaled into 0 (perfect matches) to 1000 (no match); this
evaluation function has not yet been made consistent with the
form of other KS outputs.

The second step shows the improvements obtained by the addition
of positional information to better form correspondences between
schema shutter regions and image regions. There is a good match
for five of the six shutters in the front and one of the two on
the left (Figure 17c,d). Note that the left-most shutter has
not been found and only a part of the next one has been found
because of region fragmentation.

Figure 18 demonstrates the grouping capabilities of the 2D
schema by focussing on the left two large shutters in the image.
The centroid of the schema region is used to select candidate
regions for grouping and the match function (based upon all the
features) is used to select the subset which matches best. The
right shutter of the pair is extracted by this technique, but
due to the severe fragmentation of the left shutter this
technique was not employed. The shutter on the far left was
found by moving a template, of the size of the right shutter,

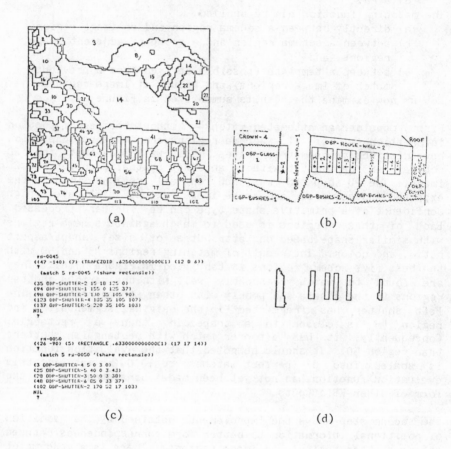

(a)

(b)

(c)

(d)

Figure 17.   Shape matching via 2D schema.   Results of matching
house shutters based upon shape matches and 2D shape KS.
(a) Portion of the original segmentation.   (b) Portion of the 2D
schema.   (c) Of the 5 image regions with high confidence of
rectangle or trapezoid, two regions, 45 and 50, are matched
against schema regions with roughly similar shapes.   The match
is based upon size, shape, and color and the best five matches
are shown.   Note that low evaluation is best.   The overall match
is shown on the left while the match factors of the features in
the order given above are shown to the right.   (d) The image
regions found to match with shutters in the schema.   Note that
the feature of position (neither schema nor region) was not
employed.

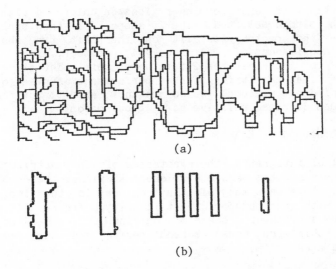

(a)

(b)

(match-groups left-shutter '(single obp-shutter-2))

```
(35 (RG-0072 RG-0064 RG-0059 RG-0052 RG-0044) 18 52)
(39 (RG-0072 RG-0064) 35 42)
(60 (RG-0064 RG-0059 RG-0052 RG-0044) 35 85)
(60 (RG-0069 RG-0064 RG-0059 RG-0052 RG-0044) 35 85)
(65 (RG-0049 RG-0040 RG-0046 RG-0045) 18 112)
(78 (RG-0050) 52 103)
(93 (RG-0075 RG-0071 RG-0047) 18 167)
(96 (RG-0075 RG-0047) 35 157)
(131 (RG-0073 RG-0069 RG-0059) 18 243)
(136 (RG-0049 RG-0046) 52 220)
(138 (RG-0053) 35 240)
(142 (RG-0048) 81 203)
(150 (RG-0074 RG-0065) 52 247)
(156 (RG-0075 RG-0071) 68 243)
(160 (RG-0063) 18 302)
(172 (RG-0070 RG-0046 RG-0045) 104 240)
(180 (RG-0095 RG-0086 RG-0070 RG-0065 RG-0045) 120 240)
(187 (RG-0055 RG-0035) 70 303)
(194 (RG-0074 RG-0070 RG-0065) 87 300)
(199 (RG-0062) 92 305)
```

(c)

Figure 18. Schema-directed grouping of image regions with simple
and distinct geometric shape. (a) Portion of original segmentation extracted from Figure 3 showing the fragmentation of the
left two shutters. (b) Grouped regions found by 2D schema. The
right-most shutter of the left pair was found using the centroid
of the schema region to select candidate regions for grouping.
(c) Results from match function when a mask formed from the right
shutter of the pair is moved to the left and matched against
groupings of candidate regions on the basis of color and size.
Regions 72, 64, 59, 52, and 44 match best. The merged collection
is shown in (b). The confidence that the second region from the
left is a rectangle has been increased from .80 (refer to Table
II) to .94.

towards the left and grouping regions on color and size. The
best match is then selected. The 2D shape KS is then applied to
determine the rectangular fit on the left two shutters,
producing confidences of 24% and 94%, respectively. It is
difficult to interpret the 24% value at this point since there
has not yet been any tuning of the performance curves of the
shape confidence measure; we do not know, as yet, how fast the
match values decrease relative to a 'good' match.

V.5.   Combination of KS Results

The result of integrating the hypotheses of the attribute KS,
line segment matches, and the 2D shape KS yields the results in
Figure 19. Note that many of the regions in the image are
labelled with the proper object identity. Figure 19(c,d) was
produced by a clean-up process of merging unlabelled adjacent
regions within the house schema region and the remaining
background area.

V.6.   Formation of a Spatial Plan Using Perspective Information

The proper use of the perspective KS requires that a set of
assumptions be generated regarding the orientation of surfaces.
In practice they would be determined via the specific 3D schema
and other information from long-term memory, but in this case
the set of assumptions necessary to drive the perspective KS are
obtained directly from the 2D schema. Thus, knowledge that a
particular region is bush, and that bushes are usually
perpendicular and attached to the ground plane, is available to
the 2D schema if it has been generated from a 3D schema. These
critical assumptions allow the perspective KS to place that
region (bush) in the 3D world model.

Let us consider the strategy for the computation of the distance
and size of an unoccluded object which is perpendicular to and
touches the ground plane; this strategy will be applied to
computing the range and height of the bushes. The strategy by
which the 2D schema controls the application of processes is
outlined in Figure 20. The spectral attribute matcher KS can be
used to validate the regions presumed to be bush and grass.
Their common boundary implies that it is unlikely that the
bottom of the bush is occluded. Next the perspective KS is
called to determine the distance and size of the bush. In this
example the range of the bushes is based upon two
assumptions: vertical orientation and the elevation of the
bottom of the bushes is 0. Then, the identity of regions 102,
110, 112, and 113 as grass implies that there is no occlusion of
these bush regions. Hence, the image coordinates of the region
can be translated into a range in the physical world. Once the
range is computed, then the image size -- region height and

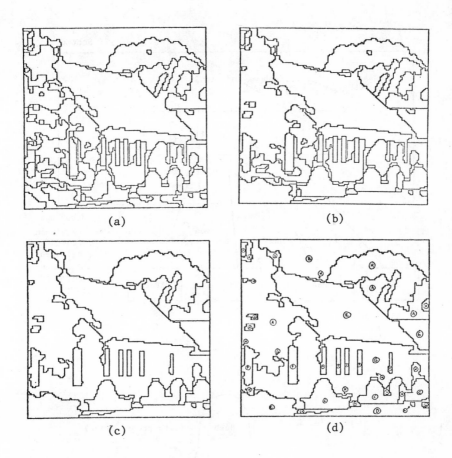

Figure 19.   Combination of schema-directed KS results.
(a) Original segmentation.   (b) Combined results of 2D schema
with attribute KS, line segment matching, and region shape KS.
(c) All regions without semantic labels are merged under guidance
of 2D schema (i.e., unlabelled house regions are kept separate
from unlabelled background regions). (d) Same as c, but labels
are provided on diagram:

Ⓐ	tree	Ⓔ	roof
Ⓑ	sky	Ⓕ	shutter
Ⓒ	bush	Ⓖ	unlabelled house
Ⓓ	grass	Ⓗ	unlabelled background

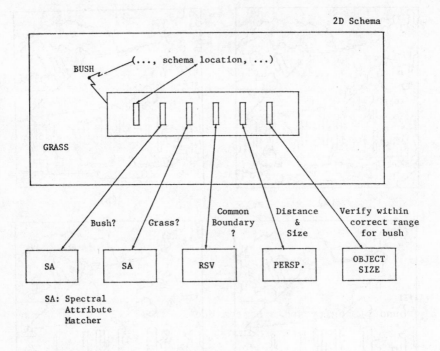

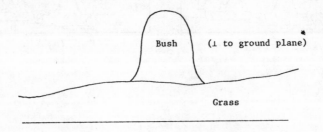

Figure 20.   Strategy for computing size and distance of
unoccluded object which touches ground. The SA KS is used to
verify that the regions expected to be bush and grass. The fact
that they have a common boundary implies that the bottom of the
bush is not occluded (assuming the ground plane is planar). The
perspective KS is used to compute the distance and size, while
the object size KS verifies that the computed size is in the
expected range of bush sizes.

width -- allows the physical size to be computed. Note that in order to carry out this analysis, the system employed the 2D schema, the spectral attribute KS, and the perspective KS. The inference drawn from this chain of hypotheses, namely that the region represents a bush, can be partially validated by noting that the computed size falls within the allowed range for bushes stored in long-term memory (see V.7).

Figure 21(a) describes the camera geometry from a bird's eye view with the image plane shown in front of the focal point for convenience. The range, offset, and elevation of a surface/object in the physical world must be computed in terms of the viewer-centered coordinate system involving the line of sight of the camera. Figure 21(b) lists results of applying the perspective KS, under control of the 2D schema, to selected regions of our test image. All the regions considered (bushes, shutters, house wall) lie roughly (particularly the bushes) in a pair of planes which are vertical to the ground plane and oriented at a diagonal to the right, away from the viewer. The location of objects are graphically portrayed in the bird's eye view of Figure 21(c).

In order to use the results in an effective manner, an error analysis should be taken into consideration. With an assumption of ground planarity and a camera model (focal length = 50 mm, elevation about 2 meters, because the person was standing on higher ground, tilt = 2 degrees upward), then the range of a physical point in the ground plane can be derived directly from the image coordinates of the point (in pixels). However, the computation is not a linear function of this image distance, and both the physical range and its associated error increase exponentially. Table III lists the absolute and relative error of a one-half pixel error for each row of pixels starting from the top of the image (i.e., row 1 in our 128x128 pixel image). The error in the range is shown superimposed on the location of objects in Figure 21(c). A one-half pixel error in width will produce an error in physical width which is relatively constant over the image unless the camera has a wide-angle lens (e.g., a fish-eye lens). Note that error in range will propagate directly into an error in physical height and width and this must be taken into consideration by the object size KS.

Even such simple perspective results as shown provide the beginnings of a 3D spatial layout. The ranges of the row of bushes in front of the house provide a range of possible orientations for region 56 (the house wall). This partial plan, shown as a bird's eye view, is illustrated in Figure 21(c). The angle of the shutters has been computed to be 24 degrees from the line of sight. The house in Figure 3(a) does not seem to be oriented at such a steep angle, but there is significant

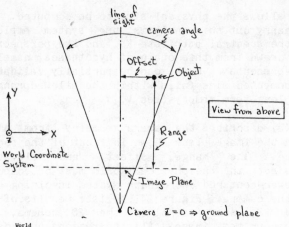

World
Coordinates

Y	Range: distance to object on a line perpendicular to image plane
X	Offset: distance of object to right of line of sight
$\|X_1-X_2\|$	Width of object: $\|$offset of right side -- offset of left side$\|$
Z	Elevation: distance of object above ground plane
Z_2-Z_1	Height of object: elevation of top -- elevation of bottom

(a)

Region	Region Identity	Range	Offset	Width	Height	Assumptions
90	Bush	32.0		2.32	1.01	vertical, attached to ground
99	Part of House front	39.1				same as 90
82 & 90	Bush	32.0	.472	2.54	1.59	same as 90; 82 vertical with same range as 90
79	Bush	37.5	3.78	2.63	1.61	same as 90
83	Bush	40.9	6.54	1.58	1.48	same as 90
56	House front	39.1		2.39	2.83	99 and 56 lie in one plane; range of 56 is same as 99; 56 is vertical
47	Shutter	33.0	.375	.300	1.50	vertical, height = 1.5 m
50	Shutter	34.8	1.22	.237	1.50	height = 1.5 m
51	Shutter	34.8	1.70	.237.	1.50	height = 1.5 m
54	Shutter	36.7	2.54	.250	1.50	height = 1.5 m
60	Shutter	44.0	5.15	.300	1.50	height = 1.5 m

(b)

Figure 21.   (caption on next page)

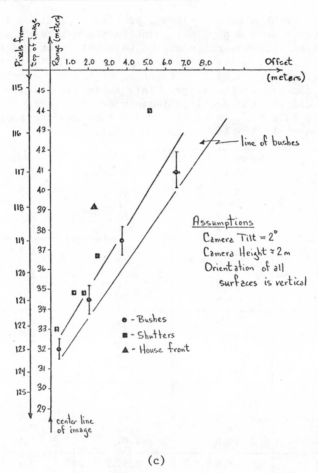

(c)

Figure 21. Results of forming a spatial plan using the Perspective KS. (a) Imaging geometry and description of terms used in
presenting the perspective results. The Z axis represents the
gravitational vertical; for the example image, the line of sight
is inclined 2° from the X,Y plane. (b) Computation of physical
location and size based upon assumptions shown in the right-hand
column. (c) Ground plan of house determined by the perspective
KS. The results of (b) in terms of range and offset fix the
locations of objects in the X,Y plane. Both range and offset
are expressed in meters. The two vertical scales show the
correlation between range and rows of pixels in the image. If a
pixel in a row is assumed to have elevation 0, then the physical
range is obtained by reading the range scale. The error range of
Table IV is superimposed as a vertical line through the location
of the bushes; the angle of the bushes is computed to be 24°
from the line of sight.

Table III. Error analysis for perspective KS. It is assumed that
a pixel represents a physical point in the ground plane (i.e.,
at elevation 0). The range of the physical point and its
associated error, under an assumption of one-half pixel error in
the image, are computed as a function of the row of pixels in
which it appears in the image. This table was derived via the
camera model for the specific image under consideration:
f = 50 mm, tilt = 2°, elevation = 2 m (because the picture was
taken from a slight rise in the terrain).

Pixel in Row	Computed Range of pixel	Absolute Error (meters)	Relative Error (%)
bottom of → 128	27.1	.416	1.5
image 127	28.0	.442	1.6
126	28.9	.472	1.6
125	29.9	.504	1.7
124	30.9	.540	1.7
123	32.0	.580	1.8
122	33.2	.624	1.9
121	34.5	.674	1.9
120	35.9	.730	2.0
119	37.5	.794	2.1
118	39.1	.865	2.2
117	40.9	.945	2.3
116	42.9	1.042	2.4
115	45.1	1.15	2.5
114	47.5	1.28	2.7
113	50.2	1.43	2.8
112	5?.3	1.61	3.0
111	56.7	1.82	3.2
110	60.6	2.01	3.4
109	65.0	2.39	3.7
108	70.2	2.79	4.0
107	76.2	3.30	4.3
106	83.4	3.95	4.7
105	92.2	4.82	5.2
104	102.9	6.01	5.8
103	116.	7.71	6.6
102	134.	10.25	7.6
101	158.	14.3	9.0
100	193.	21.3	11.0
99	247.	35.1	14.2
98	342.	69.0	20.1
97	559.	197.1	35.2
horizon 96	1529.	5294.	346.
→ 95		-6180.	296.

foreshortening. This orientation has been determined to be accurate via external physical examination of the environment.

V.7. Object Hypotheses Based on Size

Once the perspective KS has provided hypotheses about ranges of surfaces and the physical sizes of their projections, the size KS can be used to generate object hypotheses on the basis of the computed sizes. Figure 22 shows the hypotheses and their associated confidences formed by applying the size KS to selected regions from Figure 3. In each case, the default range on size (computed size $\pm 5\%$) was used, although these values can be set by the result of the perspective KS and the location of the region in the image. Also note that these results could be filtered by spatial location, much as the hypotheses formed from the spectral attribute matcher were. This results in a partial check on the assumptions used by the perspective KS during the computation of the size.

VI. CONCLUSIONS

The results cited in this paper represent the current state of development of the VISIONS system. A top-down interpretation of a scene has been successfully performed, although the conditions under which this interpretation was obtained were highly constrained. It demonstrates some degree of integration of the system, from automatic segmentation of the digitized input through symbolic output of object identities and generation of a rough plan of the three-dimensional space in the scene. An expanded version of this paper [PAR80] also contains experimental results on the use of inference networks for schema instantiation, and surface patches for 3D shape representation.

The primary emphasis of our current efforts is on the development of strategies by which the many knowledge sources can be integrated in order to interpret 2D color images. However, the ability to obtain the correct interpretation is inherently linked to the quality of information provided by these processes: without plausible hypotheses about the image, there isn't any control strategy worthy of investigation! Nevertheless, it is not feasible for us to attempt to perform extensive research in all the areas represented by the KSs. Thus, we must balance our efforts in the development of more complete knowledge sources against the development of interpretation strategies. Currently, we have implemented at least a simple version (and sometimes a complex version) of several KSs.

Each of the KSs developed can be used in different ways to

Region	Height	Width	Actual Identity	Object Size KS Hypotheses
82&90	1.59	2.54	Bush	(OBC-CAR 100) (OBC-BUSH 68) (OBC-TREE-CROWN 25)
54	1.50	.25	Building Shutter	(OBC-TREE-TRUNK 100) (OBC-BUILDING-SHUTTER 35)
41&56	4.30	5.7 to 32	House Body or Building Side	(OBC-HOUSE-BODY 100) (OBC-GARAGE 74) (OBC-ROOF 74) (OBC-HOUSE 71) (OBC-BUILDING-WALL 34) (OBC-BUILDING-SIDE 34) (OBC-TREE 18) (OBC-TREE-CROWN 14)
20	3.5 to 9.7*	3.3	Tree or Tree Crown	(OBC-GARAGE 100) (OBC-TREE 90) (OBC-TREE-CROWN 67) (OBC-ROOF 40) (OBC-GARAGE-BODY 17) (OBC-GARAGE-FRONT-SIDE 12) (OBC-BUILDING-WALL 1) (OBC-BUILDING-SIDE 1)
14	3.75	5.70	Roof	(OBC-GARAGE 100) (OBC-ROOF 53) (OBC-GARAGE-BODY 50) (OBC-TREE 33) (OBC-GARAGE-FRONT-SIDE 32) (OBC-BUILDING-WALL 26) (OBC-BUILDING-SIDE 26) (OBC-TREE-CROWN 25) (OBC-SKY 4) (OBC-HOUSE-BODY 3) (OBC-HOUSE 1)

* Elevation of top of region

Figure 22. Summary of results of object size KS for selected regions of Figure 3. The sizes shown were computed by the perspective KS using the default ± 5% error range (see Figure 17). The actual range can be set by the perspective KS on the basis of the error analysis.

produce several different kinds of hypotheses. The experiments already performed seem to indicate that there may be many mini-strategies for using the KSs in particular ways across the range of images. For example, the perspective KS can determine physical dimensions of surfaces, while the object size KS uses these results to produce a confidence measure for object hypotheses; or the horizon KS can be used to filter implausible object identities from the output of the spectral attribute matcher KS. Interesting strategies can be modelled in terms of the overlap of information related to perspective, occlusion, size, shape, junction analysis, etc. With proper design the set of local processes may be built to answer the questions that are of importance to each other, and this network of processes can be flexibly and incrementally constructed. As the strategies are understood, they can be incrementally embedded in the schemas.

The results presented in this paper were generated via top-down control of the KSs using a specific 2D schema -- in effect a plan -- for a specific house scene. The analysis was highly biased towards success because the schema is tuned to the particular situation: the case of looking at a familiar scene from a familiar point of view. It does, however, show some of the ways that the KSs are able to interact, and can also be viewed as an experiment in verifying that some stored schema is applicable to a given image. The last experiment demonstrates bottom-up interaction of the KSs in an attempt to instantiate the proper schema from a set of schemas.

The facilities now exist for actually developing to a much deeper level some of the ideas we have only been able to suggest as promising. The benefits of some of the interesting developments of our colleagues in the research community over the last few years has led to a deeper appreciation of the problems yet remaining. This is reflected somewhat in a shift of research emphasis, as we propose a highly structured research paradigm for exploring the issues we set forth. A series of increasingly more difficult experiments will provide an experimental methodology for developing schema-driven (e.g., top-down) control mechanisms; each succeeding experiment will assume a set of weaker constraints, representing image interpretation tasks where a decreasing amount of knowledge of the situation is available. It is worth noting, however, that the basic approach is not substantially different from the initial top-down approach that started the VISIONS project [HAN74, RIS74], although it is considerably richer in detail.

REFERENCES

[AGI72] G.J. Agin, "Representation and Description of Curved
         Objects," Stanford AI Memo 73, 1972.

[AGI76] G.J. Agin and T.O. Binford, "Computer Description of
         Curved Objects," IEEE Transactions on Computers, April
         1976, pp. 439-449.

[ARB77] M.A. Arbib, "Parallelism, Slides, Schemas, and Frames,"
         in Systems: Approaches, Theories, Applications (W.E.
         Hartnett, Ed.), D. Reidel Publishing Co., 1977, pp.
         27-43.

[BAJ76] R. Bajcsy and M. Tavakoli, "Computer Recognition of
         Roads from Satellite Pictures," IEEE Transactions on
         Systems, Man, and Cybernetics, Vol. SMC-6, September
         1976, pp. 623-637.

[BAL78] D.H. Ballard, C.M. Brown, and J.A. Feldman, "An
         Approach to Knowledge-Directed Image Analysis,"
         Computer Vision Systems (A. Hanson and E. Riseman,
         Eds.), Academic Press, pp. 271-281, 1978.

[BAR76] H. Barrow and J.M. Tenenbaum, "MSYS: A System for
         Reasoning About Scenes," Technical Note 121, Artificial
         Intelligence Center, Stanford Research Institute, Menlo
         Park, CA, April 1976.

[BAR78] H.G. Barrow and J.M. Tenenbaum, "Recovering Intrinsic
         Scene Characteristics from Images," Computer Vision
         Systems (A. Hanson and E. Riseman, Eds.), Academic
         Press, pp. 3-26, 1978.

[BUL78] B.L. Bullock, "The Necessity for a Theory of
         Specialized Vision," Computer Vision Systems (A. Hanson
         and E. Riseman, Eds.), Academic Press, pp. 27-35, 1978.

[DAV76] L.S. Davis, "Shape Matching Using Relaxation
         Techniques," Technical Report 480, Computer Science
         Center, University of Maryland, College Park, MD,
         September 1976.

[DUD73] R.O. Duda and P.E. Hart, Pattern Classification and
         Scene Analysis, John Wiley and Sons, 1973.

[DUD77] S.A. Dudani and A.L. Luk, "Locating Straight-Line Edge
         Segments on Outdoor Scenes," Proc. of Conf. on Pattern
         Recognition and Image Processing, Troy, NY, June, 1977,
         pp. 367-377.

[DYE80] C.R. Dyer, A. Rosenfeld, and H. Samet, "Region Representation: Boundary Codes from Quadtrees," Communications of the ACM, 3, 1980, pp. 171-179.

[ERM75] L.D. Erman and V.R. Lesser, "A Multi-Level Organization for Problem Solving Using Many Diverse Cooperating Sources of Knowledge," Proc. 4th Inter. Joint Conf. on Artificial Intelligence, Tbilisi, USSR, 1975, pp. 483-490.

[FEL74] J.A. Feldman and Y. Yakimovsky, "Decision Theory and Artificial Intelligence: I. A Semantics-Based Region Analyzer," Artificial Intelligence, Vol. 5, 1974, pp. 349-371.

[FRI69] D.P. Friedman, D.C. Dickson, J.J. Fraser, and T.W. Pratt, "GRASPE 1.5 - A Graph Processor and Its Application," Tech. Report, University of Houston, 1969.

[HAN74] A.R. Hanson and E.M. Riseman, "Preprocessing Cones: A Computational Structure for Scene Analysis," COINS TR 74C-7, Univ. of Mass., Amherst, September 1974.

[HAN78a] A.R. Hanson and E.M. Riseman (Eds.), Computer Vision Systems, Academic Press, 1978.

[HAN78b] A.R. Hanson and E.M. Riseman, "Segmentation of Natural Scenes," in Computer Vision Systems (A.R. Hanson and E.M. Riseman, Eds.), Academic Press, pp. 129-163, 1978.

[HAN78c] A.R. Hanson and E.M. Riseman, "VISIONS: A Computer System for Interpreting Scenes," in Computer Vision Systems (A.R. Hanson and E.M. Riseman, Eds.), Academic Press, pp. 303-333, 1978.

[HAN80a] A.R. Hanson, E.M. Riseman and F.C. Glazer, "Edge Relaxation and Boundary Continuity," in Consistent Labeling Problems in Pattern Recognition (R.M. Haralick, Ed.), Plenum Press, 1980.

[HAN80b] A.R. Hanson and E.M. Riseman, "Processing Cones: A Computational Structure for Image Analysis," in Structured Computer Vision (S. Tanimoto and A. Klinger, Eds.), Academic Press, 1980.

[HAR78] R. Haralick, "Using Perspective Transformations in Scene Analysis," Technical Report, Electrical Engineering Department, University of Kansas, Lawrence, May 1978.

[HAV78] W.S. Havens, "A Procedural Model of Recognition for
         Machine Perception," TR-78-3, Ph.D. Thesis, Department
         of Computer Science, University of British Columbia,
         Vancouver, Canada, 1978.

[HOR77] B.K.P. Horn, "Understanding Image Intensities,"
         Artificial Intelligence, Vol. 8, No. 2, 1977, pp.
         201-231.

[KOH79] R. Kohler, "Reference Manual for the VISIONS Low-Level
         Image Processing System," COINS Dept., Univ. of Mass.,
         Amherst, Spring 1979.

[KON75] K. Konolige, "The ALISP Manual," Univ. Computing
         Center, Univ. of Mass., August 1975.

[LES77] V.R. Lesser and L.D. Erman, "A Retrospective View of
         the Hearsay-II Architecture," Proc. Inter. Joint Conf.
         on Artificial Intelligence, Cambridge, MA, 1977, pp.
         790-800.

[LEV78] M.D. Levine, "A Knowledge-Based Computer Vision
         System," in Computer Vision Systems (A.R. Hanson and
         E.M. Riseman, Eds.), Academic Press, 1978, pp. 335-352.

[LOW76] B.T. Lowerre, "The HARPY Speech Recognition Systems,"
         Ph.D. Thesis, Department of Computer Science,
         Carnegie-Mellon University, Pittsburgh, PA, 1976.

[LOW78] J.D. Lowrance, "GRASPER 1.0 Reference Manual," COINS
         Technical Report 78-20, University of Massachusetts,
         Amherst, December 1978.

[LOW80] J.D. Lowrance, "Dependency-Graph Models of Evidential
         Support," Ph.D. Dissertation, COINS Dept., Univ. of
         Mass., Amherst, expected June 1980.

[MAC78] A.K. Mackworth, "Vision Research Strategy: Black Magic,
         Metaphors, Mechanisms, Miniworlds, and Maps," in
         Computer Vision Systems (A.R. Hanson and E.M. Riseman,
         Eds.), Academic Press, 1978, pp. 53-60.

[MAR76] D. Marr, "Early Processing of Visual Information,"
         Phil. Trans. Roy. Soc. B275, 1976, pp. 483-524.

[MAR77] D. Marr and .H.K. Nishihara, "Representation and
         Recognition of the Spatial Organization of
         Three-Dimensional Shapes," Proc. Roy. Soc. B.200, 1977,
         pp. 269-294.

[MAR78] D. Marr, "Representing Visual Information," in Computer Vision Systems (A.R. Hanson and E.M. Riseman, Eds.), Academic Press, 1978, pp. 61-80.

[MIN75] M. Minsky, "A Framework for Representing Knowledge," in The Psychology of Computer Vision (P. Winston, Ed.), McGraw-Hill, 1975, pp. 211-277.

[NAG79] P. Nagin, "Studies in Image Segmentation Algorithms Based on Histogram Clustering and Relaxation," COINS Technical Report 79-15 and Ph.D. Dissertation, Univ. of Mass., Amherst, September 1979.

[NEV76] R. Nevatia, Computer Analysis of Scenes of 3-Dimensional Curved Objects, Birkhauser-Verlag, Basel, Switzerland, 1976.

[NEV77] R. Nevatia and T.O. Binford, "Description and Recognition of Curved Objects," Artificial Intelligence, Vol. 8, 1977, pp. 77-98.

[NEV78] R. Nevatia, "Characterization and Requirements of Computer Vision Systems," in Computer Vision Systems (A.R. Hanson and E.M. Riseman, Eds.), Academic Press, 1978, pp. 81-87.

[NEW65] J.R. Newman, The Universal Encyclopedia of Mathematics, The New American Library, July 1965.

[OVE79] K.J. Overton and T.E. Weymouth, "A Noise Reducing Preprocessing Algorithm," Proc. of Pattern Recognition and Image Processing Conference, Chicago, Illinois, August 1979, pp. 498-507.

[PAR80] C.C. Parma, A.R. Hanson and E.M. Riseman, "Experiments in Schema-Driven Interpretation of a Natural Scene," COINS TR 80-10, Univ. of Mass., Amherst, April 1980.

[PRA71] T. Pratt and D. Friedman, "A Language Extension for Graph Processing and Its Formal Semantics," Communications of the ACM, 4, 1971.

[PRA79] J. Prager, "Analysis of Static and Dynamic Scenes" Ph.D. Dissertation, COINS Dept., Univ. of Mass., Amherst, March 1979.

[PRA80] J. Prager, "Extracting and Labeling Boundary Segments in Natural Scenes," IEEE Trans. Pattern Analysis and Machine Intelligence, Vol. PAMI-2, January 1980, pp. 16-27.

[RIS74] E.M. Riseman and A.R. Hanson, "Design of a Semantically
         Directed Vision Processor," COINS TR 74C-1, Univ. of
         Mass., Amherst, January 1974.

[RIS77] E.M. Riseman and M.A. Arbib, "Computational Techniques
         in the Visual Segmentation of Static Scenes," Computer
         Graphics and Image Processing, 6, 1977, pp. 221-276.

[ROB65] L.G. Roberts, "Machine Perception of Three-Dimensional
         Solids," Optical and Electro-Optical Information
         Processing (J.T. Tippet et al., Eds.), MIT Press, 1965.

[ROS76] A. Rosenfeld, R.A. Hummel and S.W. Zucker, "Scene
         Labelling by Relaxation Operations," IEEE Trans.
         Systems, Man, and Cybernetics, 6, 1976, pp. 420-433.

[RUB77] S.M. Rubin and R. Reddy, "The Locus Model of Search and
         Its Use in Image Interpretation," Proc. of Fifth IJCAI,
         Cambridge, MA, August 1977.

[SAK76] T. Sakai, T. Kanade, and Y. Ohta, "Model-Based
         Interpretation of Outdoor Scenes," Third Int. Joint
         Conf. on Pattern Recognition, Coronado, CA, November
         1976, pp. 581-585.

[SAM80] H. Samet, "Region Representation: Quadtrees from
         Boundary Codes," Communications of the ACM, 3, 1980,
         pp. 163-170.

[SCH75] R.C. Schank and R. Abelson, "Scripts, Plans, and
         Knowledge," Proc. of Fourth IJCAI, Tbilisi, 1975, pp.
         151-158.

[SCH77] R.C. Schank and R.P. Abelson, Goals, Plans, Scripts and
         Understanding: An Enquiry into Human Knowledge
         Structures, Erlbaum Press, NJ, 1977.

[SCH79] R.C. Schank, Interdisciplinary Conference, Jackson,
         Wyoming, January 1979.

[SHI78] Y. Shirai, "Recognition of Real-World Objects Using
         Edge Cues," in Computer Vision Systems (A.R. Hanson and
         E.M. Riseman, Eds.), Academic Press, 1978, pp. 353-362.

[TAN78] S.L. Tanimoto, "Regular Hierarchical Image and
         Processing Structures in Machine Vision," in Computer
         Vision Systems (A.R. Hanson and E.M. Riseman, Eds.),
         Academic Press, 1978, pp. 165-174.

[TAN80]   S. Tanimoto and A. Klinger (Eds.), _Structured_ _Computer_
          Vision, Academic Press, 1980.

[TEN77]   J.M. Tenenbaum and H.G. Barrow, "Experiments in
          Interpretation-Guided   Segmentation,"   _Artificial_
          Intelligence, 8, No. 3, 1977, pp. 241-274.

[UHR72]   L. Uhr, "Layered `Recognition Cone' Networks That
          Preprocess, Classify, and Describe," _IEEE_ _Trans._
          Computers, 1972, pp. 758-768.

[UHR78]   L. Uhr, "`Recognition Cones,' and Some Test Results;
          The Imminent Arrival of Well-Structured Parallel-Serial
          Computers; Positions, and Positons on Positions," in
          Computer _Vision_ _Systems_ (A.R. Hanson and E.M. Riseman,
          Eds.), Academic Press, 1978, pp. 363-377.

[WIL77]   T. Williams and J. Lowrance, "Model-Building in the
          VISIONS High Level System," COINS Technical Report
          77-1, Univ. of Mass., Amherst, January 1977.

[WIL80]   T. Williams, Ph.D. Dissertation (in preparation). COINS
          Dept., Univ. of Mass., expected June 1980.

[WAL75]   D. Waltz, "Understanding Line Drawings of Scenes with
          Shadows," in _The_ _Psychology_ _of_ _Computer_ _Vision_ (P.
          Winston, Ed.), McGraw-Hill, 1975, pp. 19-91.

[YAK73]   Y. Yakimovsky and J.A. Feldman, "A Semantics-Based
          Decision Theory Region Analyzer," _Proc._ _IJCAI-3_, August
          1973, pp. 580-588.

[YOR79]   B. York, "A Primer on Splines," COINS TR 79-5, Univ. of
          Mass., Amherst, Mass., March 1979.

[YOR80]   B. York, Ph.D. Dissertation (in preparation), COINS
          Dept., Univ. of Mass., expected June 1980.

[ZUC77]   S.W. Zucker, R.A. Hummel, and A. Rosenfeld, "An
          Application of Relaxation Labelling to Line and Curve
          Enhancement," _IEEE_ _Transactions_ _on_ _Computers_, Vol.
          C-26, April 1977, pp. 394-403.

FINDING CHROMOSOME CENTROMERES USING BOUNDARY AND DENSITY
INFORMATION

Jim Piper

MRC Clinical and Population Cytogenetics Unit,
Edinburgh EH4 2XU,
Scotland.

Existing algorithms for locating chromosome centromeres in
digital images are insufficiently reliable for totally automatic
analysis systems.   However, the problem of unreliability can to
some extent be overcome by the use of confidence measures at each
processing   stage.   It is therefore desirable to combine
different available algorithms in such a way as to improve both
the   reliability and the confidence in the location of the
centromere.   A combination of two methods is described, in which
chords   analysis   acts as a "plan" for the analysis of an
integrated density profile along the medial   axis   of the
chromosome.

1. INTRODUCTION.

Location   of chromosome centromeres (figure 1) is essential
in automated systems intended for karyotyping homogeneously
stained chromosomes and for chromosome aberration scoring, and is
highly desirable for karyotyping banded chromosomes (3,6,10).   A
centromere finder must firstly be reliable - it should make few
serious errors.   It should also be accurate in⁻ the positioning
of the centromere.   No method has yet been found which locates
the centromeres of a sufficient proportion of the chromosomes in
each cell to permit the construction of a totally automatic
karyotyping system, even if restricted to the 10 class problem of
homogeneously stained chromosomes.   As a result, existing
practical systems rely on operator interaction to obtain an
acceptable performance.   This increases the operating costs so
that such systems are probably uneconomic compared with
conventional manual analysis (6).

511

J. C. Simon and R. M. Haralick (eds.), Digital Image Processing, 511–518
Copyright © 1981 by D. Reidel Publishing Company.

(a)   (b)

Figure   1.   Digitised chromosomes with centromere lines marked,
(a) homogeneously stained, (b) Giemsa banded.

     A measure of confidence in the performance of an algorithm
applied to a particular chromosome can provide the means of
overcoming the reliability problem (8).   In an interactive
system, the measure can be used to select relevant machine
decisions for human review or correction (5), and in a totally
automatic system it can be used to alter the behaviour of a later
stage in the system, for example by weighting a context sensitive
or relaxation type classifier (9).   Thus a good algorithm is one
that is both reliable and accurate, and which   produces features
upon which an effective confidence measure can be based.

2. CENTROMERE LOCATION METHODS.

     There   are two basic types of centromere location algorithm.
One analyses the boundary of the chromosome.   Examples are
boundary chain code analysis (1), and "chords" analysis of the
sides or chords of the minimum enclosing convex polygon (7).
These methods are particularly prone to boundary noise problems,
and so require high image quality.   In addition, they contain a
large number of rather ad hoc rules for interpreting boundary
configurations, and are therefore difficult to train, or to
support by reliable confidence measures.   The second variety of
centromere algorithm analyses a density profile along the major
axis of the chromosome (4).   There are differences in the
methods of deriving the axis, and whether a   straight or
curvilinear axis is used in the case of bent chromosomes (6).
Algorithms of either type have proved reliable and accurate for
the larger chromosomes 1-12 and X when stained homogeneously, but
by no means so reliable for the smaller chromosomes 13-22 and Y,
or for banded chromosomes.

Given that two essentially different ways of finding centromeres are available, an obvious question is whether they can fruitfully be combined. Problems naturally arise when the methods produce results which do not agree, as a choice of one or the other has to be made. One way out of this dilemma is to choose whichever result has the higher confidence measure (8). In the approach described below the two methods are applied hierarchically, one providing a "plan" for the next (2).

3. A SYNTHESIS OF BOUNDARY AND PROFILE METHODS.

The centromere finding method described here is multipass, with a computationally cheap analysis generating a plan of plausible centromere locations and associated chromosome axes, followed by a more costly analysis to look at these possibilities in more detail. A simple chords algorithm finds between 1 and 3 candidate centromeres (occasionally more), each associated with a potential chromosome axis. Each axis is then used to derive a profile, and analysis of the profile produces a refined position for the candidate centromere. Selection of the best candidate and measurement of its confidence is based on features such as the distance between corresponding centromere candidates found by the chords and profile analysis, the depth of the major minimum in the profile, and the symmetry of the two halves of the chromosome as divided by the axis.

3.1. The chords analysis.

A minimum convex polygon is fitted to the chromosome outline (7). Each chord for which a linear combination of the length, the area between the chord and the chromosome, and the depth of the lowest chromosome boundary point below the chord exceeds some threshold is marked as "significant". In practise, the majority of chromosomes will have between 2 and 4 significant chords (figure 2).

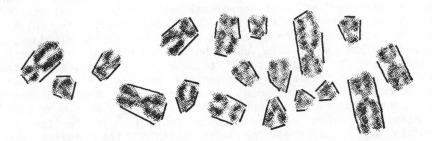

Figure 2. The "significant" chords of the minimum enclosing convex polygons.

The chords are related to each other in terms of the included angle between them, their distance apart compared to the chromosome width, and how "opposite" they are. Oppositeness can be measured by considering the distance to the chord end points from the produced chord intersection point (figure 3). Pairs of chords are selected on the basis of these measurements using some simple heuristics. For each selected pair of chords, a candidate centromere line is postulated to join the lowest points below each of the two chords, and the bisector of the angle between the two chords is the associated potential chromosome axis (figure 3). Typically, only one pair of chords is selected in a metacentric chromosome, while in an acrocentric chromosome there are three pairs, with three chords participating in two pairs each (figure 3).

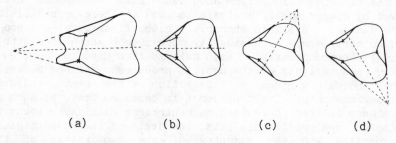

(a)   (b)   (c)   (d)

Figure 3.   "Significant" chords, their angle bisectors and the potential centromere lines obtained from each selected pair. (b), (c) and (d) show three candidates in an acrocentric chromosome.

3.2. Derivation of the profile.

Each centrome location proposed by the chords analysis is analysed further by rotating the chromosome to make the postulated axis a coordinate axis. The medial axis is derived from the set of mid-points of each coordinate row of the rotated chromosome (figure 4). The profile of the density-weighted mean square distance of each coordinate row from the medial axis of the chromosome is computed (figure 4). This profile is related to the position of the centre of density of the chromatids with respect to the medial axis, while being independent of banding intensity. It would appear to solve the problems associated with the use of the integrated density profile (4) in banded chromosomes, where in some chromosome pairs the centromere coincides with an intense optical density peak, i.e. a dark band, examples of which can be seen in figure 1. In practise, the profile is accumulated for each side of the medial axis independently, as this provides a means for measuring the symmetry of the chromosome about the axis, which is a useful feature for rejecting incorrect axis candidates.

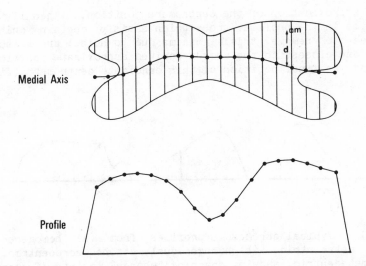

Figure 4.   A chromosome and its profile.   The profile value for
each coordinate row is computed as $\Sigma md^2 / \Sigma m$, where m is the grey
value of a pixel, and d its distance from the medial axis.

We   have   investigated a method for dealing with bent
chromosomes which in effect straightens the chromosome image
before the profile is constructed.   Such techniques are of
importance in deriving the banding density profile for band
analysis,   but   for the purpose of centromere location the
unstraightened image proves adequate even when the chromosome is
quite severely bent.

3.3. Analysis of the profile.

Again   there is more than one possible method, and it proves
to be profitable to combine two.   The first method is to look
for minima in the profile.   This method fails in the case of
acrocentric   chromosomes,   where   the   profile   increases
monotonically to the maximum and then decreases monotonically
(figure 5b).   Alternatively, applying chords analysis to the
profile itself yields significant chords above concavities in the
profile, exactly as in the chords analysis of the chromosome
outline (figure 5).

A   significance measure, similar to that used for the
chromosome boundary analysis (§3.1), when applied to these chords
leads to a choice of the most significant, and the point on the
profile whose vertical distance below the chord is greatest is
then the candidate centromere position.   In order to combine the
two methods, the chords analysis of the profile is used to obtain

a first approximation of the centromere position. Then if there
are minima in the profile, those which satisfy certain criteria
(designed to remove trivial minima due to noise) are selected,
and the one closest to the chords-derived approximate position is
chosen as the final position of this candidate centromere (figure
5).

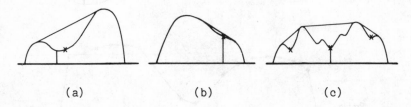

(a)   (b)   (c)

Figure 5.   Typical chromosome profiles from (a) homogeneously
stained metacentric, (b) homogeneously stained acrocentric, (c)
banded metacentric, showing spanning "chords", points of greatest
vertical distance below chords (x), and the selected centromere
position (vertical line).

3.4. Features for confidence estimation.

    To establish confidence estimates, features were measured on
a training set of chromosomes for which the correct centromere
location had been provided by a cytotechnician.   Discriminant
function analysis provided a confidence measure and a confidence
threshold for good centromeres.   This methodology is used both
to select the best of several possible candidates, and to
estimate the reliability of the single resulting centromere.
About 20 features showed some discriminatory power, the best
being (i) distance to the chords predicted centromere, (ii)
symmetry between the two half profiles, (iii) value of a
significance measure applied to the chords above the profile
(3.3), and (iv) chromosome aspect ratio (i.e. length/width).

4. PERFORMANCE AND CONCLUSIONS.

    The centromere finding method described here performs
substantially better than methods used previously in our
experimental chromosome analysis systems, whether of the chords
or profile variety.   The automatically determined centromere was
compared with a manually placed centromere on all chromosomes in
19 digitised Orcein stained cells and in 10 Giemsa banded cells.
The mean error in the centromere index (which is defined as 100
times the ratio of the area of the short arms to the total
chromosome area, and so is in the range [0, 50]) is shown in
table 1 for each "Denver" group, for both homogeneously stained

and banded chromosomes. For comparison, the standard deviation of the group centromere index obtained from the manually placed centromeres is also shown.

Table 1. The within-group standard deviation and mean error in the automatically found centromere index, by chromosome group, compared with the within-group standard deviation of the centromere index from manually placed centromeres.

chromosome group	homogeneous chromosomes			banded chromosomes		
	auto c.i. st dev	mean c.i. error	manual c.i. st dev	auto c.i. st dev	mean c.i. error	manual c.i. st dev
1	1.7	1.1	2.0	1.0	0.5	1.1
2	5.1	1.9	2.7	3.0	1.6	1.7
3	2.5	1.6	1.6	2.6	2.3	2.3
4,5	5.7	2.8	2.3	4.9	2.5	2.5
6-12,X	4.8	2.1	4.8	4.8	1.7	5.3
13-15	8.5	5.2	3.6	3.4	2.9	3.6
16	6.7	3.1	2.9	5.8	3.9	4.0
17,18	4.6	2.1	4.6	5.3	3.0	4.0
19,20	4.7	3.0	3.6	4.2	2.3	4.1
21,22,Y	7.8	4.9	6.4	9.7	5.9	6.4

For homogeneously stained chromosomes, these errors, taken with the other elements of our chromosome analysis system, result in overall error rates of about 9% chromosome mis-identifications and 3% chromosome losses in a completely automatic analysis with no human interaction. Confidence measures proved to be satisfactory, and this coupled with the improved accuracy of the algorithm has resulted in a halving of the operator work load in an interactive version of the system compared to that previously reported (5).

However, confidence measures were less good for banded chromosomes. This seems to be due to two factors. Firstly, the outline of digitised banded chromosomes has features due to stain distribution (figure 1) as well as due to the centromere, and as a result the significant chords are not such good predictors of the centromere. Secondly, the centromere region of a homogeneous chromosome is characteristically paler than elsewhere, and testing this density in a candidate centromere provides a useful confidence feature. Banded chromosome centromeres can have a density ranging from very pale to very bright (figure 1) and so this particular feature has no ability to discriminate good from bad candidate centromeres.

The main advantages of the method of centromere finding described here are (i) it is not necessary to have a complicated chords algorithm, (ii) the precision of the profile method is retained, (iii) the problem of how to combine two centromere finding methods is solved, (iv) the method appears to work as well for banded as for homogeneously stained chromosomes, and (v) there are a considerable number of features available for confidence measures, which makes subsequent automatic or interactive processing more productive.

5. REFERENCES.

1. Gallus, G., and Regoliosi, G.: 1974, J. Histochem. Cytochem. 22, pp. 546-553.

2. Kelly, M.D.: 1971, "Machine Intelligence 6" (eds. Meltzer, B., and Michie, D.), Edinburgh University Press, pp. 397-409.

3. Ledley, R.S., Lubs, H.A., and Ruddle, F.H.: 1972, Comput. Biol. Med. 2, pp. 107-128.

4. Mendelsohn, M.L., Hungerford, D.A., Mayall, B.H., Perry, B.H., Conway, T.J., and Prewitt, J.M.S.: 1969, Ann. N.Y. Acad. Sci. 157, pp. 376-392.

5. Piper, J., Mason, D., Rutovitz, D., Ruttledge, H., and Smith, L.: 1979, J. Histochem. Cytochem. 27, pp. 432-435.

6. Piper, J., Granum, E., Rutovitz, D., and Ruttledge, H.: 1980, Signal Processing 2, pp. 203-221.

7. Rutovitz, D.: 1969, "Machine Intelligence 5" (eds. Meltzer, B., and Michie, D.), Edinburgh University Press, pp. 435-462.

8. Rutovitz, D., Cameron, J., Farrow, A.S.J., Goldberg, R., Green, D.K., and Hilditch, C.J.: 1969, "Human Population Cytogenetics" (eds. Jacobs, P.A., Price, W.H., and Law, P.), Edinburgh University Press, pp. 281-296.

9. Rutovitz, D., Green, D.K., Farrow, A.S.J., and Mason, D.C.: 1978, "Pattern Recognition" (ed. Batchelor, B.G.), Plenum, New York, pp. 303-329.

10. Wald, N., Li, C.C., Herron, J.M., Davis, L., and Fatora, S.R.: 1976, "Automation of Cytogenetics - Asilomar Workshop" (ed. Mendelsohn, M.L.), National Technical Information Service, Springfield, Va., pp. 39-45.

FORENSIC WRITER RECOGNITION

Volker Klement

Forensic Science Institute, Bundeskriminalamt
Wiesbaden, Federal Republic of Germany

Abstract. The application of image processing and
pattern recognition techniques to support the
comparative handwriting analysis is reported. The
dominating problem proved to be the selection and
formulation of suited task-oriented features. Adequate
solutions have been obtained by heuristic approaches,
they have to be verified statistically on large test
data sets.

1. Introduction

The forensic comparison of handwritings is a
well-introduced and generally appreciated method of
high probative force (1). Considering the large number
of incoming samples per year, one expects advantages
from computer-oriented methods by objectivation,
quantification and automation. Computer-oriented image
processing and pattern recognition techniques for
writer recognition are being developed in a current
research project "Objectivation and automation of the
handwriting comparison" (2,3,4).

The objective is to recognize the writer of a text. As
such, it should be distinguished from work to
recognize the text itself (character recognition), to
perform graphological studies or to analyze the
dynamics of the writing process. There are two aspects
of writer recognition:

J. C. Simon and R. M. Haralick (eds.), Digital Image Processing, 519–524.

  - identification
identifying a writer (by features of his handwriting)
out of a limited set of known writers,

  - verification
stating a degree of certainty, that two handwritings
originate from the same writer (based on a similarity
measure).

For illustration, Fig. 1 shows a typical example of a
discretized and binarized handwriting image stored in
the computer.

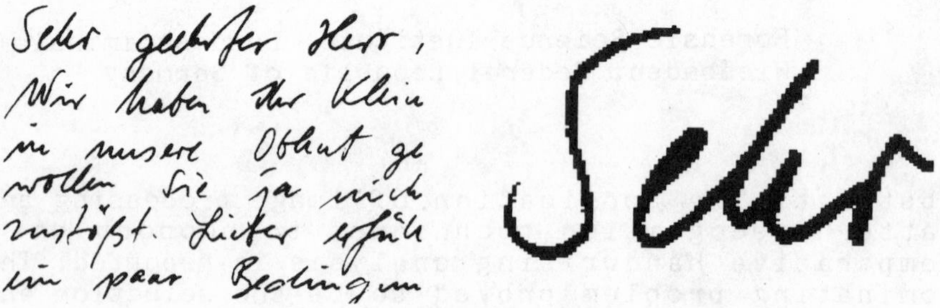

a) entire image   b) magnified segment

Fig. 1 Typical example of a stored handwriting image

2. The specification of writer-characteristic features

The handwritten text to be analyzed is represented by
a static image; thus the recognition of the writer can
be seen as a problem of pictorial pattern recognition
(4). When chosing a typical approach with a subdivision into image preprocessing and segmentation, feature extraction and classification, it turns out that
the specification of an adequate feature set will be
the dominating problem. Adequate in this sense means
describing the image in a sufficient manner for the
given task. As this task is the recognition of the
writer, one has to specify features which characterize
the writer and which do not characterize other properties like for example the written text.

The solution of this problem is not very obvious because it is not statable a priori, which pictorial features will be writer-characteristic. Some clues may be derived from the criteria used in conventional forensic handwriting comparison, especially those in the MALLY system (5). Based on the experiences gained, these criteria seem to be relatively writer-characteristic. Unfortunately, they are of a qualitative-descriptive type and therefore useless for direct application on a computer. As the meaning of the picture content is not clearly defined as far as writer-characteristic properties are concerned, any approach will have to start with the heuristic selection of features and will continue testing them on large learning sets.

A list of postulates can a priori be drawn up for the selection of features:

- writer-specifity
This means small variations for different patterns of one writer (small within-class variances) and large variations for the patterns of different writers (large between-class variances).

- completeness
This means all possible writers should be distinguishable. (This contradicts the requirements of economy, which means a reduction of redundancy in the set of features.)

- invariance.
The features should be invariant against the written text, the writing material and utensils used (blackness, contrast, width of stroke), disguised writing, psychological conditions, natural and situation-specific long-term changes in writing.

Two basic concepts for the extraction of writer-related information from a handwriting image may be distinguished, both concepts are promising and complement each other:

- textinsensitive features
global statistical features derived from the entire image,

- textsensitive features
individual features of special characters or groups of characters.

The textinsensitive approach leads to different models
of the image content. One model considers the image as
a primitive "texture" and attempts to describe this
texture numerically in either the space or the spectral domain (2). The other considers the handwriting
as a meaningful "line curve" (3) and attempts to
describe structural and statistical properties of this
line curve. The latter is computationally much more
extensive, requiring operations like skeletonization
and intelligent line following.

The textsensitive features require the localization of
characters or text segments, whose contour or reconstructed line curve will be analyzed. For special
applications regarding repetitive texts (e.g. checks),
the interactive localization is realistic, as the
problem of automatic character-recognition within a
handwritten text has not yet been solved in practice.

3. An example of textinsensitive feature extraction

In the following, an approach supposing a region-
-oriented texture and using fast parallel image
operations will be demonstrated as an example of
textinsensitive feature extraction. Fig. 2 shows the
process of feature extraction and classification as a
multi-grade process of information reduction.

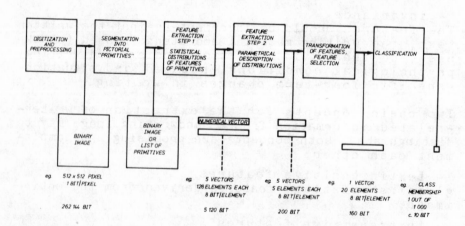

Fig. 2   Textinsensitive writer recognition as a process
         of information reduction

A segmentation into pictorial primitives is understood
as a segmentation into meaningful components of limited size. Very simple algorithms combining shifting in
one of four directions and logical operations result
in new images consisting of separated line curve
elements which are parts of the inner contour. Fig. 3
shows one of these images, it contains only those
pixels of the original (Fig. 1) which have no left
neighbour. The statistical distribution of properties
of these pixel-chains over the entire image is a first
step of feature extraction. Such properties may be the
length, the horizontal or vertical extension, the
direction angle or the curvature. Fig. 4 shows the
statistical distribution of direction angles for a
handwriting slanting to the right (Figs. 1, 3). In a
second step of feature extraction, each of the
distribution histograms obtained is described by
suitable parameters like the coefficients of a
Gaussian fit or the coefficients of a high-order
polynomial fit. In the next steps of feature
transformation and selection as well as classification
rather general statistical techniques are applied
(e.g. 6).

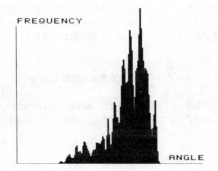

Fig.3 Segmentation into   Fig. 4 Distribution of
          primitive elements   direction angles

4. The application-oriented suitability of features

The suitability of features to solve the given problem
is determined by two factors:

   - the capability of class discrimination,

   - the computational effort.

To assess the class discrimination capability of a
feature, heuristic criteria, especially such based on
obviousness, are only of limited validity. However in
this practical multivariate statistical problem, which
is additionally characterized by being a multiclass-
-problem, an exhaustive analytical judgement will be
impossible too. The primary reason for this is the
impossibility to get a well-founded knowledge about
the feature distributions in all possible classes.
Thus all the quantifying criteria for feature selection and the methods to eliminate relations between
features (e.g. 6) are meaningful, but in a limited
degree only. In addition to such methods, a somehow
pragmatic solution for the practice will be attained
from the recognition rates obtained on a learning set.
By this one may extend the suitability considerations
from single features to feature combinations.

5. Conclusions

With a limited data set of 20 writers and 40 samples
per writer, the research project reached a recognition
rate of 99.6 % with 15 textinsensitive features. There
is no doubt that this data base has to be enlarged in
order to obtain statistically proved statements. This
again is an absolutely necessary prerequisite for any
application in practice.

6. Acknowledgements

This research was supported by the Bundesministerium
fuer Forschung und Technologie, contract DV 5.800.

7. References

(1) M.R.HECKER : Kriminalistik 25, pp.316-318, 1971
(2) W.KUCKUCK, B.RIEGER, K.STEINKE : Proc. 1979
Carnahan Conf. on Crime Countermeasures, pp.57-64,
Univ. of Kentucky, Lexington 1979
(3) K.STEINKE : in "Angewandte Szenenanalyse", ed.
J.P.FOITH, pp.180-189, Springer, Berlin 1979
(4) V. KLEMENT, R.-D.NASKE, K.STEINKE : Third Int.
Conf. on Security through Science and Engineering,
Berlin 1980
(5) R.MALLY : BKA-Schriftenreihe 2, pp.145-150, 1955
(6) K.FUKUNAGA : Introduction to Statistical Pattern
Recognition, Academic Press, New York 1972

EVALUATION OF IMAGE SEQUENCES: A LOOK BEYOND APPLICATIONS

B. Radig and H.-H. Nagel

Fachbereich Informatik
Universität Hamburg
D-2000 Hamburg 13

0. ABSTRACT

Rastering of a digital image usually reflects the spatial coordinates of the optical image which is projected onto the sensor. Stacking images to a sequence allows to record a third dimension which in most applications corresponds to time. Selected applications are classified according to the techniques they use in order to extract or handle the temporal variations of phenomena recorded in the sequence. The intensity variations of a translating object image are approximately described by equations which are used to estimate the frame-to-frame displacement. Recently reported refinements to this estimation approach are discussed.

1.0 INTRODUCTION

A digital image is obtained by quantizing the sensor signal from one of possibly several spectral channels at each grid node of a two-dimensional raster. By organizing two-dimensional samples as an image sequence, the sampling process is extended to a third dimension. This may be the spectral dimension, the third spatial dimension, or the time dimension. One may even think of a sensor dimension as in the case of binocular or multiocular stereo.

J. C. Simon and R. M. Haralick (eds.), Digital Image Processing, 525–560.
Copyright © 1981 by D. Reidel Publishing Company.

There are applications where sequences of sequences are formed to sample simultaneously along the time and three spatial dimensions in order to explore, e.g., the time variation of a 3-D density distribution of a beating heart. Motion is an essential action in our world. If not otherwise stated we describe applications where time is the third dimension of a sequence.

Motion of objects leads to systematic variations between consecutive images from a sequence. Here the sampling rate is assumed to be high enough to capture the systematics of physical motion. If snapshots from a scene with action are recorded at random time intervals, more scene specific knowledge will be required in general to evaluate the resulting series of images. Two basic problems are encountered by all investigations of such systematic variations:

a) How to isolate and describe image parts (namely projections of moving objects) which remain (quasi-)invariant from one frame to the next although not necessarily throughout an entire sequence? The isolation of such image parts has to discriminate between significant changes, e.g. variation of relative configurations, and insignificant ones, i.e. noise.

b) How to find a transformation (the physical law) which uses the smallest number of parameters to describe the observed changes throughout an image sequence?

The search for suitable transformations is based on two assumptions:

b1) A scene depicted in the image sequence may be modeled in terms of separate objects which are more or less rigid or rigid parts of compound objects.

b2) The observed variations in time can be explained by a relative motion between one or more of such objects and the sensor.

Modeling of the original scene is thus recognized to play an important role for the analysis of image sequences. Progress towards an abstract description of variations in an analysed image sequence requires improved models of all components involved. These comprise the sensors, the physics of the imaging process, moving objects, the stationary background, the admissible configuration between objects, and their motion relative to the sensor. By measuring how consistently and concisely the input data from the image sequence can be described, the selection and adaptation of models can be evaluated within the image sequence analysis system. This is possible to a larger degree than in the case of single images, since the analysis of a single image offers less redundancy which can be exploited for system internal evaluation. Proper choice of models based on the analysis of a subsequence may even allow to predict individual image frames and frame-to-frame variations in the remaining image sequence. Sequence analysis does not only contribute to the growth of means and techniques developed for single image analysis, rather it is expected to contribute towards improved modeling for image analysis in general.

The development of more widely applicable systems for the analysis of image sequences is forced away from ad-hoc assumptions towards the incorporation of more general concepts. 3-D structure may serve as an example. The concept of a surface patch in 3-D space can be based on the coordinated displacement of a connected set of pixels. To account in general for changes due to rotation around an axis which has a component parallel to the image plane, the concept of a 3-D volume has to be introduced. From there it is one step to the concept of a single rigid object as a hypothesis to explain a variety of changes. As a next step one may introduce a collection of opaque rigid objects moving independently from each other along 3-D trajectories. Even if modeling has been restricted to convex bodies, at this point one has to cope with occlusion. A further evolution would attempt to model a non-rigid object as a flexible assembly of rigid subobjects.

The concepts mentioned so far are related to modeling the appearance of visible scene components. A different category of concepts describing the action in a scene and the interactions between objects has to be incorporated into advanced analysis systems. Concepts related to certain motion verbs may be taken as examples. The next higher level of complexity is to explore the meaning of an object in terms of its functions which are observed from time variations in spatial structures. Interactive dialog systems are employed to facilitate the formation of new concepts by analysis of image sequences. The trend towards explicit modeling of increasingly complex concepts is demonstrated by an overview of actual applications. Common developments according to the considerations outlined above can already be recognized.

2.0 APPLICATIONS

Among the numerous publications which can be considered as relevant in this context, a few examples are selected to demonstrate the variety of aspects that may have to be taken into account. Further references can be found in the book by Ullman 79 (109) and in the survey articles by MARTIN and AGGARWAL 78 (66), NAGEL 78b (73), and SCACCHI 79 (96). A more comprehensive discussion is presented in NAGEL 80 (74).

2.1 Coding Techniques For Image Sequences

A large body of well organized knowledge has been accumulated about the coding of image sequences for bandwidth compression during transmission. Such coding techniques exploit the following premises:
- The similarity between successive frames.
- Spatial resolution is automatically reduced proportional to the velocity of moving objects by the integrating effect of current TV-camera targets.
- Resolution loss in rapidly moving areas is tolerated by human perception as long as low to moderate motion is portrayed with good resolution.
- The camera is not moved very often - if at all.

Three considerations suggest that scene analysis may profit from paying closer attention to the work on interframe coding:
- The variety of techniques for comparing image frames and for segmenting them into nonstationary and stationary areas.
- The fairly general assumption about the contents of image sequences that form the basis for these approaches.
- The efforts to establish objective quality criteria for judging the results of different approaches, i.e. LIMB 79 (58).

2.1.1 Transformation Coding Techniques –

HELLER 74 (38) used a 4x4 Hadamard transform and truncated the resulting coefficients to a fixed precision which had been selected for each coefficient according to its significance. The coded coefficients were transmitted for every fourth frame. At the three intervening frames only the first three transform coefficients were updated. They correspond to the average intensity, a horizontal and a vertical edge within the 4x4 segment. Updating was done by transmitting only the coarsely quantized differences between the current values and those most recently transmitted in full precision. Image quality appeared acceptable for stationary scenes at transmission rates down to 1 bit per pixel. Nonstationary image areas showed disturbing effects at rates below 2 bits per pixel. Sending only coarse updates for three coefficients turned out to be insufficient for segments in nonstationary areas. Some other faults could be traced to the fact that the fixed precision did not allow to transmit some occasionally large transform coefficients which are important despite of their infrequent occurrence, as READER 75 (90) pointed out, because they involve prominent features like sharp high contrast edges. Adaptive image coding – see PRATT 79 (85) or HABIBI 77 (32) – is required to cope with such situations adequately.

The approaches to hybrid interframe coding investigated by ROESE and others differ from the work of HELLER 74 (38) and JONES 76 (46) in several aspects. The discrete cosine transform (DCT) or alternatively the Fourier transform (FT) on 16x16 segments was used. Differential pulse code modulation (DPCM) techniques were employed to code the differences between corresponding transform coefficients from consecutive frames. Rather than implementing the coder electronics, computer simulation experiments were performed. With a DCT hybrid interframe coder at a transmission rate of 0.25 bit/pixel, ROESE and ROBINSON 75 (94) obtained for a typical videotelephone "head and shoulder" scene an image quality which was comparable to the result of an intraframe coder requiring 2 bit/pixel. Reduced performance has to be anticipated for image sequences which exhibit a larger amount of changes between corresponding pixels from consecutive frames, e.g. due to camera motion. In this context ROESE and ROBINSON 75 (94) pointed out that certain kinds of camera motion like panning let successive frames differ essentially by a translation. Here FT appears theoretically attractive because the magnitude of Fourier coefficients should remain approximately constant, the change

being reflected by a systematic variation of their phase factors.
The same is true - see HASKELL 74 (35) - for the subimage
representing a translating object. In this case, the problem of
segmenting images into stationary and nonstationary parts arises.

An example of extending 2-D transformations to the time
dimension is the work of KNAUER 76 (50). He described a real-time
digital video processor to perform Hadamard transformations on
4x4x4 pixel blocks from video frames digitized with 512x525x6 bit
resolution. The first coefficient is determined as the average
intensity in the block whereas the remaing coefficients measure
the relative variations within such a block along the horizontal,
vertical, and temporal axis. The coding step was adapted to yield
high temporal and low spatial fidelity if temporal variations
dominated. Vice versa, in stationary areas high spatial fidelity
was emphasized at the expense of low temporal fidelity.

2.1.2 Spatial Segmentation Based On Temporal Characteristics -

PEASE and LIMB 71 (82) sampled every other pixel within a
scanline (changing the sampling pattern every other frame). They
aggregated pixel positions into a 1-D nonstationary segment
within a scanline where significant deviations from the intensity
observed in the previous frame had been detected. At the
receiver, pixels at positions sampled in the current frame would
be displayed at the updated intensity value. Intensity values for
pixels in stationary segments not transmitted in the current
frame would be copied from the last frame (lower temporal
resolution). Pixels not sampled in nonstationary segments would
be linearly interpolated from their updated neighbors within
their scanline (lower spatial resolution). A discussion of four
approaches of this type can be found in HASKELL et al. 72 (37).
CONNOR et al. 73 (18) exploited the spatial and temporal
correlation between interframe differences in nonstationary image
areas. They computed at each rasterposition three dissimilarity
predicates, a moving edge predicate, a conservative and a
sensitive low texture nonstationarity predicate. Interframe
differences within nonstationary image components would only be
transmitted for every other field, if the transmission buffer
filled up. Linear interpolation between corresponding pixels of
the two neighboring updated fields would be used to display
fields not updated. Techniques of this kind are called
'conditional vertical subsampling'. A theoretical analysis of

interframe differences in stationary and nonstationary areas has
been presented by CONNOR and LIMB 74 (17). BROFFERIO and ROCCA 77
(10) proposed to classify the pixels within a scanline into three
segment types, based on how best to predict their greyvalues from
those previously encountered. Pixels in stationary segments were
predicted from the same rasterposition. Pixels belonging to an
object with known velocity were predicted from earlier segments
attributed to this object. Remaining pixels were predicted from
the neighbor in the scanline above. Simulation results - with the
assumption of a single displacement vector - seem to offer the
possibility to halve the bit rate required for coding.

These and similar approaches make no explicit distinction
between a nonstationary image component and the image of a moving
object. However, these notions are not equivalent. Apart from
other reasons (noise, changing illumination) which may generate a
nonstationary component, the nonstationary component obtained by
comparing two consecutive frames using change detection
techniques comprises not only the image of a moving object but in
addition the background uncovered by the displacement of an
object. To separate these two contributions to the nonstationary
image component KLIE 78 (49) analysed the effects of a temporal
low-pass filter. He showed that this filter affected, too, the
spatial frequency distribution within each frame. By appropriate
choice of the filter parameters he could arrange that this
"crosstalk" effect would be restricted to image areas
representing objects in motion whereas the stationary background
area would not be affected - even where it was just uncovered by
a moving object. By comparing a signal that had been subjected to
a temporal low-pass filter with an appropriately delayed version
of the unfiltered signal, Klie was able to derive a signal which
deviated from zero only in the image area corresponding to a
moving object. Converting this signal into a mask he thus was
able to separate the image of the moving object from the
uncovered background. Therefore, his "motion" detector differed
from his "change" detector. This approach assumed that the entire
projection of a moving object could be assigned to the
nonstationary category by evaluation of interframe differences.
Problems should be expected in the following constellation: (i)
The image of the moving object contains homogeneous segments
whose extensions along a scanline are significantly greater than
the displacement of the object projection along the scanline.
Such a situation has been discussed in a slightly different
context by JAIN and NAGEL 79 (44). (ii) The displacement of a
moving object projection is an integral multiple of a dominating

spatial wavelength in this subimage (stroboscopic effects).

2.1.3 Interframe Coding Based On Movement Compensation -

Assume a moving object image is translated from frame to frame by a 2-D displacement vector (dx,dy). Within the interior of this subimage a pixel at position (x+dx,y+dy) in the current frame may be predicted from a pixel at (x,y) in the previous frame. Larger deviations from this prediction would only be expected for pixels near the subimage boundary. ROCCA 69 (93) suggested to exploit this consideration in order to reduce the transmission rate by transmitting first the displacement vector and subsequently only the differences between the greyvalues predicted on this basis and the actual ones. To accommodate more general situations (camera panning, several objects) Rocca proposed to subdivide each frame into square segments of fixed size. The displacement vector for each segment had to be determined by finding the displacement of maximum crosscorrelation between the segments from the current and the previous frame. Unless the exact position of the crosscorrelation maximum would be interpolated, such an - computationally still too expensive - approach yields only integer values for the components of the displacement vector.

LIMB and MURPHY 75a (59), 75b (60) were the first to report about an implementation which determined the absolute frame-to-frame displacement of a single object image translated with constant velocity parallel to the TV-lines. A slight modification of this circuit enabled them even to determine the sign of the velocity. CAFFORIO and ROCCA 76 (11) generalized this approach in order to determine both velocity components - parallel as well as perpendicular to the scanlines. They could do this even for several sufficiently large object images moving with different velocities. They demonstrated by computer simulations that spatial lowpass filtering of interframe differences improved the velocity estimate. FUKINUKI 78 (25) applied the same technique - to compute the displacement from the intraframe greyvalue gradient and the interframe difference - to fixed 16x16 pixel windows, with a mean square error approach to estimate the individual displacement vectors.

NETRAVALI and ROBBINS 79 (76), ROBBINS and NETRAVALI 79 (92), as well as STULLER and NETRAVALI 79 (101) employed elaborate prediction methods in order to compensate movement.

2.2 Meteorological Measurements

Aerial and satellite image sequences are routinely evaluated for a wide variety of applications. Typical is the measurement of cloud displacements in order to estimate horizontal and vertical wind velocities.

2.2.1 Visual Channel -

First attempts to determine cloud displacements from satellite images are characterized by the needs to rely on manual measurements, i.e. FUJITA 69 (24). As a next step, an operator interactively selected a window in the image which was dominated by a single cloud layer with a reasonably unique greyvalue pattern. The displacement was computed by a crosscovariance match (LEESE et al. 70 (52), 71 (53)). ENDLICH et al. 71 (21) applied the ISODATA algorithm to cluster interactively the 2-D intensity distribution within 120x120 pixel subwindows. Cluster descriptors were matched with those from the same subwindow in a satellite image recorded 30 to 90 minutes later. Differences between x- and y-components from tentative cluster pairings were histogrammed. In an iterative process, peaks in this histogram were selected as indicative for a possible cloud displacement vector in the subwindow studied.

The quality of such approaches depends on a careful selection of the cloud segments and a correction of all systematic errors. The most significant uncertainty was due to lack of knowledge about cloud height. Wind velocity determination became unreliable if a mixture of two or more cloud layers at different altitudes and speeds existed in the selected windows.

2.2.2 Infrared Channel -

Estimates of cloud temperature obtained by an infrared
sensor considerably improved the separation of different cloud
layers. SHENK and KREINS 70 (98) selected cloud segments with the
lowest temperature to estimate wind velocities which correlated
reasonably well with observed values. In the approach of HUGHES
et al. 78 (43), routine calibration of infrared intensity reading
versus height was used to select an infrared intensity range
which corresponded to cloud top heights not exceeding 3000 m.
16x16 pixel windows centered at predefined grid locations were
only accepted for further processing if a minimum number of
pixels within this infrared intensity range was exceeded. For
such windows a crosscorrelation search was performed for the
optimum match in a satellite picture recorded 30 minutes later.
However, only pixels in the accepted intensity range participated
in the crosscorrelation. Windows with insufficient low level
clouds were thus excluded and spots of higher level clouds could
not influence the displacement result. HAASS and BRUBACKER 79
(31) suggested to save time in crosscorrelation search by
establishing a set of predictors comprising not only the centroid
coordinates but in addition area size, brightness, and other
descriptors for cloudsegments. Size, center coordinates together
with their standard deviations, and average intensities in the
visual and infrared channel describe cloud segments in the
approach of WOLF et al. 77 (113), 79 (114). An empirical
dissimilarity function based on these descriptors was used to
score a possible pairing of cloud segments within a 120x70 pixel
search window from two consecutive frames, similar to the
approach of Endlich et al.; however, Wolf et al. abandoned the
ISODATA-algorithm.

2.2.3 Combined Information -

In order to obtain a more reliable conversion of cloud
displacements into wind velocities, the selected cloud formations
should exhibit sufficiently uniform meteorological properties.
The question how these properties are reflected in the recorded
data can only be answered in cooperation with meteorologists.
PARIKH and ROSENFELD 78 (81) segmented windows in infrared images
using a histogram technique. A multistage algorithm evaluated
textural and tonal features which characterize meteorological
properties of desired cloud patterns in order to classify and

merge segments. Further progress requires the incorporation of
more knowledge about both the static and dynamic aspects of cloud
formation, growth, and dissipation in order to fully exploit the
information offered by multispectral satellite image sequences.

This is especially required for the study and prediction of
severe bad weather constellations. ADLER and FENN 76 (1)
investigated the development of thunderstorms, based on infrared
satellite data recorded within a restricted picture area at 5
minute intervals. In order to determine thermodynamic
instabilities leading to severe storms, KANAL and PARIKH 77 (47)
employed a pattern recognition approach based on classification
of cloud types and especially the evolution of their features
with time. Another indicator for certain severe weather
situations is cloud rotation about a vertical axis. HAASS and
BRUBACKER 79 (31) estimated it by describing the contour of a
cloud segment in a polar coordinate system around an
interactively selected origin and by subsequent crosscorrelation
of such contours from two consecutive frames.

Sequences of infrared images allow to observe the
temperature at a fixed geophysical location as a function of
time. This opened the possibility to estimate vertical cloud
motion (LO 75 (65)) and clear air motion (HALL et al. 72 (34)) as
well as time variations in the temperature of large sea surface
areas (WATERS 78 (111)). YATES and BANDEEN 75 (119) mentioned in
this context measurements of sea ice coverage. They pointed out
that clouds and ice which are often indistinguishable in a single
image might nevertheless be separated by an interframe comparison
of a picture sequence recorded on successive days: cloud patterns
are dynamic whereas the ice features are relatively stable within
a day.

2.3 Medical And Biomedical Applications

Beyond pure medical applications, the evaluation of image
sequences spreads into biomedicine, biophysics, and biology.
Analysis of sequences allows to study such phenomena as
stimulus-response, transportation, growth, and transformation.

2.3.1 Medical Applications -

Usually the sequence data come from digitized x-ray images or from TV-recordings. An application is image enhancement - after injection of a roentgen-opaque material into the blood (angiography). If an average of several digitized pre-injection images is subtracted from a post-injection image, components not affected by the injection can be suppressed (CHOW and KANEKO 72 (15)); modern TV-techniques facilitate this approach in real-time (BRENNECKE et al. 76 (6)). The signal-to-noise ratio can be improved by averaging only phase registered images (BRENNECKE et al. 78 (8)). In order to detect abnormalities in ventricular wall motion CHOW and KANEKO 72 (15) searched a suitably enhanced x-ray film sequence of the left ventricle for subregions with larger greyvalue variance. There, greyvalues were histogrammed in order to determine a threshold. Pixels in such regions were classified as either interior or exterior to the ventricle by fitting two normal distributions to the greyvalue histogram. HACHIMURA et al. 78 (33) were able to determine the left ventricular contour in scintigraphic image sequences. They estimated the left ventricular volume and plotted it versus time. The idea behind the approach of TASTO 73 (103) was to save computing time by a guided search. He sampled the motion of the heart in a x-ray sequence with a rate high enough to obtain only slight displacements from frame to frame. After having located the left ventricular contour by a first guess, search space in subsequent frames could be reduced. TSOTSOS 80 (107) developed a general system for motion description and applied it to left ventricular wall motion.

Information about dynamic changes in size, shape, and position of intact working organs is of great interest for a detailed understanding of their function (WOOD 76 (116)). HOEHNE et al. 77 (41), e.g., studied the blood flow in the kidney. LEDLEY 73 (51) explored the blood flow in the eye by recording the passage of fluorescent material from the retinal arterioles through the capillary mesh out through the retinal vein. To extend this approach for routine applications, a television ophthalmoscope image processor was developed by READ et al. 77 (89) which digitized and recorded images in real time at a rate of 5 to 15 frames/sec.

2.3.2 Biomedical Applications -

In order to study the stimulus-response behavior of microorganisms moving in a wet-slide preparation under a microscope GREAVES 75 (29) recorded image sequences on videotape. He obtained binary pictures by interactive thresholding and thinned each organism to a single point. Sequences of point coordinates were evaluated to aggregate quantitative descriptions like velocity or mean direction from which the stimulus-response behavior could be inferred. In order to obtain statistically reliable quantitative observations about the so-called chemotaxis, a large number of lymphocytes had to be tracked in time-lapse cinemicrophotographs recorded at 20 second intervals over a period of 1 to 2 hours. GREENE and BARNES 77 (30) mechanically tracked one cell at a time by positioning the microscope stage according to a feedback from a five unit photodiode array. LEVINE et al. 79 (55) designed a computer controlled system to achieve the same goal by evaluating time-lapse TV-images. This experimental device might be employed to scan a proliferating cell in order to trace the genealogies of selected cells. LIPKIN et al. 79 (63) started an approach not only to trace cells but moreover to measure morphologically discernible cell activities. ARIKI et al. 78 (3) implemented a system for constructing interactively a model of an image pattern, for tracing this pattern, for measuring features, and for detecting changes in the measured features. One application was to analyse the morphogenetic movement of a dissociated cell of Xenopus laevis, a protozoon. YACHIDA et al. 78 (117) observed fishes swimming in a vat by a TV-camera connected to a videotape recorder to study their behavior under a variety of light and acoustic stimuli. A sophisticated image analysis allowed them to identify the often motion-blurred images of fishes, to model their shape, and to employ shape prediction in order to disambiguate situations where one fish occluded another. The systems of Ariki et al. as well as of YACHIDA et al. attempted to design larger parts independent of the specific application. This trend can be observed in other fields, too.

2.4 Object Tracking In Outdoor Scenes

Actors and action could become rather complex and their environment is not controllable in outdoor scenes. Even lighting conditions are hard to influence by the experimentalist. Therefore, such a domain represents a challenge for an automatic analysis system.

2.4.1 Traffic Monitoring –

WOLFERTS 74 (115) described an operational semi-automatic image processing system for the determination of vehicle velocities from time-lapsed aerial film recordings of street scenes. The image of a vehicle had to be selected by an operator. Crosscorrelation was used to track this vehicle image in subsequent frames. The basic idea to subtract two images in order to obtain locations where some changes occured had already been implemented by ONOE et al. 73 (80). They subtracted the current video signal of a TV-camera from a reference frame stored on a solid-state storage tube. The difference signal was large in regions where cars had been displaced. Onoe et al. demonstrated velocity estimation in a two car scene. To reduce processing time ONOE and OHBA 76 (79) restricted the analysis to pixels selectively digitized along the image of lanes. Cars were assumed to be present at positions where the difference between pixel intensities and the corrected average background values exceeded a threshold. GERLACH 79 (28) successfully combined several techniques in order to achieve a more robust system for tracking single vehicles in road scenes recorded by a ground based TV-camera. A window around the vehicle image was selected interactively to become the reference subimage. It was crosscorrelated with a suitably selected search area in subsequent frames. As long as the crosscorrelation maximum exceeded a threshold the reference subimage was smoothly updated from the current frame using weighting factors for contributions from the reference and the current frame. In this way the reference subimage could be adapted to the changing appearance of the object whereas the background part in the reference window was blurred. Therefore larger greyvalue gradients could be expected predominantly within the object image. This effect was exploited to obtain an object mask. Then only pixels covered by the mask contributed to the correlations, thus reducing the influence of background variation on the crosscorrelation maximum. Moreover, the analysis of greyvalue differences in the reference subimage - which remained centered on the object - allowed to detect foreground components as structures shifting between frames with respect to the window boundary. Such components could be excluded from the object mask.

DRESCHLER and NAGEL 78 (20) as well as JAIN and NAGEL 79 (44) investigated TV-image sequences from downtown street intersection scenes as a way to study principle problems of image sequence analysis. DRESCHLER clustered corresponding regions - extracted by an image subtraction technique - by a minimal spanning tree algorithm. JAIN and NAGEL developed an approach based on comparison between frames from a sequence and a fixed frame of this sequence as a reference. They could successfully isolate the image of a moving object and determine its displacement velocity.

2.4.2 Tracking -

SCHALKOFF and MCVEY 79 (97) developed an algorithm to determine the 2-D displacement, rotation, and dilatation of an object image between consecutive video frames. Their approach is closely related to methods developed for movement compensation in interframe coding. It is based on the assumptions that the object image could be reliably separated from the background, the object image is large and compact so that boundary effects are negligible, and that the greyvalue distribution of the object would exhibit a significant variation although restricted to lower spatial frequency components.

AGGARWAL and DUDA 75 (2) studied the tracking of simulated clouds, especially the problem of occlusion. They described their objects as binary masks with polygonal boundaries. CHOW and AGGARWAL 77 (14) admitted non-polygonal boundaries subject to the constraint that object images were observed unoccluded for at least two frames in order to estimate their contour and their displacement velocities. MARTIN and AGGARWAL 79 (67) described object contours in terms of straight and curved line segments. This enabled them to track individual boundary segments in order to decompose the contour of two overlapping object images during the tracking process into boundary parts related to the component objects. The study of ROACH and AGGARWAL 79 (91) extended the investigations from planar uniformly colored objects to the modeling of objects in a 3-D blocksworld, taking internal edges into account.

MILGRAM 77 (69) applied dynamic programming to organize the search for a consistent description of regions that had to be tracked throughout a sequence of image frames. The idea common to this approach and the ones of DRESCHLER and NAGEL 78 (20) as well as RADIG 78 (88) is the consideration to base the tracking decisions on more global information about the sequences rather than on a single pair of descriptor sets derived from two images.

2.4.3 Industrial Automation And Robotics -

Application of visual sensors to industrial automation appears to be dominated still by the evaluation of single images. NEUMANN 78 (77) studied the tracking of an object on a simulated conveyor belt by instantiating 2-D relational models based on straight line contour approximation to images of the sequence. DAVIES and SMITH 79 (19) determined relative displacements between parts of a gasturbine engine as a function of various operating conditions. Windows in high-energy x-ray images were selected and crosscorrelated with subsequent frames. HERNAN and JIMENEZ 79 (40) accumulated statistic measurements on the formation and disappearance of blobs in the turbulent mixing layer between high speed parallel streams of different gases. Image processing techniques were applied to isolate and classify blob boundaries. Ellipses were fitted to blobs and subsequently used to track a blob from frame to frame until it disappeared by coalescence with other ones.

Tracking under real-time constraints is required for applications such as the visual control of a mechanical manipulator. UNO et al. 76 (110) reported on using a TV-camera to detect the position of bolts and reenforcement ribs on moving steel moulds in order to control a manipulator. They inspected every video frame only within one or a few windows at fixed locations and measured how well a feature of an object fitted into the window when the object passed through the field of view. The displacement between the window center and a reliably detected object feature in two frames was used to determine position values. BIRK et al. 77 (9), in order to correct workpiece orientation by an industrial manipulator, derived values for simple features like centroid coordinates, areas, and

principle axes, from binary images of the workpiece recorded by
two TV-cameras. The feature values were compared to reference
values obtained during a learning phase in order to estimate
initial position and orientation of the workpiece. Corrections to
the orientation were derived from changes of feature values in
subsequent frames.

The research on navigable robots forced the development of
environment models to plan and control the robot trajectory.
Moreover, the requirement of real-time feedback pressed towards
reasonable fast heuristics and implementations. Both the JPL and
the Stanford robot are equipped with stereo TV-sensing, in
addition the JPL navigable robot has a laser range finder. See
GENNERY 77a (26), 77b (27), MORAVEC 77 (70) for the Stanford
robot and LEVINE et al. 73 (54), O'HANDLEY 73 (78), LEWIS and
JOHNSTON 77 (56), THOMPSON 77 (106), YAKIMOVSKY and CUNNINGHAM 78
(118) for the JPL robot.

3.0 COMPARING IMAGES FROM A SEQUENCE

Examples have been presented from different fields of
application. They were grouped according to these fields. Another
way to look at them is to classify the method they apply in order
to extract that information which can only be obtained from a
sequence and not from a single image. We discuss in the sequel
the categories proposed by NAGEL 78b (73). Examples for each
category are given by tables of references to the literature,
predominantly about publications discussed in the previous
chapter.

3.1 No Interframe Comparison

In this category approaches are listed where the frames of a
sequence are not compared with each other. Only intraframe image
processing techniques are applied in order to derive a sequence
description which is intended to be evaluated by external means
such as human perception or non-pictorial data processing.

TABLE 1: No interframe comparison

CHIEN and JONES 75 (13),	HACHIMURA et al. 78 (33),
LEDLEY 73 (51),	McCORMICK et al. 76 (68),
NATARAJAN and AHMED 77 (75),	ROESE et al. 77 (95),
SMALLING et al. 76 (99),	TASTO 73 (103),
TASTO et al. 78 (105),	WINKLER and VATTRODT 78 (112)

3.2 Indirect Interframe Comparison

This class contains such approaches where an explicit comparison between images is not performed but information from several frames is combined to achieve the desired result. An example is the processing of pixels within a cube from the three dimensional space which is spanned by the sequence coordinates as e.g., in the case of transform coding. In this case 2 to 16 frames are combined. Another example is the guidance of image processing on the current frame by results obtained in previous frames. A third example is the selection of a certain image for further processing by detecting a special feature in the sequence.

TABLE 2: Indirect interframe comparison

HELLER 74 (38),	HERMAN and LIU 78 (39),
JONES 76 (46),	KNAUER 76 (50),
LIU 77 (64),	TASTO 74 (104),
UNO et al. 76 (110)	

3.3 Dissimilarity Grading

A comparison algorithm is classified as "dissimilarity grading" if image descriptors of two different frames are compared and the descriptors are evaluated on the same raster. No positional search should be necessary to match these descriptors. A prior manual registration of the two frames is not regarded as such a search. NAGEL 78 (73) assigns an "order"-attribute to the descriptors depending on the number of pixels which have to be looked at to compute the descriptors. The simplest form of dissimilarity grading is the comparison of greyvalues from two pixels at the same position in two frames.

TABLE 3: Dissimilarity grading

BRENNECKE et al. 76 (6), 77 (7), 78 (8),	
BROFFERIO and ROCCA 77 (10),	CAFFORIO and ROCCA 76 (11),
CANDY et al. 71 (12),	CHOW and KANEKO 72 (15),
CONNOR et al. 73 (18),	FENNEMA and THOMPSON 79 (23),
HASKELL 75 (36),	HASKELL et al. 72 (37),
HOGG 77 (42),	KLIE 78 (49),
LIMB and MURPHY 75a (59), 75b (60),	
LIMB and PEASE 71 (61),	LIMB et al. 74 (62),
MOUNTS 69 (71),	PEASE and LIMB 71 (82),
ROESE et al. 77 (95)	

3.4 Similarity Search, Followed By Dissimilarity Grading

Substructures in one image are selected and for each the best match in another image is determined by a search. Once sufficient correspondences are established, both images can be registered on a single raster in order to determine to what extend they differ at identical positions. Typical change detection approaches - which were omitted from the applications overview if they process only two frames - could be categorised in this class, too, if they require preregistration as in the case of satellite images.

TABLE 4: Similarity search - dissimilarity grading

CHOW et al. 73 (16),	KAWAMURA 71 (48),
LILLESTRAND 72 (57),	PRICE 76 (86),
PRICE and REDDY 79 (87),	ULSTADT 73 (108)

3.5 Dissimilarity Grading, Followed By Similarity Search

If the images from a sequence are registered with respect to each other - as, e.g., TV-frames are - dissimilarity grading techniques can be used to detect image areas where changes occur. Such areas may be classified as nonstationary under the hypothesis that they are generated by projections of moving objects or related phenomena, i.e. moving shadows. The nonstationary image components may then be tracked from frame to frame - performing a similarity search - in order to learn more about the objects which gave rise to the nonstationary image

components. A recognition system could explore the temporal
behavior of such an object or derive a more complete description
of it by combining information from its different aspects
(frames). The latter task leads to "motion-stereo" techniques.

TABLE 5: Dissimilarity grading - similarity search

DRESCHLER and NAGEL 78 (20),	JAIN et al. 77 (45),
JAIN and NAGEL 79 (44),	NAGEL 78a (72),
ONOE and OHBA 76 (79),	ONOE et al. 73 (80),
TAKAGI and SAKAUE 78 (102),	YACHIDA et al. 78 (117)

3.6 Similarity Search

The approaches to be assigned to this category are the ones most
easily associated with the idea of tracking images of moving
objects. They usually apply techniques from the repertoire of
pattern classification or static scene analysis in order to
compute descriptors for the interesting parts of the images. A
similarity function of these descriptors is used to solve the
correspondence problem. If descriptors are employed which allow
to identify the image of the same real object throughout a
sequence of frames a temporal description of such an object may
be derived. This could lead to a development of a computer
internal model of the dynamics in the observed scene.

TABLE 6: Similarity search

AGGARWAL and DUDA 75 (2),	ARIKI et al. 78 (3),
ARKING et al. 75 (4),	BARNARD and THOMPSON 78 (5),
CHOW and AGGARWAL 77 (14),	DAVIES and SMITH 79 (19),
ENDLICH et al. 71 (21),	ESKENAZI and CUNNINGHAM 78 (22),
GERLACH 79 (28),	HAASS and BRUBACKER 79 (31),
HALL et al. 72 (34),	LEESE et al. 70 (52),
LEVINE et al. 79 (55),	MARTIN and AGGARWAL 79 (67),
MILGRAM 77 (69),	NEUMANN 78 (77),
POTTER 75 (83), 77 (84),	RADIG 78 (88),
ROACH and AGGARWAL 79 (91),	SMITH and PHILLIPS 72 (100),
WOLFERTS 74 (115)	

4.0 DISPLACEMENT ESTIMATION

If an object moves in front of a stationary observer, e.g. a TV-camera, its moving projection gives rise to a systematic variation of the recorded image sequence. Under some simplified assumptions these variations can be described by equations which allow to estimate the velocity of an object image or even its rotation and dilatation. The results are not only of importance for transmission coding approaches (movement compensation) but also for tracking of real-world objects. In the latter application, a poor approximation of object velocity - e.g. by the displacement of a region center - could be replaced by such a more exact and more reliable technique.

4.1 Translation

Let us assume that the image intensity is given as a continuous function $I(x,y,t)$. Its Taylor series starts as

$$I(x+\Delta x, y+\Delta y, t+\Delta t) = I(x,y,t) + \frac{dI}{dx}\Delta x + \frac{dI}{dy}\Delta y + \frac{dI}{dt}\Delta t$$

Let $O(x,y,t)$ denote the intensity function within the subimage which represents the projection of a moving object. If the object image is translated during the time interval Δt by a distance $(\Delta x, \Delta y)$ then $O(x+\Delta x, y+\Delta y, t+\Delta t) = O(x,y,t)$ If we restrict our interest to the subimage O of I, we obtain, neglecting higher terms of the Taylor series:

$$-\frac{dI}{dt} = \frac{dI}{dx}V_x + \frac{dI}{dy}V_y$$

with velocity estimates $V_x = \frac{\Delta x}{\Delta t}$ and $V_y = \frac{\Delta y}{\Delta t}$.

For a stationary object (background) this reduces to:
$I(x,y,t+\Delta t) = I(x,y,t)$ ==> $V_x = V_y = 0$.

LIMB and MURPHY 75a (59) measured the absolute value of the velocity for an object image which was displaced parallel to the scanlines ($V_y = 0$)

$$\left|\frac{dI}{dt}\right| = \left|\frac{dI}{dx}\right| * |V_x|.$$

In a digital picture function, the time derivative is approximated by the interframe difference and the horizontal gradient component by the intraframe difference of two adjacent pixels within a scanline. Limb and Murphy estimated

$$|\hat{V}_x| = \frac{\sum_{x,y}|O(x,y,t)-O(x,y,t+\Delta t)|}{\sum_y\sum_x|O(x,y,t)-O(x+\Delta x,y,t)|}.$$

The sign of the velocity could be obtained by

$$\hat{V}_x = \frac{\sum_{x,y}(O(x,y,t)-O(x,y,t+\Delta t))}{\sum_y\sum_x|O(x,y,t)-O(x+\Delta x,y,t)|}.$$

This estimation is even justified if the object image velocity has an y-component, under the assumption that the interframe differences due to the y-component are not correlated with the differences due to the horizontal motion and thus cancel out (LIMB and MURPHY 75b (60)).

4.2 Rotation And Dilatation

SCHALKOFF and MCVEY 79 (97) extended the velocity estimation of a translating object to the case of additional rotation and dilatation. The scale factor and the rotation angle have to be determined, too, from the image sequence. The object image O is a "snapshot" of the object at a time t in a fixed coordinate system Q. The movement of the object image relative to the sensor coordinate system $S = (x,y)^T$ is described by a time dependent affine transformation of Q to S (note that Q and S are vectors, A is a matrix)
$$Q(S,t) = A(t) * (x,y)^T - (\Delta x(t), \Delta y(t))^T .$$
If we introduce the time dependent translation vector
$\Delta S(t) = (\Delta x(t), \Delta y(t))^T$ we may write $Q(S,t) = A(t)*S - \Delta S(t)$
The object image O is expressed in a frame at time t by I(S,t) = O(Q(S,t)). After a time intervall Δt the greyvalue at a fixed location (x,y) in the image has changed due to the object motion.

$$\Delta I = I(S,t+\Delta t)-I(S,t) = \frac{dI(S,t)}{dt} * \Delta t$$

$$= \frac{dO(Q(S,t))}{dt} * \Delta t$$

$$= GRAD^T\ O(Q(S,t)) * \frac{dQ(S,t)}{dt} * \Delta t$$

Since the object image O in the Q-system is by definition time invariant the temporal dependency can be expressed by

$$\frac{dQ(S,t)}{dt} * \Delta t = Q(S, t+\Delta t) - Q(S,t)$$
$$= [A(t+\Delta t) - A(t)] * S - [\Delta S(t+\Delta t) - \Delta S(t)]$$
$$= \Delta A(t, \Delta t) * S - \Delta \Delta S(t, \Delta t)$$

Let the pixels of the object image in the frame be enumerated by i, then the greyvalue difference after a time Δt is

$$\Delta I_i = I(S_i, t+\Delta t) - I(S_i, t) \text{ with } S_i = (x_i, y_i)^T.$$

Suppressing momentarily the explicit notation of time dependency:

$$\Delta I_i = GRAD^T O(A * S_i - \Delta S) * [\Delta A * S_i - \Delta \Delta S]$$

We restrict the general affine transformation to a rotation and dilatation in the image plane in order to reduce the number of unknowns from 6 to 4: angle, scale, x- and y-displacement.

$$A = s * \begin{pmatrix} \cos\Theta & -\sin\Theta \\ \sin\Theta & \cos\Theta \end{pmatrix}$$

At the time t when the "snapshot" was taken we define s=1 and $\Theta=0$. Then ΔA becomes

$$\Delta A = \begin{pmatrix} s*\cos\Theta-1 & -s*\sin\Theta \\ s*\sin\Theta & s*\cos\Theta-1 \end{pmatrix} = \begin{pmatrix} a & -c \\ c & a \end{pmatrix}$$

Substituting $O_i = O(A(t) * S_i - \Delta S(t))$ we obtain

$$\Delta I_i = F_i 1 * a + F_i 2 * c + G_i 1 * \Delta \Delta S_x + G_i 2 * \Delta \Delta S_y \quad \text{with}$$

$$F_i 1 = x_i * GRAD_x O_i + y_i * GRAD_y O_i$$
$$F_i 2 = x_i * GRAD_y O_i + y_i * GRAD_x O_i$$
$$G_i 1 = -GRAD_x O_i \text{ and } G_i 2 = -GRAD_y O_i.$$

Combining the matrix elements a and c to a vector $E=(E1, E2)^T$ (with E1=a and E2=c) we rewrite

$$\Delta I_i = (F_i 1, F_i 2, G_i 1, G_i 2) * (E1, E2, \Delta \Delta S_x, \Delta \Delta S_y)^T$$

In order to estimate the unknowns we minimize

$$\sum_i [F_i * E + G_i * \triangle\triangle S - \triangle I_i]^2 = \text{minimum}.$$

This results in the equations:

$$F^T * F * E + F^T * G * \triangle\triangle S = F^T * \triangle I$$

$$G^T * F * E + G^T * G * \triangle\triangle S = G^T * \triangle I$$

Here F and G are regarded as two-column matrices with i as row index. The solution of this system (see NAGEL 80 (74) for a more complete discussion) is

$$\hat{E} = [F^T * [1 - G*G^{\#}] * F]^{-1} * F^T Z [1 - G*G^{\#}] * \triangle I$$

with the pseudo-inverse $G^{\#} = [G^T * G]^{-1} * G^T$ and 1 as the unity matrix.

$$\triangle\triangle\hat{S} = G^{\#} * [\triangle I - F * \hat{E}]$$

The estimation of E can be rewritten as

$$\hat{E} = \hat{F} * [\triangle I - G * \triangle\triangle\hat{S}]$$

These results are equivalent to those of Schalkoff and McVey.

Two aspects of these solutions will be discussed. The last expression for $\hat{E}$ shows that the parameters specifying an incremental rotation and scale change of the object image will vanish to the extend that all observable interframe intensities $\triangle I$ (interior to the object image) can be computed from a common shift vector $\triangle\triangle S$.

On the other hand, let the object movement be dominated by the frame-to-frame translation $\triangle\triangle S$. Then the components of $\hat{E}$ will be negligible and we obtain the estimate $\triangle\triangle S = G^{\#} * \triangle I$. Since the matrix $G^{\#} * G$ can be explicitly inverted, one obtains

$$\triangle\triangle\hat{S}_x = \frac{1}{\det|G^T * G|} *$$

$$[\quad -\sum_i (\text{GRAD}_y O_i * \text{GRAD}_y O_i) * \sum_i (\text{GRAD}_x O_i * \triangle I_i)$$

$$+\sum_i (\text{GRAD}_x O_i * \text{GRAD}_y O_i) * \sum_i (\text{GRAD}_y O_i * \triangle I_i) \quad]$$

$$\Delta \Delta \hat{S}_y = \frac{1}{\det |G^T * G|} *$$

$$[\quad -\sum_i (GRAD_x O_i * GRAD_y O_i) * \sum_i (GRAD_x O_i * \Delta I_i)$$

$$+\sum_i (GRAD_x O_i * GRAD_x O_i) * \sum_i (GRAD_y O_i * \Delta I_i) \quad]$$

This is exactly the result derived by FUKINUKI 78 (25) as well as NETRAVALI and ROBBINS 79 (76).

It will have to be verified using actual data from TV or satellite sensors wether the introduction of rotation and dilatation parameters represents a consistent improvement for the description of moving object images. Furthermore, it will have to be investigated to what extend the use of approximate digital operators for determination of $GRAD\ O_i(Q(S,t))$ influences the reliability of the results for $\Delta \Delta S$ and E. Nevertheless, this model offers the opportunity to at least discuss specific assumptions and to point out the relations between the different approaches to velocity estimation.

5.0 CONCLUSION

The field of image sequence analysis passes through a period of rapid development, especially as far as modeling of 2-D image variations is concerned. A trend to more rigorous mathematical formulations is clearly discernible despite the fact that simplifying assumptions still have to be made. Progress in this direction demonstrates the transition to a state where this field should no longer be viewed as a mere agglomeration of applications.

Most recently, there are first results which extend modeling of temporal image sequences into three dimensions, explaining the temporal variation as being caused by rigid body motion in three dimensional space. Further knowledge has to be collected in order to cope with situations such as occlusion or moving shadows in a real-world scene domain. We feel that sequence processing will stimulate single image analysis, thus in future the flow of expertise might be from dynamic to static scene analysis.

ACKNOWLEDGEMENTS
The authors are grateful to Mrs. R. Jancke who helped to prepare this manuscript with an admirable patience and expertise. G.

Hille's suggestions helped considerably to improve the manuscript. We couldn't have done without them, but considering the nice weather outside, they could perhaps have done better without us.

6.0 REFERENCES

WCATVI = Workshop on Computer Analysis of Time-Varying Imagery, Philadelphia/PA, April 5-6, 1979

(1) R.F. Adler and D.D. Fenn
"Thunderstorm Monitoring from a Geosynchronous Satellite", 7th Conference on Aerospace and Aeronautical Meteorology (1976) pp. 307-311

(2) J.K. Aggarwal and R.O. Duda
"Computer Analysis of Moving Polygonal Images", IEEE Trans. Computers C-24 (1975) 966-976

(3) Y. Ariki, T. Kanade, and T. Sakai
"An Interactive Image Modeling and Tracing System for Moving Pictures", IJCPR-78 Nov. 7-10, 1978 Kyoto/Japan, pp. 681-685

(4) A.A. Arking, R.C. Lo, and A. Rosenfeld
"An Evaluation of Fourier Transform Techniques for Cloud Motion Estimation", TR-351 (January 1975) Computer Science Department, University of Maryland, College Park/MD

(5) S.T. Barnard and W.B. Thompson
"Visual Disparity as a Cue for Depth and Velocity in Real World Scenes", IEEE Conference on Pattern Recognition and Image Processing Chicago/Ill., May 31 - June 2, 1978, pp. 402-404

(6) R. Brennecke, T.K. Brown, J. Buersch, and P.H. Heintzen
"Digital Processing of Video angio-cardiographic Image Series Using a Minicomputer", Proc. IEEE Conference on Computers in Cardiology, St. Louis 1976

(7) R. Brennecke, T.K. Brown, J. Buersch, and P.H. Heintzen
"Computerized Video-Image Preprocessing with Applications to Cardio-Angiographic Roentgen-Image-Series", GI/NTG Fachtagung Digitale Bildverarbeitung Muenchen, 28.-30. Maerz 1977, (H.-H. Nagel, ed.) Informatik Fachberichte 8 (1977) pp. 244-262, Springer Verlag Berlin-Heidelberg-New York, 1977

(8) R. Brennecke, T.K. Brown, J. Buersch, and P.H. Heintzen
    "A Digital System for Roentgen Video Image Processing", in:
    Roentgen-Video-Techniques for Dynamic Studies of Structure
    and Function of the Heart and Circulation P.H. Heintzen and
    J.H. Buersch (eds.) Georg Thieme Verlag, Stuttgart 1978, pp.
    150-157
(9) J.R. Birk, R.B. Kelley, and V.V. Badami
    "Workpiece Orientation Correction with a Robot Arm Using
    Visual Information", IJCAI-77, p. 758
(10) S. Brofferio and F. Rocca
    "Interframe Redundancy Reduction of Video Signals Generated
    by Translating Objects", IEEE Trans. Communications COM-25
    (1977) 448-455
(11) C. Cafforio and F. Rocca
    "Methods for Measuring Small Displacements of Television
    Images", IEEE Trans. Information Theory IT-22 (1976) 573-579
(12) J.C. Candy, M.A. Franke, B.G. Haskell, and F.W. Mounts
    "Transmitting Television as Clusters of Frame-to-Frame
    Differences", Bell System Techn. J. 50 (1971) 1889-1917
(13) R.T. Chien and V.C. Jones
    "Acquisition of Moving Objects and Hand-Eye Coordination",
    IJCAI-75, pp. 737-741, Tbilisi, Georgia/USSR, September 3-8,
    1975
(14) W.K. Chow and J.K. Aggarwal
    "Computer Analysis of Planar Curvilinear Moving Images",
    IEEE Trans. Computers C-26 (1977) 179-185
(15) C.K. Chow and T. Kaneko
    "Automatic Boundary Detection of the Left Ventricle from
    Cineangiograms", Computers and Biomedical Research 5 (1972)
    388-410
(16) C.K. Chow, S.K. Hilal, and K.E. Niebuhr
    "X-ray Image Subtraction by Digital Means", IBM J. Res.
    Develop. 17 (1973) 206 -218
(17) D.J. Connor and J.O. Limb
    "Properties of Frame-Difference Signals Generated by Moving
    Images", IEEE Trans. on Communications COM-22 (1974)
    1564-1575
(18) D.J. Connor, B.G. Haskell, F.W. Mounts
    "A Frame-to-Frame Picturephone Coder for Signals Containing
    Differential Quantizing Noise", Bell System Techn. J. 52
    (1973) 35-51

(19) D.L. Davies and P.H. Smith
"Digital Analysis of Internal Machine Part Motion in a Time
Series of Radiographs", WCATVI-79, pp. 128-130
(20) L. Dreschler and H.-H. Nagel
"Using 'Affinity' for Extracting Images of Moving Objects
from TV-Frame Sequences", IfI-HH-B-44/78 (February 1978)
(21) R.M.Endlich, D.E. Wolf, D.J. Hall, and A.E. Brain
"Use of a Pattern Recognition Technique for Determining
Cloud Motions from Sequences of Satellite Photographs", J.
of Applied Meteorology 10 (1971) 105-117
(22) R. Eskenazi and R.T. Cunningham
"Real-Time Tracking of Moving Objects in TV Images", IEEE
Workshop on Pattern Recognition and Artificial Intelligence,
Princeton/NJ, April 12-14, 1978, pp. 4-6
(23) C.L. Fennema and W.B. Thompson
"Velocity Determination in Scenes Containing Several Moving
Objects", Computer Graphics and Image Processing 9 (1979)
301-315
(24) T.T. Fujita
"Present Status of Cloud Velocity Computations from the ATS
I and ATS III Satellites", Space Research IX, pp. 557-570,
North Holland Publ. Co., Amsterdam 1969
(25) T. Fukinuki
"Measurement of Movement and Velocity of Moving Objects with
Picture Signals", Record of Technical Group on Image
Technology of the Institute of Electronics and
Communications Engineers (IECE) of Japan October 29, 1978
(IE 78-67, pp. 35-41)
(26) D.B. Gennery
"A Stereo Vision System for an Autonomous Vehicle",
IJCAI-77, pp. 576-582
(27) D.B. Gennery
"A Stereo Vision System", Proc. Image Understanding Workshop
L.S. Baumann (ed.) pp. 31-46 Palo Alto/CA, Oct. 20-21, 1977
Science Applications, Inc., Arlington/VA
(28) H. Gerlach
"Digitale Bildfolgenauswertung zum Wiederfinden von Objekten
in natuerlicher Umgebung", in: Angewandte Szenenanalyse,
J.P. Foith (ed.) Informatik Fachberichte 20, pp. 199-207
Springer Verlag, Berlin-Heidelberg-New York 1979
(29) J.O.B. Greaves
"The Bugsystem: The Software Structure for the Reduction of
Quantized Video Data of Moving Organisms", Proc. IEEE 63
(1975) 1415-1425

(30) F.M. Greene, Jr., and F.S. Barnes
      "System for Automatically Tracking White Blood Cells", Rev.
      Sci. Instruments 48 (1977) 602-604
(31) U. Haass and T.A. Brubaker
      "Estimation of Lateral and Rotational Cloud Displacement
      from Satellite Pictures", WCATVI-79, pp. 103-104
(32) A. Habibi
      "Survey of Adaptive Image Coding Techniques", IEEE Trans.
      Communications COM-25 (1977) 1275-1284
(33) K. Hachimura, M. Kuwahara, and M. Kinoshita
      "Left Ventricular Contour Extraction from Radioisotope
      Angiocardiograms and Classification of Left Ventricular Wall
      Motion", IJCPR-78 Nov. 7-10, 1978 Kyoto/Japan, pp. 911-913
(34) D.J. Hall, R.M. Endlich, D.E. Wolf, and A.E. Brain
      "Objective Methods for Registering Landmarks and Determining
      Cloud Motions from Satellite Data", IEEE Trans. Computers
      C-21 (1972) 768-776
(35) B.G. Haskell
      "Frame-to-Frame Coding of Television Pictures Using
      Two-Dimensional Fourier Transforms", IEEE Trans. Information
      Theory IT-20 (1974) 119-120
(36) B.G. Haskell
      "Entropy Measurements for Nonadaptive and Adaptive,
      Frame-to-Frame, Linear Predictive Coding of Videotelephone
      Signals", Bell System Techn. J. 54 (1975) 1155-1174
(37) B.G. Haskell, F.W. Mounts, and J.C. Candy
      "Interframe Coding of Videotelephone Pictures", Proc. IEEE
      60 (1972) 792-800
(38) J.A. Heller
      "A Real Time Hadamard Transform Video Compression System
      Using Frame-to-Frame Differencing", Proc. Nat.
      Telecommunications Conference, San Diego, 1974, pp. 77-82
(39) G.T. Herman and H.K.Liu
      "Dynamic Boundary Surface Detection", Computer Graphics and
      Image Processing 7 (1978) 130-138 see also: Proc. Symposium
      on Computer-Aided Diagnosis of Medical Images, J. Sklansky
      (ed.) Coronado/CA, November 11, 1976, pp. 27-32
(40) M.A. Hernan and J. Jimenez
      "Automatic Analysis of Movies in Fluid Mechanics",
      WCATVI-79, pp. 134-135
(41) K.H. Hoehne, G. Nicolae, G. Pfeiffer, R.-W. Dix, W.
      Ebenritter, D. Novak, M. Boehm, B. Sonne, and E. Buecheler
      "An Interactive System for Clinical Application of
      Angiodensitometry", GI/NTG Fachtagung Digitale
      Bildverarbeitung Muenchen, 28.-30. Maerz 1977 (H.-H. Nagel,

ed.) Informatik Fachberichte 8 (1977) pp. 232-243, Springer Verlag, Berlin-Heidelberg-New York 1977

(42) D.C. Hogg
"A Methodology for Real Time Scene Analysis", IJCAI-77, p. 627

(43) G. Hughes, C. Novak, and R. Schreitz
"Automated Techniques for the Detection and Displacement Measurement of Selected Cloud Imagery Observed in Geostationary Satellite Data", Proc. 8th Annual Automatic Imagery Pattern Recognition Symposium, R.A. Kirsch and R.N. Nagel (eds.) Gaithersburg/MD, April 3-4, 1978, pp. 81-90 Electronic Industries Association, Washington/DC 1978

(44) R. Jain and H.-H. Nagel
"On the Analysis of Accumulative Difference Pictures from Image Sequences of Real World Scenes", IEEE Trans. Pattern Analysis and Machine Intelligence PAMI-1 (1979) 206-214

(45) R. Jain, D. Militzer, and H.-H. Nagel
"Separating Non-Stationary from Stationary Scene Components in a Sequence of Real World TV-Images", IJCAI-77, 612-618 and IfI-HH-B-32/77 (March 1977) Institut fuer Informatik, Universitaet Hamburg

(46) H.W. Jones
"A Real-Time Adaptive Hadamard Transform Video Compressor", SPIE 87 (1976) 2-9

(47) L.N. Kanal and J.A. Parikh
"Severe Storm Pattern Recognition from Meteorological Satellite Data: a Report on Current Status and Prospects", Research and Development Technical Report ECOM-77-3, Atmospheric Sciences Lab., US Army Electronics Command, White Sands Missile Range/NM (March 1, 1977)

(48) J.G. Kawamura
"Automatic Recognition of Changes in Urban Development from Aerial Photographs", IEEE Trans. Systems, Man, and Cyb. SMC-1 (1971) 230-239

(49) J. Klie
"Codierung von Fernsehsignalen fuer niedrige Uebertragungsbitraten", Dissertation, Juni 1978 Lehrstuhl fuer Theoretische Nachrichtentechnik und Informationsverarbeitung, Techn. Universitaet Hannover, D-3000 Hannover 1

(50) S.C. Knauer
"Real-Time Video Compression Algorithm for Hadamard Transform Processing", IEEE Trans. Electromagnetic Compatibility EMC-18 (1976) 28-36

(51) R.S. Ledley
"Some Clinical Applications of Pattern Recognition",
IJCPR-73, pp. 89-112
(52) J.A. Leese, C.S. Novak, and V.R. Taylor
"The Determination of Cloud Pattern Motions from
Geosynchronous Satellite Image Data", Pattern Recognition 2
(1970) 279-292
(53) J.A. Leese, C.S. Novak, and B.B. Clark
"An Automated Technique for Obtaining Cloud Motion from
Geosynchronous Satellite Data Using Cross-Correlation", J.
of Applied Meteorology 10 (1971) 118-132
(54) M.D. Levine, D.A. O'Handley, G.M. Yagi
"Computer Determination of Depth Maps", Computer Graphics
and Image Processing 2 (1973) 131-150
(55) M.D. Levine, Y.M. Youssef, and F. Ferrie
"Cell Movements: Its Characterization and Analysis",
WCATVI-79, pp. 93-95
(56) R.A. Lewis and A.R. Johnston
"A Scanning Laser Rangefinder for a Robot Vehicle",
IJCAI-77, pp. 762-768
(57) R.L. Lillestrand
"Techniques for Change Detection", IEEE Trans. Computers
C-21 (1972) 654-659
(58) J.O. Limb
"Distortion Criteria of the Human Viewer", IEEE Trans.
Systems, Man, and Cybernetics SCM-9 (1979) 778-793
(59) J.O. Limb and J.A. Murphy
"Measuring the Speed of Moving Objects from Television
Signals", IEEE Trans. Communications COM-23 (1975) 474-478
(60) J.O. Limb and J.A. Murphy
"Estimating the Velocity of Moving Images in Television
Signals", Computer Graphics and Image Processing 4 (1975)
311-327
(61) J.O. Limb and R.F.W. Pease
"A Simple Interframe Coder for Video Telephony", Bell System
Techn. J. 50 (1971) 1877-1888
(62) J.O. Limb, R.F.W. Pease, and K.A. Walsh
"Combining Intra-Frame and Frame-to-Frame Coding for
Television", Bell System Techn. J. 53 (1974) 1137-1173
(63) L.E. Lipkin, P. Lemkin, and M. Wade
"Digital Analysis of Living Cell Image Sequences in Support
of Cytotoxicity-Carcinogenesis Research", WCATVI-79, p. 25

(64) H.K. Liu
   "Two- and Three-Dimensional Boundary Detection", Computer
   Graphics and Image Processing 6 (1977) 123-134
(65) R.C. Lo
   "The Application of a Thresholding Technique in Cloud Motion
   Estimation from Satellite Observations", TR-357 (February
   1975) Computer Science Department, University of Maryland,
   College Park/MD
(66) W.N. Martin and J.K. Aggarwal
   "Survey: Dynamic Scene Analysis", Computer Graphics and
   Image Processing 7 (1978) 356-374
(67) W.N. Martin and J.K. Aggarwal
   "Computer Analysis of Dynamic Scenes Containing Curvilinear
   Figures", Pattern Recognition 11 (1979) 169-178
(68) B.H. McCormick, J.S. Read, R.T. Borovec, and R.C. Amendola
   "Image Processing in Television Ophthalmoscopy", Digital
   Processing of Biomedical Images K. Preston, jr., and M. Onoe
   (eds.) Plenum Publ. Co., New York 1976
(69) D.L. Milgram
   "Region Tracking Using Dynamic Programming", TR-539 (May
   1977), Computer Science Center, University of Maryland,
   College Park/MD see, too, WCATVI-79, p. 13
(70) H.P. Moravec
   "Towards Automatic Visual Obstacle Avoidance", IJCAI-77, p.
   584
(71) F.W. Mounts
   "A Video Encoding System with Conditional Picture-Element
   Replenishment", Bell System Techn. J. 48 (1969) 2545-2554
(72) H.-H. Nagel
   "Formation of an Object Concept by Analysis of Systematic
   Time Variations in the Optically Perceptible Environment",
   Computer Graphics and Image Processing 7 (1978) 149-194
(73) H.-H. Nagel
   "Analysis Techniques for Image Sequences", IJCPR-78, pp.
   186-211
(74) H.-H. Nagel
   "Analysis of Image Sequences : What Can We Learn from
   Applications ? ", to appear in: Image Sequence Analysis,
   T.S. Huang (ed.), Springer Verlag Berlin-Heidelberg-New York
   1980
(75) T.R. Natarajan and N. Ahmed
   "On Interframe Transform Coding", IEEE Trans. Communications
   COM-25 (1977) 1323-1329

(76) A.N. Netravali and J.D. Robbins
      "Motion Compensated Television Coding: Part 1", Bell System
      Technical J. 58 (1979) 631-670
(77) B. Neumann
      "Interpretation   of   Imperfect   Object   Contours   for
      Identification and Tracking", IJCPR-78 Nov. 7-10, 1978
      Kyoto/Japan, pp. 691-693
(78) D.A. O'Handley
      "Scene Analysis in Support of a Mars Rover", Computer
      Graphics and Image Processing 2 (1973) 281-297
(79) M. Onoe and K. Ohba
      "Digital Image Analysis of Traffic Flow", IJCPR-76, pp.
      803-808
(80) M. Onoe, N. Hamano, K. Ohba
      "Computer Analysis of Traffic Flow Observed by Subtractive
      Television", Computer Graphics and Image Processing 2 (1973)
      377-392
(81) J.A. Parikh and A. Rosenfeld
      "Automatic Segmentation and Classification of   Infrared
      Meteorological Satellite Data", IEEE Trans. Systems, Man,
      and Cybernetics SMC-8 (1978) 736-743
(82) R.F.W. Pease and J.O.Limb
      "Exchange of Spatial and Temporal Resolution in Television
      Coding", Bell System Techn. J. 50 (1971) 191-200
(83) J. Potter
      "Velocity as a Cue to Segmentation", IEEE Trans. Systems,
      Man and Cybernetics, SCM-5 (1975) 390-394
(84) J.L. Potter
      "Scene Segmentation Using Motion Information", Computer
      Graphics and Image Processing 6 (1977) 558-581
(85) W.K. Pratt (ed.)
      "Image Transmission Techniques", Academic Press New York/NY
      1979
(86) K.E. Price
      "Change Detection and Analysis in Multi-Spectral Images",
      Ph.D. Thesis, December 18, 1976 Department of Computer
      Science, Carnegie-Mellon University, Pittsburgh/PA
(87) K. Price and R. Reddy
      "Matching Segments of Images", IEEE Trans. Pattern Analysis
      and Machine Intelligence PAMI-1 (1979) 110-116
(88) B. Radig
      "Description of Moving Objects Based on Parameterized Region
      Extracting", IJCPR-78 Nov. 7-10, 1978 Kyoto/Japan, pp.
      723-725

(89) J.S. Read, R.T. Borovec, R.C. Amendola, A.C. Petersen, M.H.
      Goldbaum, M. Kottow, B.H. McCormick, and M.F. Goldberg
      "The Television Ophthalmoscope Image Processor", Proc. IEEE
      Workshop on Picture Data Description and Management,
      Chicao/Ill., April 21-22, 1977, pp. 64-67
(90) C. Reader
      "Intraframe and Interframe Adaptive Transform Coding", Proc.
      SPIE 66 (1975) 108-117
(91) J. Roach and J.K. Aggarwal
      "Computer Tracking of Objects Moving in Space", IEEE Trans.
      Pattern Analysis and Machine Intelligence PAMI-1 (1979)
      127-135
(92) J.D. Robbins and A.N. Netravali
      "Interframe Television Coding Using Movement Compensation",
      International Conference on Communications June 1979, IEEE
      publication CH 1435-7/79 pp. 23.4.1 - 23.4.5
(93) F. Rocca
      "Television Bandwidth Compression Utilizing Frame-to-Frame
      Correlation and Movement Compensation", Symposium on Picture
      Bandwidth Compression MIT, Cambridge/MA, April 1969 (T.S.
      Huang and O.J. Tretiak, eds.) Gordon and Breach, New York
      1972, pp. 675-693
(94) J.A. Roese and G. S. Robinson
      "Combined Spatial and Temporal Coding of Digital Image
      Sequences", Proc. SPIE 66 (1975) 172-180
(95) J.A. Roese, W.V. Pratt, and G.S. Robinson
      "Interframe Cosine Transform Image Coding", IEEE Trans.
      Communications COM-25 (1977) 1329-1339
(96) W. Scacchi
      "Visual Motion Perception by Intelligent Systems", IEEE
      Conference on Pattern Recognition and Image Processing,
      Chicago/IL, August 6-8, 1979, pp. 646-652
(97) R.J. Schalkoff and E.S. McVey
      "Algorithm Development for Real-Time Automatic Video
      Tracking System", preprint 1979
(98) W.E. Shenk and E.R. Kreins
      "A Comparison Between Observed Winds and Cloud Motions
      Derived from Satellite Infrared Measurements", J. of Applied
      Meteorology 9 (1970) 702-710
(99) R.W. Smalling, M.H.Skolnick, D. Myers, R. Shabetai, J.C.
      Cole, and D. Johnston
      "Digital Boundary Detection, Volumetric and Wall Motion
      Analysis of Left Ventricular Cine Angiograms", Comput. Biol.
      Med. 6 (1976) 78-85

(100) E.A. Smith and D.R. Phillips
"Automated Cloud Tracking Using Precisely Aligned Digital ATS Pictures", IEEE Trans. Computers C-21 (1972) 715-729

(101) J.A. Stuller and A.N. Netravali
"Transform Domain Motion Estimation", WCATVI-79, p. 82

(102) M. Takagi and K. Sakaue
"The Analysis of Moving Granules in a Pancreatic Cell by Digital Moving Image Processing", IJCPR-78 Nov. 7-10, 1978 Kyoto/Japan, pp. 735-739

(103) M. Tasto
"Guided Boundary Detection for Left Ventricular Volume Measurements", IJCPR-73, pp. 119-124

(104) M. Tasto
"Motion Extraction for Left-Ventricular Volume Measurement", IEEE Trans. Biomedical Engineering BME-21 (1974) 207-213

(105) M. Tasto, M. Felgendreher, W. Spiesberger, and P. Spiller
"Comparison of Manual versus Computer Determination of Left Ventricular Boundaries from X-Ray Cineangiocardiograms", in: Roentgen-Video-Techniques for Dynamic Studies of Structure and Function of the Heart and Circulation P.H. Heintzen and J.H. Buersch (eds.) Georg Thieme Verlag, Stuttgart 1978, pp. 168-183

(106) A.M. Thompson
"The Navigation System of the JPL Robot", IJCAI-77, pp. 749-757

(107) J.K. Tsotsos
"A Framework for Visual Motion Understanding", Department of Computer Science, University of Toronto Toronto/Canada 1980

(108) M.S. Ulstad
"An Algorithm for Estimating Small Scale Differences between Two Digital Images", Pattern Recognition 5 (1973) 323-333

(109) S. Ullman
"The Interpretation of Visual Motion", The MIT Press, Cambridge/Mass., 1979

(110) T. Uno, M. Ejiri, and T. Tokunaga
"A Method of Real-Time Recognition of Moving Objects and its Application", Pattern Recognition 8 (1976) 201-208

(111) M.P. Waters, III
"Time Compositing of Meteorological Satellite Data for Ocean Current Identification", Proc. 8th Annual Automatic Imagery Pattern Recognition Symposium, R.A. Kirsch and R.N. Nagel (eds.) Gaithersburg/MD, April 3-4, 1978, Electronics Industries Association, Washington/DC 1978, pp. 59-65

(112) G. Winkler and K. Vattrodt
"Measures for Conspicuousness of Images", Computer Graphics and Image Processing 8 (1978) 355-368

(113) D.E. Wolf, D.J. Hall, and R.M. Endlich
"Experiments in Automatic Cloud Tracking Using SMS-GOES Data", J. Applied Meteorology 16 (1977) 1219-1230

(114) D.E. Wolf, D.J. Hall, and R.M. Endlich
"An Automatic Method for Determining Cloud Motions from Pictures Taken by Geosynchronous Weather Satellites", WCATVI-79, pp. 101-102

(115) K. Wolferts
"Special Problems in Interactive Image Processing for Traffic Analysis", IJCPR-74, pp. 1-2 Copenhagen, August 13-15, 1974

(116) E.H. Wood
"New Horizons for Study of the Cardiopulmonary and Circulatory Systems", Chest 69 (1976) 394-408

(117) M. Yachida, M. Asada, and S. Tsuji
"Automatic Motion Analysis System of Moving Objects from the Records of Natural Processes", IJCPR-78 Nov. 7-10, 1978 Kyoto/Japan, pp. 726-730

(118) Y. Yakimovsky and R.T. Cunningham
"A System for Extracting Three-Dimensional Measurements from a Stereo-Pair of TV-Cameras", Computer Graphics and Image Processing 7 (1978) 195-210

(119) H.W. Yates and W.R. Bandeen
"Meteorological Applications of Remote Sensing from Satellites", Proc. IEEE 63 (1975) 148-163

AUTOMATED IMAGE-TO-IMAGE REGISTRATION , A WAY TO MULTI-TEMPORAL ANALYSIS OF REMOTELY SENSED DATA

Robert JEANSOULIN

Laboratoire "Langages et Systèmes Informatiques". Centre National de la Recherche Scientifique. TOULOUSE. FRANCE. (*)

ABSTRACT. The digital overlaying of images is the only way to the multidimensional analysis of data, but a qualitative assessment. In Remote Sensing, the multi-temporal analysis is useful for studying the evolution of many physical phenomena, and now the merging of Visible, near or thermal Infra-Red and Microwave data, opens a new field of investigations.
A set of several methods - some of them are now classic ones - is shown in this chapter, and we try to point out the troubles met in automated registration : what happens to the physical content of the remotely sensed information with respect to the processing applied?

KEY WORDS. Correlation, Edge Detection, Multitemporal Analysis, Rectification, Registration, Overlaying, Remote Sensing.

(*) Work sponsored by a research contract between the group of Professor BRUEL (Université Paul Sabatier and ENSEEIHT) and the Centre National d'Etudes Spatiales, "Etudes Thématiques" group of Mr. SAINT.

J. C. Simon and R. M. Haralick (eds.), Digital Image Processing, 561–577.

1. THE IMAGE OVERLAYING AND THE "MULTI" CONCEPT IN REMOTE SENSING.

The image-to-image registration is not really a new problem, but it is concerned by automatical processing for only about ten years.
Indeed, since the launch of the first Earth Ressources satellite, the ability of getting multiple data has notably increased in the Remote Sensing domain.
It is what   is called the "Multi" Concept by COLWELL in (1): multispectral,multitemporal,multistage,multistation,multisensor.
Let us write the Multi Concept as a function :

$$(x,y) \in \text{geographical domain} \longrightarrow f(x,y) \in \mathbb{R}^n$$

Most of the work published on Registration and Overlaying has been done on LANDSAT-MSS images:(2) to(8).
The Table 1.1 summarizes the overlaying difficulties with respect to the Multi Concept, and the Table 1.2 describes the registration and the rectification in the overlaying processing.

2. THE REGISTRATION AND ITS PROCESSING.

In the registration processing -i.e. the determination of couples of homologous pixels (or "control points", or "amers") - several questions need answer :
- what to register for getting a location of pixel?
- how to register?
- can we trust the registration result?

WHAT ?   the processing cannot be punctual but local.
         Two ways can be followed, which handle:
         - the radiometric levels on a neighborhood
         (information of continuity)
         - the result of some local edge operator
         (information of discontinuity): table 2.1.

HOW ?   several similarity functions can be used for
         computing a registration index: table 2.2.

TRUST?   two ways for selecting trusty indexes:
         - a-priori selecting: a criterion discrimina-
         tes the control points which are the ablest
         for succeeding in the registration (t. 2.3(a))
         - a-posteriori selecting: several criteria
         discriminate reliable indexes (t. 2.3(b))

Table 1.1. The Multi-Concept and the Image Overlaying.

MULTI- .	AMOUNT OF DATA (examples) .	OVERLAYING ABILITIES
SPECTRAL	LANDSAT-MSS: 4 bands METEOSAT: 3 bands SPOT: 3 bands (20meters) 1 band (10meters)	Easy (in general) Offset or slight rota- tion due to the instru ment

TEMPORAL
(same
sensor)
	LANDSAT: 1 scene/18 days
	METEOSAT: 48 . / 1 .
	SPOT: 1 . /26 .
	    (vertical mode)
	    1 . / 5 .
	    (non vertical)
		More or less difficult
		(more for airborne
		data than for satel-
		lite data)
		Model of the movement
		to be estimated

SENSOR	LANDSAT(80m.) on Radar VIGIE(15m.) or on DAEDALUS (3m.)	Various and complex modeling Intermediate cartogra- phy is necessary

Table 1.2. The REGISTRATION and the RECTIFICATION.

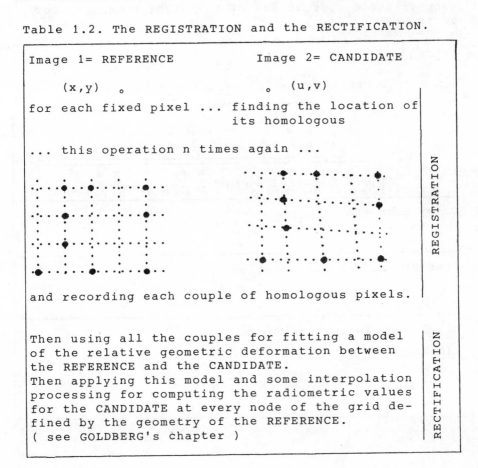

Table 2.1. Example of discontinuity detection.

Original image

↓

Improve   the local homogeneity by using a non-linear
median smoothing:
on each 3 by 3 neighborhood, let us compare the central
pixel with the others. If it has the highest or the lowest value, replace it by the median value.

↓

Detect   the edge by using a local derivating like:

$$d_{i,j} = 1/4 \, (|x_{i-1,j-1} - x_{i+1,j+1}| + |x_{i,j-1} - x_{i,j+1}| + |x_{i-1,j} - x_{i+1,j}| + |x_{i-1,j+1} - x_{i+1,j-1}|)$$

↓

Get   a binary information by thresholding the histogram of the $d_{i,j}$'s at 20% of the highest values, for
example.

Table 2.2. Registration index (or Similarity)

Correlation coefficient

$$\rho_{lk} = \frac{[N^2 \sum_{i=1}^{N} \sum_{j=1}^{N} x_{ij} y_{i+l,\,j+k} - (\sum_{i=1}^{N} \sum_{j=1}^{N} x_{ij})(\sum_{i=1}^{N} \sum_{j=1}^{N} y_{i+l,\,j+k})]}{\sqrt{[N^2 \sum_{i=1}^{N} \sum_{j=1}^{N} x_{ij}^2 - (\sum_{i=1}^{N} \sum_{j=1}^{N} x_{ij})^2]\,[N^2 \sum_{i=1}^{N} \sum_{j=1}^{N} y_{i+l,\,j+k} - (\sum_{i=1}^{N} \sum_{j=1}^{N} y_{i+l,\,j+k})^2]}}$$

Covariance function

$$r_{lk} = \sum_{i=1}^{N} \sum_{j=1}^{N} x_{ij} \, y_{i+l,\,j+k}$$

Absolute difference function

$$a_{lk} = \sum_{i=1}^{N} \sum_{j=1}^{N} |x_{ij} - y_{i+l,\,j+k}|$$

Graph correlation coefficient
(binary values)

$$g_{lk} = \frac{\sum_{i=1}^{N} \sum_{j=1}^{N} x_{ij} \, y_{i+l,\,j+k}}{\sum_{i=1}^{N} \sum_{j=1}^{N} x_{ij}}$$

Table 2.3(a) A-priori selecting.

-Among N windows around a fixed reference window, let us choose the one which maximizes the number of "chained edge pixels", for a given edge detector and for a given chain length (9).

-If the type of edges to be detected is well known, let us unite the information of discontinuity and the radiometric information: an example with METEO-SAT images is given in (10).

Table 2.3(b) A-posteriori validation.

The Registration Index always has a maximum value on the surface on which it is computed. The question is : is this maximum a peak of correlation which indicates the good position of the homologous pixel on the CW ?
The answer will be positive if the three following criteria are positive :

$$\text{let } \rho(max) \text{ be the maximum of the set } \{ \rho_{1,k} \}$$

1. EXISTENCE $\rho(max) > \bar{\rho} + 3\ \sigma(\rho)$

2. UNIQUENESS the i highest values below the max. :
$\rho(max-1), \rho(max-2), \ldots \rho(max-i)$
belong to some neighborhood of $\rho(max)$

3. SHARPNESS let N be the cardinal of the subset
$\{ (1,k)\ ;\ \rho(max)/\sqrt{e} < \rho_{1,k} < o(max) \}$
and N_{tot} the cardinal of the complete set $\{ (1,k) \}$
then : $\quad p = N_{tot}/N < p_{max}$

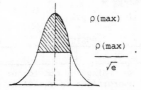

$\rho(max)$

$\dfrac{\rho(max)}{\sqrt{e}}$

for estimating p_{max} let us see the normal Gaussian distribution : the hatched part contains 19.8 % of the total surface.

Several combinations of these processing have been tried by several authors, comparative studies have been done (11)(12).

The Flow-Chart of the table 2.4.(below) shows the processing sequence used in (12) which works with discontinuity detection, graph-correlation (or FFT-correlation) and a-posteriori validation.

Let us notice that the similarity -or registration-index can measure only the fitness of a registration by local translation. If the deformation is too strong, it cannot be locally approximated by a translation and then it is necessary first to rectify the Reference window with a model (coarse or updated): this part is written in *script*.

An example of the Previous Rectification is shown by the table 2.5.

Table 2.4. Flow-Chart of the Registration Processing.

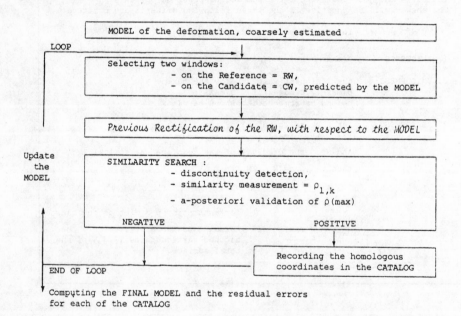

Table 2.5. Example of the application of the previous rectification

The images are from an airborne DAEDALUS, taken from two perpendicular flights (high resolution = 3 m. along track, wide field of view = 78°)

Part of the Reference,   of the Candidate

1. Edges detected on the Reference window.

2. Edges detected on the Candidate window.

3. Edges detected on the RW after the previous rectification.

3. HEALTH OR PATHOLOGY OF SOME EXAMPLES WITH LANDSAT

Let us watch the behaviour of the former algorithm,
with respect to several physical parameters, by inves-
tigating some regions in FRANCE (see list below),
which are rather different from most of the agricultural regions in the USA.

Regions	Characteristics
BEAUCE	Agricultural area, surface of fields: from several hectares to several tens of hectares
MARAIS POITEVIN	Coastal area, presence of water/land discontinuity
GIRONDE	Heterogeneous agricultural area (vineyards) Wide river (Dordogne)

3.1. Relationship between the Resolution (R) of the
sensor and the Surface (S) of the topological elements.
The optimal ratio for getting discontinuities which fit
the similarity search, should verify:

$$30 < S / R^2 < 300$$

for an algorithm using RW of 32 by 32 pixels and an
edge operator of 3 by 3 pixels.
The figures 3.1(a) and (b) show examples where S is too
large or too small for the LANDSAT resolution = 80 m.

The Figure 3.1(c) shows a pathologic case, where the
main element of the countryside (the river) has a size which fits the resolution quite well along one direction and badly along the perpendicular one. The
final result is a success because of some other small
structures being in the same area.
The three main ridges which appear on the correlation
surface mean we can register the riverbanks well, or
the right one over the left one, and vice versa.

Figures 3.1.   Influence of the ratio S/R^2

(a) the "Côte Vendéenne"   (b) the "Marais Poitevin"
    S too large   S too small

Two unsuccessful correlation surfaces.

Figure 3.1(c)   The Dordogne river

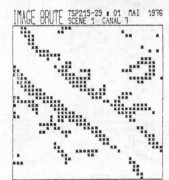

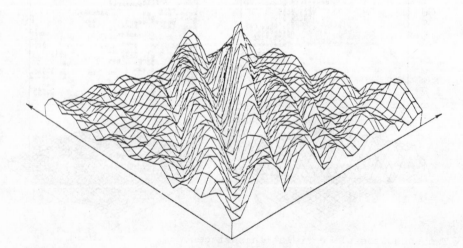

3.2. Relationship between the Date, the Spectral Band, and the Type of countryside.
Over an agricultural area like the BEAUCE, we can observe the following results (over 24 tests) :

BAND	March/ April	April/ April	April/ July	July/ July	July/ August	% of
5	0	75	92	83	96	TP
	0	21	8	17	4	FN
	75	4	0	0	0	TN
	25	0	0	0	0	FP
7	33	96	8	0	71	TP
	58	4	38	0	29	FN
	8	0	54	75	0	TN
	0	0	0	25	0	FP

Comments: TP=True Positive (the diagnosis given by the algorithm is "Positive", and the visual observation says it is "True"), FN=False Negative (a good registration is not seen by the algorithm), TN=True Negative, FP=False Positive.
Obviously, the best is maximizing the TP rate and eliminating the FP, which is the worst case because a false location is recorded on the CATALOG.
We can notice that the band 5 yields the best contrasts and then the best discontinuities, at the end of Spring and in Summer (prevalence of chlorophylian activity).
On the other hand, the band 7 is better in Winter and at the beginning of Spring (prevalence of bared soil).

Over a coastal area and, generally, every time there is a presence of water in the countryside, the band 7, which clearly discriminates land and water, is more efficient (see Figures 3.2(a) and (b)) .

All these results are in (12),tome II.

Figures 3.2.   Influence of the Spectral Band

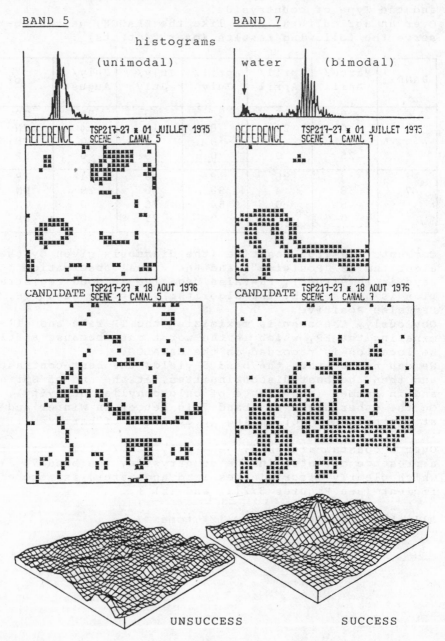

BAND 5   BAND 7

histograms

(unimodal)   water   (bimodal)

UNSUCCESS   SUCCESS

4. THE "LAURAGAIS PROJECT":A LIVE EXAMPLE OF THE MULTI-CONCEPT.

The "Lauragais project" (13) is intended to vegetation monitoring by using both a data bank of ground truth measurements and a set of remotely sensed data taken by an airborne scanner: the DAEDALUS.

Ground data = canopy cartography, biological and structural characteristics of the different crops.
Remotely sensed data = eight spectral bands, ranging from 400 nm. to 1000 nm.,at each date. Two flights in 1978, five in 1979 and five planned in 1980.

Potentially, the Multi-Concept can produce a function:

for each pixel $(i,j) \longrightarrow f(i,j) \in \mathbb{R}^{96}$
of a size of about
3 by 3 meters
96=8 bands by 12 dates!

The problem is to get the 12 images overlayed with a good accuracy.

4.1. Rectification by "Sliding Models":

Owing to the severe deformation, the rectification model cannot be computed by a polynomial over the whole image: even with a five degree, the residuals errors remain high (4 to 5 pixels RMSE, 20 pixels max.)
A better way is fitting local models along the plane track, according with the zones of flight stability (skew variations are the main causes of error).
The local models are linear panoramic ones (8 parameters, see table 4.1.), and the rectification is done by "sliding" from one model to the following, over an overlaying zone of 10% or 20% of each model (14).
The residual errors are notably lower: 1 to 2 pixels RMSE, 10 pixels max., but the Sliding Model computing needs so many homologous pixels that only the automated registration is able to do it (minimum number of pixels/local model 10, number of models/image 10, then about 1200 locating are requested for the 12 images)
The table 4.1. shows the different operations to work out for giving reality to the Multi Concept in this experiment.

Table 4.1.   Operating Chart

- Choosing a Reference (image of best quality)

- Coarse estimating of the relative deformations between the Reference and each of the eleven other images: visual selection of 4 pixels on the Reference and their homologous = 48 visual locating (vs. 1200)

- Eleven automated registrations and recording of catalogs of control points

- Eleven computing of Sliding Models.

Description of the Sliding Model:

1. compute a linear regression line for all the pixels in the CW, which are the homologous of pixels of a same scan-line in the RW

it gives a simulated scan-line

2. estimating the parameters of a panoramic model between two simulated scan-lines

$$u = P_1(x,y) = x_0 + \alpha x + \beta tg((y-p_0)\Delta\omega)$$
$$v = P_2(x,y) = y_0 + \gamma y + \delta tg((y-p_0)\Delta\omega)$$

$\alpha, \beta, \gamma, \delta$ are computed from estimations of:
- a relative skew,
-      "   angle,   between the two flights
-      "   speed,
-      "   altitude.

3. linear smoothing of the parameters between every couple of adjacent models

10% or 20% of overlaying $\longrightarrow$

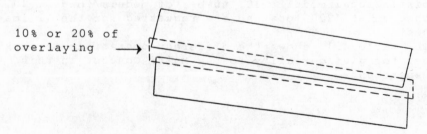

4.2. Using the Sliding Models:

- field by field analysis:
the ground data are recorded field by field and not
pixel by pixel, then the misregistration is not really
a problem as well it does not exceed 2 pixels (there
are few of greater misregistration).
Then we can merge the information of the ground data
bank and the radiometric values of any field, at any
date, in any spectral band, only by knowing each Sliding Model.
The Rectification itself is not necessary, we can get
access to requested informations by the method of Self
Defining Data Sets (SDDS = original radiometric information + geometric information, see (15)).

- SPOT-images simulation (16) :
the radiometric simulation is obtained from the eight
bands of DAEDALUS and gives the three bands of SPOT,
or the panchromatic band;
the geometric simulation is obtained by combining the
spatial filter of the SPOT-radiometer, its flight model
and the Sliding Model previously computed for every
image.
Then the "LAURAGAIS Project" can be used as a performance valuer of the capabilities of SPOT for vegetation
monitoring.

5. CONCLUDING REMARKS.

Recent work on image-to-image Registration is presented
in this paper, with reference to the Multi Concept.
Further studies should be done on several aspects of
this problem, particularly on the a-priori selecting of
reference windows, on the adaptability of discontinuity
detection (automated selection of the best detector),
on the computation of the model from the control points.

REFERENCES.

(1).American Society of Photogrammetry.
    "Manual of Remote Sensing"
    Chapter I (Introduction) -1975

(2).P.E.Anuta.
    "Spatial Registration of Multispectral and Multi-
    temporal Digital Imagery Using Fast Fourier Trans-
    form Techniques"
    IEEE, Vol.GE-8, n°4,october 1970

(3).D.I.Barnea, H.F.Silverman.
    "A Class of Algorithms for Fast Digital Image Re-
    gistration"
    IEEE, Vol.C-21, n°2,february 1972

(4).P.Van Wie, M.Stein.
    "A LANDSAT Digital Image Rectification System"
    Goddard Space Flight Center report,may 1976

(5).W.A.Davis, S.K.Kenue.
    "Automatic Selection of Control Points for the Re-
    gistration of Digital Images"
    Proceeding of the 4th IJCPR, KYOTO,nov. 1978

(6).M.L.Nack.
    "Rectification and Registration of Digital Images,
    and the effect of Cloud Detection"
    1977 Machine Processing of Remotely Sensed Data
    Symposium, PURDUE,june 1977

(7).D.G.Goodenough.
    "L'Analyseur d'Images du Centre Canadien de Télé-
    détection"
    4th Canadian Symposium on Remote Sensing, QUEBEC,
    may 1977

(8).R.Jeansoulin.
    "Le Recalage Automatique d'Images de Télédétection"
    2ème Congrés AFCET-IRIA, Reconnaissance des Formes
    et Intelligence Artificielle, TOULOUSE,sept. 1979

(9).F.Salgé.
    "Etude de la Corrélation Automatique pour la Recti-
    fication des Images SPOT (niveaux 2 et 3)"
    Rapport OT/TI/CRIS, IGN-CNES, TOULOUSE,july 1980

(10).M.Adelantado, R.Jeansoulin.
    "Le Recalage Automatique des Côtes pour la Recti-
    fication des Images Météorologiques METEOSAT"
    Rapport LSI, Université P.Sabatier, TOULOUSE,1980

(11).M.Svedlow, C.D.McGillem, P.E.Anuta.
    "Image Registration: Similarity Measure and Pre-
    processing Method Comparisons"
    IEEE, Vol.AES-14, n°1,january 1978

(12).R.Jeansoulin.
    "Le Recalage Automatique d'Images de Télédétection
    Tome I= Méthodes et Algorithmes (mars 1980),
    Tome II= Images LANDSAT-MSS (juin 1980)"
    Rapports LSI, Université P.Sabatier, TOULOUSE

(13).G.Saint, A.Killmayer, G.Guyot, J.Riom.
    "Relations between Biological and Phenological
    Characteristics of Vegetation and its Radiome-
    tric Properties: examples in Agriculture and Fo-
    restry"
    14th International Symposium on Remote Sensing of
    Environment, SAN JOSE-COSTA RICA,april 1980

(14).R.Jeansoulin.
    "Influence des paramètres physiques de prise de
    vue, dans la Corrélation Automatique d'images"
    14th International Congress of the ISP, HAMBURG,
    july 1980

(15).P.E.Anuta.
    "Geometric Representation Methods for Multi-type
    Self Defining Remote Sensing Data Sets"
    14th International Congress of the ISP, HAMBURG,
    july 1980

(16).C.Mouvier.
    "Expérimentation LAURAGAIS: simulation de données
    SPOT"
    Rapport OT/TI/TH, IGN-CNES, TOULOUSE,august 1979

OCCLUSION IN DYNAMIC SCENE ANALYSIS

W.N. Martin and J.K. Aggarwal

Departments of Electrical Engineering
and Computer Sciences
The University of Texas at Austin
Austin, Texas 78712

Abstract: This paper presents several fundamental concepts necessary for the successful analysis of dynamic scenes containing occluding objects by discussing various systems which have been developed to perform this analysis. The dynamic scenes are represented by time ordered sequences of images. Data must be extracted from each of these images and then integrated into coherent information about the sequence as a whole.

1. INTRODUCTION

The term "dynamic scene analysis" has been used to refer to the process of analyzing time ordered sequences of images, and in an earlier survey [1] we have observed that for a computer vision system to recognize and understand the motion apparent in the image sequences several levels of analysis are required. Peripheral and attentive processes constitute two of the more important levels. The peripheral processes detect motion and direct the attentive processes to it, while the attentive processes track the movement and attend to the details of the objects in motion. Cognitive processes form a level in which the specfic information acquired through the other levels is integrated with the goals and expectations of the vision system. In this paper we will discuss two systems [2,3] incorporating attentive level processes that have been developed at the University of Texas at Austin. Several other systems have been developed here and reproted in [4,5,6,7,8], while additional surveys of the area can be found in [9,10].

J. C. Simon and R. M. Haralick (eds.), Digital Image Processing, 579–590.

2. DYNAMIC SCENE ANALYSIS WITHIN A DOMAIN OF POLYGONAL FIGURES

Our initial system [2,11] analyzed dynamic scenes containing arbitrarily complex rigid polygonal figures, possibly having holes. These figures were allowed to move with various rotational and translational velocities in planes parallel to the image plane. For this domain the polygons were required to be rigid, but were not restricted in their movement. In this way the polygons could occlude one another, combining their projections in the image plane to form apparent objects which changed shape as the occluding polygons changed their relative positions. Thus the main task of the system was to decompose the apparent polygons into constituents that corresponded to the visible portions of the actual polygons in the scene. This task was accomplished by tracking the rigid parts (called "real" vertices) of the apparent polygons and using the acquired information to interpret the changes in the remaining parts.

Here we should make the important point that dynamic scene analysis systems must have access to image features which remain constant through the image sequence or at least change so slowly as to be constant through subsequences of images. These image feature may be descriptors associated with various "tokens" in the image. Tokens are groupings of descriptors considered to be indicative of salient components of the scene. In [2] the tokens were the vertices of the apparent polygons and had descriptors for the spatial location of the vertex point, the length of the two polygonal sides incident on the vertex, and the interior angular measure of the vertex. Clearly, for moving polygons that can occlude one another only the latter descriptor would remain constant and thus it was used as the primary tracking feature.

The tracking, however, was mostly concerned with forming a correspondence between given scene components and the tokens in each image of the sequence. The angular measure at a vertex was not a completely sufficient tracking feature for several reasons. First, the angular measure was not a uniquely identifying property. Second, the analysis of the apparent polygons required the identification of the vertices formed by occlusion (called "false" vertices) even though their angular measure was changing. The tracking was accomplished with the aid of object models, created and continually updated by the motion analysis system. These models provided the additional constraints for identification by grouping the vertex tokens into ordered sets, i.e., objects, indicating the system's interpretation of the apparent polygons at any given time. The interaction between the tracking process and the interpretation structure, i.e., the object models, is another important feature of systems which attempt to analyze the motions of complex objects.

The example scene displayed in Figure 1 contained one large polygon rotating in a clockwise direction and initially obscuring two smaller polygons. A vertex of one of the obscured polygons appeared in Scene 2 while the remaining polygon first appeared in Scene 4. The interpretation of this scene derived by the system after five frames is shown in Model 5. Three separate objects were represented with two objects having only partial descriptions and the large polygon having its occluded parts restored by information from previous frames. These models provided the additional capability of forming complete object descriptions for polygons that were only partially visible in any given image.

3. DYNAMIC SCENE ANALYSIS WITHIN A DOMAIN OF CURVILINEAR FIGURES

The two subsequent studies [12,3] removed the constraint on polygonal objects. The first [12] made extensive use of a predictive model. The primary tracking was performed on a token created for each apparent object. The descriptors used were the spatial location of the centroid of the figure, the area, the direction of the major axis, and the size of the enclosing rectangle oriented to the major axis. For tracking of non-occluding figures the match between tokens in consecutive images was established on the basis of the latter three descriptors. If the figures occluded, however, the predictive model incorporated information from previous images to form expectations for the descriptor values. These expectations were then matched against the current token values. A disadvantage of the predictive scheme was that it required occluding objects to maintain constant motions. The reason for this restriction was that the descriptors retained in the tokens did not allow the updating of the motion estimates of the figures while they were occluding.

The second study [3] removed this restriction by modeling the figures with ordered sets of tokens. An edge detecting preprocessor [13] extracted the object boundaries from graylevel images of curvilinear planar figures which exhibited the following properties :

1. the edges in the images were the boundaries of figures;

2. when two figures overlapped the boundary between them was not discernible, thus occluding figures appeared as a single figure; and

3. the figures moved in planes parallel to the image plane, so that the scale of the images did not change.

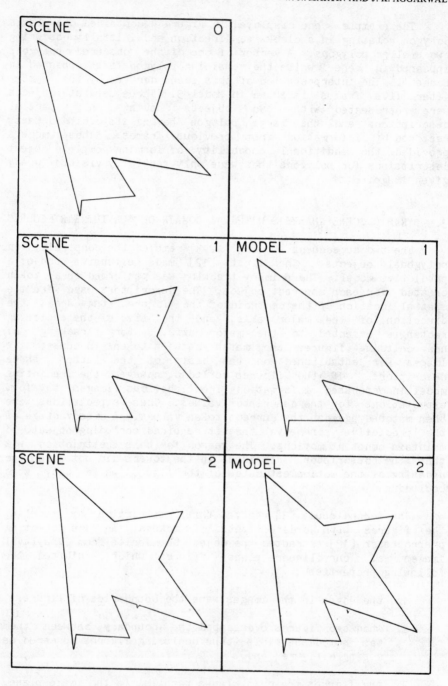

Figure 1.   The input frames and generated model interpretations
of a scene containing three polygonal figures.

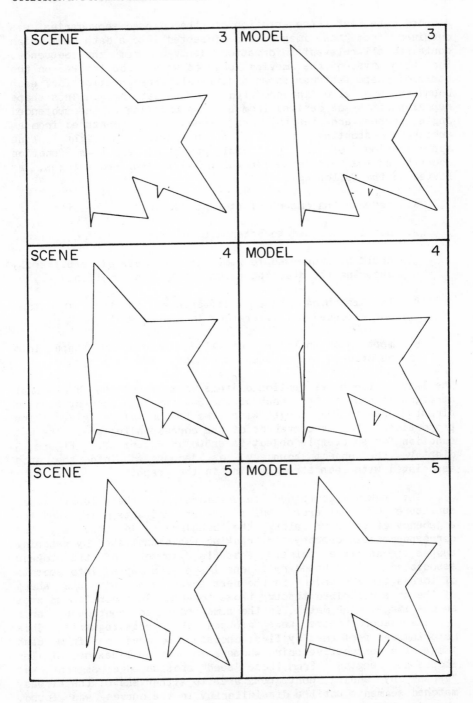

Figure 1. cont.

A coordinate list representation of the closed boundaries was obtained from the images and inserted into a data base which contained all relevant information derived from the sequence. For this system the moving objects were to be tracked on the basis of shape features so an additional representation for each boundary was also incorporated into the data base. This shape representation was derived from a chain encoding of the subtended angle versus arc length, ψ -s, function as measured from an arbitrary starting point. A piecewise straight line approximation of the pictorial graph of the ψ -s function constituted the representation which had several advantages, as listed in the following :

1. reduced the amount of storage required;

2. easily smoothed to attenuate noise;

3. could be processed to remove all effects of arbitrarily choosing the starting point of the ψ-s function;

4. for matching shapes, differences in the functions corresponded to differences in shape; and

5. most importantly, effectively decomposed the shape into an ordered set of arcs, the tokens for this system.

The last of the above mentioned items arose from the fact that straight lines in the graph of the ψ-s function corresponded to circular arcs in the image with the slope of a given line proportional to the curvature of the corresponding arc. The ψ-s function for an example object is shown in Figure 2. Figure 3 displays the object boundary, as decomposed into the arcs associated with each straight line in the graph.

The tokens contained descriptors for the length and curvature of the arcs, while the ordered sets maintained the adjacency of the arcs along the boundary. In this way the correspondence necessary for tracking was established by matching shapes across image pairs. That is, subsets of the tokens associated with a boundary in one image were mapped into subsets of tokens for a boundary in the next image by matching the shape of the arcs described through those tokens. Thus contiguous arcs in one image which match, in the same order, to contiguous arcs in the second image were grouped into edge segments. This matching was performed by first choosing two arcs, one from each image of a consecutive pair, whose ψ-s function lines had similar slopes and lengths. From these "seed" arcs an edge segment was "grown" by adding contiguous arcs to either end of the already matched segments until a dissimilarity in the curves was found. The dissimilarity of two curves was measured by the area between

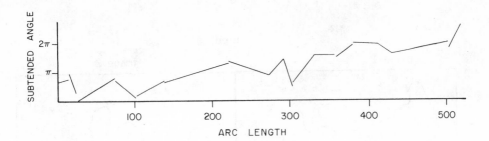

Figure 2. A straight line approximation to the graph of the ψ -s function for an example figure.

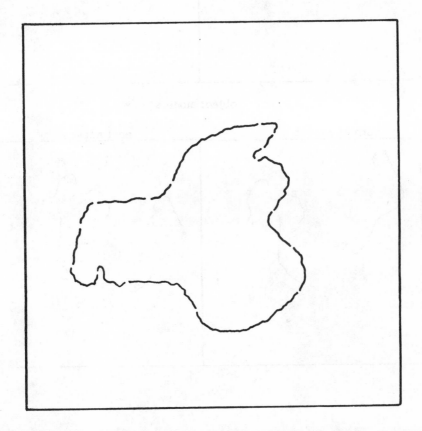

Figure 3. The figure boundary as segmented into arcs by the ψ -s function shown in Figure 2.

input

image 1

image 2

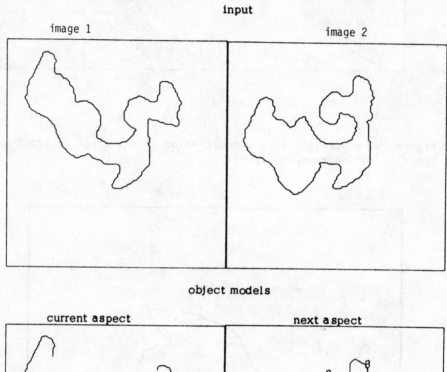

object models

current aspect

next aspect

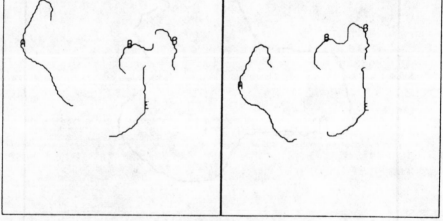

Figure 4.   The input frames and generated model   interpretations
of a scene containing three curvilinear figures.

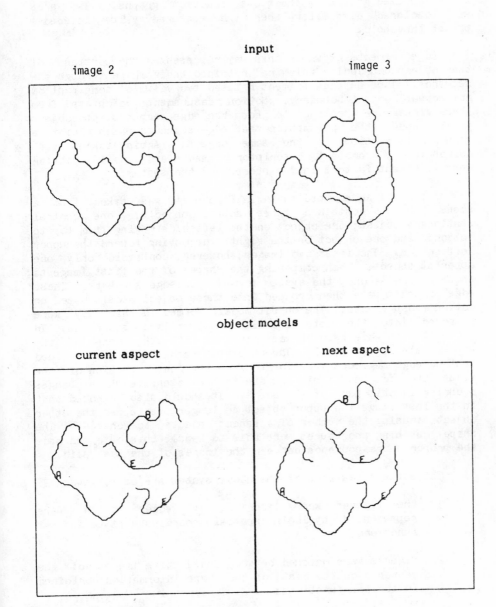

Figure 4. cont.

the normalized graphs of their ψ-s function graphs. Two arcs were declared dissimilar when the measured value exceeded a preset threshold.

Edge segments grown in this way represented the portions of the object boundaries which had retained their shape through the sequence. Thus an edge segment related two views of some part of an actual object boundary. Motion measurements calculated from these views were then used to group the edge segments into object models under the assumption that edge segments which exhibit a common motion belong to the same object. Again tokens with appropriately constant descriptor elements and dynamic modeling were important factors in the success of the system.

A part of an example shown in Figure 4 was taken from a scene which contained three actual objects: one central stationary object; one object on the left side moving from top to bottom; and one object on the right side moving toward the upper left corner. The first two images, however, contained only one apparent object. When comparing the shapes of the first image to those of the second, the system formed four edge segments. These edge segments were then grouped into three object models based on motion measurements. The object models formed in this way were inserted into the data base with arbitrarily chosen names. In Figure 4 each edge segment was labeled with the name of the appropriate object model. The observant reader will have noticed that no edge segments were formed from the center section of the first two images. This was due to the extensive shape changes occuring in that part of the scene. It should also be noted that in the last image the upper object no longer overlapped the other object, causing the number of apparent objects to change. The shape matching procedures were able to handle this case and made the proper correspondence between the images of the last pair.

The salient features of the above system are as follows:

1. the figures were described by edges which were represented by both spatial coordinate lists and ψ-s functions;

2. figures were matched between consecutive images of the sequence on the basis of the shape information contained in the ψ-s functions;

3. the matched figures were used to calculate motion measurements; and

4. motion was used to form object models under the rigid body assumption.

The system used these features to analyze successfully several sequences, such as the example above, which were generated by our image dissector camera.

4. CONCLUSION

In this paper we have attempted to elucidate several important aspects of the problem of analyzing occlusion in dynamic scenes. These included the following : system considerations such as multilevel design with the peripheral, attentive, and cognitive functions implemented as separate yet cooperating processes; the use of constant or slowly changing object attributes in forming the correspondence between components, i.e., tokens, in consecutive images of the sequence; the necessity of having decomposable object representations for the identification of partially occluded objects; object model creation and maintenance for interpretation of the scene; and the fundamental importance of the interaction between component tracking and scene interpretation.

Acknowledgement: We wish to express our gratitude to Professor Simon and Professor Haralick for organizing and hosting the NATO ASI. The research presented here was supported by the Air Force Office of Scientific Research under Grant AFOSR 77-3190.

REFERENCES

1. W.N. Martin and J.K. Aggarwal, "Dynamic scene analysis: A survey," Computer Graphics and Image Processing, vol. 7, no. 3, June 1978, pp. 356-374.

2. J.K. Aggarwal and R.O. Duda, "Computer analysis of moving polygonal images," IEEE Trans. on Comp., vol. C-24, no. 10, October 1975, pp. 966-976.

3. W.N. Martin and J.K. Aggarwal, "Computer analysis of dynamic scenes containing curvilinear figures," Pattern Recognition, vol. 11, 1979, pp. 169-178.

4. R. Jain, W.N. Martin, and J.K. Aggarwal, "Segmentation through the detection of changes due to motion," Computer Graphics and Image Processing, vol. 11, no. 3, 1979, pp. 13-34.

5. J.W. Roach and J.K. Aggarwal, "Computer tracking of objects moving in space," IEEE Trans. on Pattern Analysis and Machine Intelligence, vol. PAMI-1, no. 2, April 1979, pp. 127-135.

6. J.W. Roach and J.K. Aggarwal, "On the ambiguity of three-dimensional analysis of a moving object from its images," WCATVI, 1979, pp. 46–47.

7. J. Webb, "Static analysis of moving jointed objects," to appear in the First National Artificial Intelligence Conference, Stanford, California, 1980.

8. S. Yalamanchili, W.N. Martin, and J.K. Aggarwal, "Differencing operations for the segmentation of moving objects in dynamic scenes," to appear in the Fifth International Conference on Pattern Recognition, Miami, Florida, 1980.

9. H.-H. Nagel, "Analysis techniques for image sequences," International Joint Conference on Pattern Recognition, 1978, pp. 186–211.

10. Abstracts of the Workshop on Computer Analysis of Time-Varing Imagery (WCATVI), J.K. Aggarwal and N.I. Badler, eds., 1979.

11. R. Petermann, "Computer analysis of planar motion of polygons," Master's Thesis, University of Texas at Austin, Jan. 1975.

12. W.K. Chow and J.K. Aggarwal, "Computer analysis of planar curvilinear moving images," IEEE Trans. on Comp., vol. C-26, no. 2, Feb. 1977, pp. 179–185.

13. J.W. McKee and J.K. Aggarwal, "Finding the edges of surfaces of three-dimensional curved objects by computer," Pattern Recognition, vol. 7, 1975, pp. 25–52.

LIST OF PARTICIPANTS

M. ALMEIDA, Lisboa, Portugal.
B. APOLLONI, Arcavacato de Rende, Italy.
J.K. AGGARWALL, Austin, USA.
D. BARBA, Rennes, France.
M. BELLISANT, Grenoble France.
C.D. BENCHIMOL, Paris, France.
M. BERTHOD, Rocquencourt-Le Chesnay, France.
P.A. BIRD, Hatfield, U.K.
M. BISHOP, Glasgow, U.K.
H. BLEY, Erlangen, RFA.
B.Y. BRETAGNOLLE, Grenoble, France.
K.D. BRINCKMANN, Wiedbaden, RFA.
M. BOHM, Hamburg, RFA.
M. CANTONI, Pavia, Italy.
T.T. CAO, Orléans, France.
S. CASTAN, Toulouse, France.
E. CATANZARITI, Napoli, Italy.
M. CEBULA, Huntingdon Valley, USA.
M. CHASSERY, Grenoble, France.
Mme CHOUCQ-BRUSTON, Verrières-le-Buisson, France.
J.P. CRETTEZ, Paris, France.
P. DALGAARD, Lyngby, Denmark.
M. DAVIS, Austin, Texas, USA.
A. DELLA-VENTURA, Milano, Italy.
J. DESACHY, Toulouse, France
M. DIDAY, Le Chesnay, France.
J. FLEURET, Ivry, France.
M. FREEMAN, New York, USA.
J.P. GAMBOTTO, Arcueil, France.
D. GAMBART, Toulouse, France.
M. GOLDBERG, Ottawa, Ontario, Canada.
M. GONCALVES, Manchester, U.K.
M. GUEDJ, Orsay, France.
C. GUERRA, Roma, Italy.
M. GUGLIELMO, Torino, Italy.
M. HAAS, Delft, The Netherlands.
H. HAPPEL, Neubiberg, RFA.
M. HARALICK, Blacksburg, USA.
J.P. HATON, Nancy, France.
T.C.H. HENDERSON, Wessling, RFA.
R. HOLBEN, Huntington Beach, USA.
A. IKONOMOPOULOS, Orsay, France.
R. JEANSOULIN, Toulouse, France.
M. JUTIER, Paris, France.
E. KAHN, Villejuif, France.
N. KHOCHTINAT, Fontenay-aux-Roses, France.
 V. KLEMENT, Wiesbaden, RFA.
 G. KRUSE, Paris-la-Défense, France.

P. LAGO, Porto, Portugal.
M.T.Y. LAI, Montréal, Canada.
R. LE GO, Fontenay-aux-Roses, France.
M. LEVIALDI, Napoli, Italy.
B. LLEWELLYN, Oxford, U.K.
G. LORETTE, Créteil, France.
M. MA, Pékin, China.
S. MAITRA, Palo Alto, USA.
P. MARTHON, Toulouse, France.
W. MARTIN, Austin, USA.
H. MASTROYANNAKIS, Athens, Greece.
M. MITICHE, Austin, USA.
M. MOHAJERI, London, U.K.
M. MUSSO, Genova, Italy.
N. NISLEN, Tromso, Norway.
E. ØSTEVOLD, Kjeller, Norway.
L. OUDIN, Saint-Louis, France.
H. OZDEMIR, Istambul, Turkey.
S.D. PASS, London, U.K.
J. PIPER, Edinburg, U.K.
T.G. PUN, Lausanne, Suisse.
M. RADIG, Hamburg, RFA.
N. REHFELD, Karlsruhe, FRG
M. RISEMAN, Amherst, USA.
Y. ROBERTSON, London, U.K.
P. ROUX, Palaiseau, France.
F. SCHMITT, Paris, France.
E. SCHUSTER, Wien, Austria.
S. SCLOVE, Highland Park, USA.
L.G. SHAPIRO, Blacksburg, USA.
J.A. SORENSEN, Lyngby, Denmark.
G. STAMON, Belfort, France.
S.L. TANIMOTO, Seattle, USA.
S. TATARI, Karlsruhe, RFA.
W.K. TAYLOR, London, U.K.
R. TILNEY, Oxford, U.K.
A.E. TODD POKROPEK, London U.K.
A. VOSSEPOEL, Al Leiden, The Netherlands.
H.M. WECHSLER, Milwaukee, USA.
H. WESTPHAL, Hamburg, RFA.
P. WETTA, Villeurbanne, France.
M. WINOGRAD, New York, USA.
S.K. WU, London, U.K.
S. ZUCKER, Montréal, Canada.

INDEX